THE FUTURE X NETWORK

A BELL LABS PERSPECTIVE

THE FUTURE
X NETWORK
A BELL LABS
PERSPECTIVE

Marcus K. Weldon

CRC Press is an imprint of the
Taylor & Francis Group, an **informa** business

CRC Press
Taylor & Francis Group
6000 Broken Sound Parkway NW, Suite 300
Boca Raton, FL 33487-2742

© 2016 by Taylor & Francis Group, LLC
CRC Press is an imprint of Taylor & Francis Group, an Informa business

No claim to original U.S. Government works

Printed on acid-free paper
Version Date: 20150910

International Standard Book Number-13: 978-1-4987-5926-7 (Hardback)

Visit the Taylor & Francis Web site at
http://www.taylorandfrancis.com

and the CRC Press Web site at
http://www.crcpress.com

Table of contents

Introduction

Bell Labs research has never written a book. Bell Labs researchers have, of course, written or edited thousands of books, and tens of books have been written about Bell Labs and its unique history of innovation. But, in the 90 years that Bell Labs has been in existence, Bell Labs research as an organization has never written a book. So, why start now? Well, in short, it seemed that the time was right. We are at the nexus of a human technological revolution that will be different than any prior era, as it will simultaneously be both global and local, with innovation occurring everywhere. Therefore, almost by definition, this technological revolution will be less coherent in execution, as myriad different perspectives collide and compete. The collision of ideas often distills the optimum solution to a problem, which is the classic modus operandi of Bell Labs and is the basis of the phenomenal breakthroughs that have come from the Labs over its history. The coming global-local collision of ideas will make this technological revolution one of the most fruitful and far-reaching in human history.

Inside Bell Labs we have been exploring the limits of current technologies, architectures and solutions with a view to 2020 and beyond. From our explorations, we have created a vision for the future of networking for the new digital era, including its associated systems, applications and services. Rather than keep this vision internal to the Labs, we are sharing it with the industry at large, with the goal of fostering an informed discussion of the future network, a common understanding of its potential to transform human existence and the essential investments needed to bring it about.

Any technological revolution requires three things to exist, the introduction of a new technology being the obvious starting point. It also requires a network that connects the users of this technology in new ways, and it must result in a new social and economic reality shaped by the adoption of the technology. We are entering an era when everything and everyone will potentially have a digital identity and be connected by a global network. This will lead to the virtualization of devices, homes, enterprises and the networks that connect them to the services they consume. This in turn will give rise to a new global-local socio-economic order, where digital services are offered globally and tailored and delivered locally. Even manifestly physical goods will be affected by the spread of technologies, such as 3D printing, with goods being distributed digitally around the globe and created and delivered locally.

In this book we consider the key human needs and the driving forces for this revolution. We investigate in depth the critical enabling technologies, the architectural shifts and the new business models that could result. But this book is just a starting point for a discussion; it is inherently incomplete. It is intended to be a reference to be debated, refuted, expanded and adapted by anyone. In many ways it should be viewed as an open – and rather long – "letter" to industrial, academic and individual innovators and start-ups to help drive distributed and disruptive thinking and innovation, but with a measure of coherence of vision and purpose.

As a forward-looking, visionary work, it is intended to reach a broader audience; yet, it is grounded in and provides a clear technical basis for the conjecture. As such, it tries to find a reasonable middle ground, inevitably making choices and compromises along the way. It is consequently like most human efforts, imperfect. It is technical in places, but not technical enough for those interested in a deep understanding of a topic. It is forward-looking, but likely not sufficiently visionary for the most futuristic thinkers. It attempts to provide a relatively complete end-to-end view, but inevitably fails to capture important evolutions or innovations in key areas or cover some domains or topics in sufficient depth. However imperfect, it is nonetheless a starting point, effectively an introductory chapter to a much larger work that will be the task of the world's innovators in the coming decades. Only time will tell whether this book achieves the intended value of increasing the "signal strength" for The Future X Network – as we are calling it – or whether it simply adds to the "noise". But the goal is, I believe, a worthy one.

Marcus Weldon
President of Bell Labs and Chief Technology Officer of Alcatel-Lucent

Acknowledgements

Marcus Weldon would like to acknowledge the support provided by Lisa Ciangiulli, Charlie Bahr, Kevin Sparks and Jean-Pierre Hamaide in the production of this book on an aggressive schedule. The authors wish to acknowledge the significant contributions of many Bell Labs researchers, Bell Labs Consultants, as well as members of the Alcatel-Lucent Corporate Chief Technology Office team. The creation of this book was in many ways a classic Bell Labs undertaking involving collaboration, analysis, editing, and production support from multiple people with different perspectives and expertise. In alphabetical order these are:

Paul Adams
Patrice Rondao Alface
Prasanth Anthapadmanabhan
Gary Atkinson
Mathieu Boussard
Dominique Chiaroni
Mark Cloughtery
Phil Cole
Roberto Di Pietro
Dave Eckard
Edward Eckert
Bob Ensor
Steve Fortune
Barry Freedman
Ronan Frizzell
Adiseshu Hari
Martin Havemann
Fahim Kawsar
Vlad Kolesnikov
Bartek Kozicki
Anne Lee

Nicolas Le Sauze
Leigh Long
Huilan Lu
Jean-Francois Macq
Kevin McNamee
Joe Morabito
Rebecca Nadolny
David Neilson
Brian O'Donnell
Sanjay Patel
Frank Ploumen
Todd Salamon
Sameer Sharma
Caroline Siurna
Erwin Six
Bart Theetan
Michael Timmers
Lieven Trappeniers
Rao Vasireddy
Chris White
Paul Wilford
Peter Winzer

Experts from Bell Labs Consulting:

Ednny Aguilar
Furquan Ansari
Jim Borger
Jim Davenport
Vikas Dhingra
Paul Gagen
Martin Glapa
Ashok Gupta
Alina Ionescu-Graff
Viby Jacob
Anand Kagalkar
Sanjay Kamat
Anish Kelkar
Olivier Kermin
Bill Krogfoss
Samrat Kulkarni
Luc Ongena
Subra Prakash
Narayan Raman
Abdol Saleh
Gina Shih
Andre Trocheris
Carlos Urrutia-Valdes
Rajesh Vale
Thomas Weishaeupl

CHAPTER 1

THE FUTURE X NETWORK

Enabling a new digital era

Marcus Weldon

The essential vision

We are at the dawn of an era in networking that has the potential to define a new phase of human existence. This era will be shaped by the digitization and connection of everything and everyone with the goal of automating much of life, effectively creating time by maximizing the efficiency of everything we do, augmenting our intelligence with knowledge that expedites and optimizes decision-making and everyday routines and processes.

This futuristic vision would, on the surface, seem to be the stuff of science fiction, rather than rational technological thinking and analysis; however, we are at a unique moment in time. There is a confluence of common technological elements and human needs (consumer and enterprise) that is driving a convergence of players and market segments (for example, IT/web and telecom/networking), as well as the emergence of new business models. In essence this is a classic supply-demand equation but one that is about to be rebalanced as the demand becomes increasingly common between consumer and enterprise, and will be satisfied using a set of standard, scalable technologies, so that an optimization of supply makes it possible to meet this demand with innovative new value propositions.

We believe that this is a technology-driven transformation similar to the first Internet era (1995-2005) and the era of the cloud and the device

(2005-2015). Just as they were driven primarily by technological forces that supported new connectivity, computing and consumption paradigms, this coming transformation will also be driven by technology shifts that cause significant changes in human behavior and massive business disruption. We call this the *Future X Network transformation*, with X denoting the multiple factors of 10 that will characterize this shift. It will be a 10-year journey, resulting in multiples of 10 times the scale in devices, cloud and network infrastructure, as well as also representing the unknown variable x that characterizes the inherently variable nature of this future, which depends on successfully solving for multiple technological and economic factors.

We believe that the catalytic technological drivers for this Future X Network will be threefold:

1. A new dynamic approach to networks that creates the sense of seemingly infinite capacity by pushing beyond current scientific, informatics and engineering limits to create a new *cloud-integrated network* (CIN) that not only provides the essential input and output mechanisms, but also, intelligence.

2. The rise of Internet-connected machines and devices that will send and receive a massive amount of new digital information, as well as reshape the manufacturing landscape with a diverse array of 3D printers – reversing the current trend in developed countries of off-shoring manufacturing to lower-cost countries.

This era will be defined by the digitization and connection of everything and everyone with the goal of automating much of life, effectively creating time, by maximizing the efficiency of everything we do...

3. New data analysis techniques based on *inference* of needs and information, instead of seeking "perfect knowledge". New *augmented* intelligence systems will use the smallest amount of digital information, derived from the massive amount of data collected, to infer what is needed in each situation and context, to assist – not replace – human intelligence.[1]

Through these disruptions, human existence will be transformed. We will live with greater efficiency: enabled by data, but not overwhelmed. We will return to local production paradigms, managed and connected by the new global-local, cloud-integrated network of virtual and physical locations and workers. This will require shifts in control, collaboration and communications systems. Operations systems will need to allow accelerated onboarding, scaling and seamless interaction between new digital services – for both consumers and enterprises.

In this book, we outline how Bell Labs sees this future unfolding and the key technological breakthroughs needed at both an architectural and systems level. Each chapter of the book is dedicated to a major area of change. As with any attempt to predict or foresee the future, we no doubt exhibit a level of certainty about a range of phenomena that are inherently uncertain and shifting. But the goal is to frame the discussion in an informed way, to speculate on the future based on a clear set of assumptions, and identify a set of "game-changing" technologies and models that are markedly superior than those today. Where possible, we will reach into the past to understand the present and predict the future, but in many cases, the future we foresee is radically different from either past or present.

If there is one overarching summary statement that describes the thesis of this book, it is perhaps that by Marshall McLuhan, "as technology advances, it reverses the characteristics of every situation again and again. The age of automation is going to be the age of *do it yourself*" (McLuhan 1997). However, "do it yourself" in 2020 will mean "do it *for* yourself" in that any service or process will be personalized and contextualized, with the virtualization and software definition of networks and services allowing connectivity of anyone to anything, anywhere, on demand. It is this "perpetual optimization" that will drive the next phase of our human technological evolution.

---

1 - *Augmented* is to be contrasted with *artificial* intelligence, which aims to create "perfect knowledge" in artificial systems. Augmented intelligence, in contrast, is intended merely to assist – not replace – human beings.

The past and present

In our exploration of this technology-driven revolution, it is worthwhile clarifying what we mean by the term. The *Webster's Dictionary* definition of technology is: *a manner of accomplishing a task especially using technical processes, methods or knowledge* (Webster 2015). The earliest technological revolutions can then be understood as the invention of simple tools (accomplishing a physical task based on understanding mechanical action); language and the Phoenician alphabet (accomplishing the task of communicating more effectively and more extensively, based on the concepts of phonetic word formation and grammar); and the new engineering and construction methods of the great civilizations (accomplishing the task of building and managing large infrastructures based on understanding of mathematics and physics). In the modern era, Daniel Šmihula has identified several universal technological revolutions that occurred in western culture (Šmihula 2011):

1. *(1600–1740) Financial-agricultural revolution* – centered on the creation of banking infrastructure and stock markets to form early financial networks, as well as the development of more efficient agricultural practices that drove unprecedented population growth and commerce.

2. *(1780–1840) Industrial revolution* – centered on the use of iron and steam engine technologies, fueled by coal, to gain mechanical advantage, enabling the automated production of textiles and the first steam-based transportation networks for the delivery of mail and printed goods (also produced by steam presses).

3. *(1880–1920) Second industrial revolution* – centered on the mass production of steel and chemicals fueled by a new energy source – oil – and the creation of extended networks for transportation and electricity, and the first telephone networks for communication.

4. *(1940–1970) Scientific-technical revolution* – centered on the creation of analog and then digital electronic signal processing and computing systems, and the first digital information communications networks required to connect and control these systems at a distance.

5. *(1985–2000) Information and telecommunications revolution* – centered on creation of the Internet and the development of broadband access networks to allow high-speed access to the Internet and its associated digital information and services, as well as the first phase of mobile communications, which shifted focus from places to people.

6. *(2000-2015) Cloud and mobile revolution* – centered on the development of global cloud infrastructure (data centers and interconnection networks) and the concomitant development of smart mobile devices and mobile broadband digital access networks, which, acting together, are accelerating the transition to a new phase of nomadic, distributed human existence.

David Brown argues that the latter two actually comprise the beginning of a third industrial revolution initially defined by a new source of fuel (solar and wind), a new communication paradigm (the Internet) and new financial systems (crowdsourcing and mobile payment), but with more radical evolutions in each dimension still to come (De Vasconcelos 2015).

In each case, one could argue that there is an initial technological revolution (invention of something new) that has a concomitant networking revolution (to connect this new something), which then leads to an economic revolution. Indeed this is essentially the premise of the Šmihula theory of waves of technological revolution (Šmihula 2009). In the same vein, Carlota Perez argues that: "What distinguishes a technological revolution from a random collection of technology systems, and justifies conceptualizing it as a revolution are two basic features:

1. The strong interconnectedness and interdependence of the participating systems in their technologies and markets.

2. The capacity to transform profoundly the rest of the economy (and eventually society)" (Perez 2009).

In other words, networks that connect new technologies and result in the transformation of economics and social interactions are, by definition, *technological* revolutions. By this definition, we are indeed at the beginning of a new (6th?) social and economic upheaval driven by technology. The massive build-out of the Internet and ultra-broadband networks will connect not only people, but as well, digitized "things", which will, in turn, lead to unprecedented levels of automation. This will transform business around a new global-local paradigm that will also effectively "create time" by optimizing the execution of mundane or repetitive tasks, and so change human existence as we know it.

Steve Case has recently argued that this is actually the third Internet era, with the first defined by the building of the Internet (1985-2000) and the second by building new services on top of the Internet (2000-2015) – mostly with a consumer focus. The third era or wave will be defined by

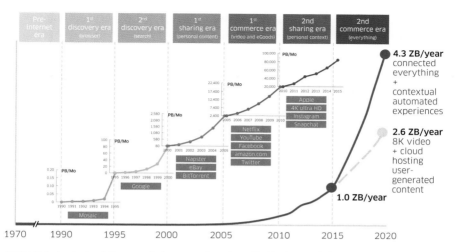

FIGURE 1: Analysis of the growth in core network traffic (dark curve) since the dawn of the Internet era in terms of the constituent five-year trend segments (data shown with expanded scales) (Bell Labs Consulting)[2]

building the Internet into everything (2015+), resulting in disruption of many industry sectors (O'Brien 2015).

We have mapped these three waves into smaller, five-year increments, which show the underlying human behavioral trends (figure 1). The Internet era started with the ability to discover content on the Internet by entering known URLs into a browser (1990–1995), which was then quickly optimized by adding the ability to discover unknown content using search functions (1995–2000). This in turn led to the ability to share personal content (2000–2005), often without the requisite permissions for media content. Next came the first commercial digital era (2005–2010), with the availability of multimedia content and associated digital rights; the beginning of eCommerce, with the ability to acquire physical goods over the Internet; and the sharing of personal images on social media. The last five years (2010–2015) have seen a continuation of this commercial growth, but have been dominated by a sharing of personal content delivered from and to mobile devices.

From this simple analysis, it is clear that another phase of Internet usage is likely upon us, which we believe will be led by another phase of commercial usage, this time driven by the digitization of everything, and the associated services that will be enabled. This new era will see unprecedented growth in traffic, as human behavior and businesses are automated to an extent that was never possible before.

2 - Unless otherwise noted, all figures come from Bell Labs Consulting analysis.

In summary, whether we are entering the sixth technological revolution (Šmihula), or the third industrial technological revolution (Brown), or the third wave of the Internet (Case), or all of the above, the conclusion is that a human technological revolution is upon us. In the following section, we look at the key business, architectural and technological elements of this revolution in more detail or to describe the essence of 2020 as we see it.

The Future X Network

As outlined in the preceding analysis, the formation of new networks is a driving force for a new technological revolution. Looking ahead to 2020, we believe the same will again be true with the creation of a new cloud-integrated network being the primary driver of what will come. The second novel technological element – arguably the catalyst for the revolution – will be the *Internet of Things* (IoT), the vast array of networked machines that will digitize our world in ways that have previously only been imagined. The third related technological element, which is also essential in order for this digitization to produce value in the form of efficiency, is the ability to capture and analyze the data streaming from these devices and effectively turn big data into the smallest amounts of actionable knowledge, what we refer to as *augmented intelligence*. This new knowledge paradigm, and the automation which will result, will underlie the economic transformation to come. We will summarize each of these three key elements – the new network, the role of IoT and the augmented intelligence paradigm – in the rest of this chapter, and throughout the other chapters of this book.

The cloud-integrated network

The evolution in communications networks since the earliest telephone and telegraph networks has been simply astounding (figure 2). We have moved from all-copper-based, wired telephony networks designed for 56 kb/s voice channels and similar dial-up modem speeds in 1995, to a few 10s of Mb/s for ADSL in the late 1990s, reaching 100 Mb/s for VDSL in 2010 (using advanced signal processing) and today anticipate the arrival of G.fast at 1 Gb/s in the near future. A similar trend has been seen in the evolution of optical access, where passive optical networks (PON) have seen approximately 10-100 times higher speeds on average during the same period. In the core network, rates of 2.5 Gb/s were possible in 1990 using simple modulation formats, but have advanced over this period to rates of 100 Gb/s in 2010, and 400 Gb/s in 2015 (using advanced signal processing with complex modulation and two polarizations of light).

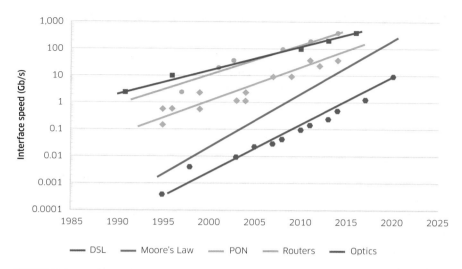

FIGURE 2: Comparison of access network and core network interface rate evolution, compared to Moore's Law exponential growth (Bell Labs data)

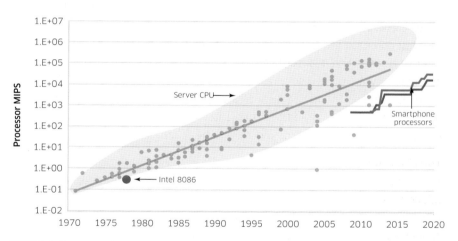

FIGURE 3: Exponential Moore's Law-driven evolution of processor performance over time (Mobile processor performance is measured in DMIPS and converted to MIPS for comparison.) (Bell Labs Consulting)

This represents an increase in the core network of almost 160 times in 25 years, and more than 250 times in the access network over 20 years. This is essentially driven by the evolution in transistor technology described by Moore's Law, which has enabled more dense, higher speed processors of different types (packet processors, digital signal processors, general purpose processors), as illustrated in figure 3.

There has also been a wholesale migration from TDM and ATM data over SDH/SONET networks, to IP over IP/MPLS networks, allowing the lowest cost and highest performance networks to be built by leveraging the much higher volumes of Ethernet and IP networking components compared to the more bespoke and expensive telecommunications components. There is ample evidence that this increase in device performance will continue at rates predicted by Moore's Law (figure 3), but a number of different approaches are required to achieve this, due to the extremely high data rates being contemplated:

1. Use of new multiple-gate, CMOS transistor designs (for example, finFET and TriGate) to allow better control of current leakage at very small (<20 nm) device sizes

2. Use of new integrated analog and digital circuitry using BiCMOS technology and 3D component integration

3. Use of new device materials, such as silicon-germanium (SiGe) in order to support higher clock frequencies and higher electron mobilities

4. Use of silicon photonics — the direct integration of photonic components into and onto CMOS driver circuitry for improved optical component performance and economics

Using all these approaches, we will be able to follow the "Moore's Law" rate of a doubling in rate every 18 months, or an increase of 10x in the five years between 2015 and 2020. But if we are entering an era where potentially 100x more capacity is required, we will also have to take a different approach to networking, by exploring new architectural dimensions. These new dimensions are described in this book, and can broadly be summarized as follows (and illustrated in figure 4):

- The embedding of the cloud in the network to form a new *edge* cloud to provide the optimum performance (throughput and latency) and economics for both virtualized networking functions and any other performance-critical enterprise or web service

- The emergence of an end-to-end, software-defined networking layer that dynamically connects distributed and diverse workloads, networks and devices, and creates end-to-end virtual network paths or slices

- The creation of a new ultra-high capacity and continuously reconfigurable network fabric, with the reimagining of the core, metro and access layers and architecture

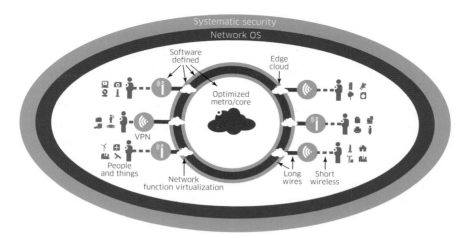

FIGURE 4: Schematic depiction of the cloud-integrated network architecture.

- The creation of a new converged, ultra-narrowband and ultra-broadband access architecture, simultaneously optimized for people, places and things

- The emergence of a *network* OS that abstracts all network resources and manages the utilization to create exactly the needed capacity from the source to the destination

- The emergence of a systematic approach to security, using endpoint-based (for example, device, server), perimeter-based, and network-based security with new, self-learning autonomics that will allow identification of *zero-day* attack vectors in real time

This combination of capabilities we refer to as the new cloud-integrated network (CIN), and is the first element of the Future X Network architecture. The CIN will blur the lines between application, device and network functionality by linking together device software functions with application and network functions in dynamic service chains, on demand. This is how a network of seemingly infinite capacity will be created, that will optimally serve the digital needs of all people, processes and things. It will also blur the line between the cloud and the network, as the cloud becomes distributed within the network for the required performance.

The chapters of this book will describe these aspects in more detail, but we will highlight two of the key performance drivers here:

1. Ultra-low, cloud-connectivity latency

2. Ultra-high access capacity to these cloud locations

The two elements – ultra-high capacity and ultra-low-latency – are related. The primary cause of latency (other than the speed of light, which induces 4.5 ms of latency for every 1,000 km), is the delay induced by network hops, when packets are queued for delivery over an interface that has lower capacity than the sum of the input flows. This *queuing delay* is generally less than a millisecond on average, but in times of severe congestion can amount to 10s of milliseconds, which will significantly compromise the performance of latency-sensitive services. In order to offer low-latency service guarantees, one must minimize the number of network hops and maximize the available bandwidth. These dual requirements essentially mandate the creation of edge cloud nodes and ultra-high-capacity access and metropolitan aggregation networks providing the required "on-ramp" connectivity to these nodes.

To understand this evolution, consider the evolution of computing resources over time. As depicted in figure 5, in the 1960s, at the beginning of the first digital computing revolution, computing resources were the bottleneck, as they were extremely expensive and scarce, so users were required to share these resources by using a sequential batch job system (based on punch cards). Then, with the development of multi-user operating systems like Unix, it was possible to share the computing resources in real time from multiple terminals directly connected to the mainframe in a computing center.

At this stage in the evolution, the network was the bottleneck, or more correctly the absence of the network was the bottleneck. However, with the advent of TCP/IP networking, the network bottleneck was removed and remote communication with mainframes over local area networks (LANs) was possible, from anywhere within a few hundred meters. Then, as computing became significantly cheaper in the 1980s and 1990s, a profound shift occurred and processing shifted to the end device – the PC. And with the creation of the Internet that allowed LANs to be connected to each other and to centralized servers that hosted applications, neither computing nor the network were the bottleneck, rather the ready access to useful information was the bottleneck.

This led to creation of the first web service platforms and browsers that allowed information to be discovered and searched with ease, and the creation of the first large-scale data centers to host these platforms. Broadband access connectivity to homes and then mobile devices led to the ability to connect to this information and content from anywhere, not just from within enterprise LANs. New services started appearing that were

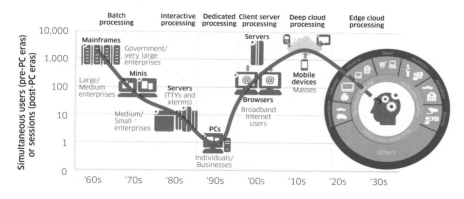

FIGURE 5: Evolution of computing resources from centralized mainframes to personal computers to the new edge cloud nodes.

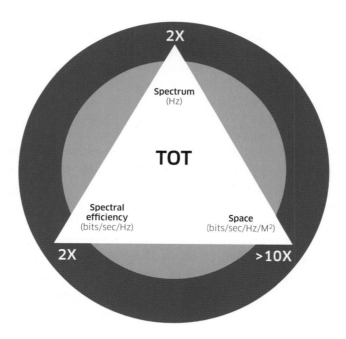

FIGURE 6: "Triangle of truth" (TOT) about transmission over copper, fiber or air. The numbers denote the additional improvement possible in each dimension relative to the centralized architecture deployed today.

"mash-ups" of individual web services using open application programming interfaces (APIs) between systems to create more sophisticated services, resulting in the massive scale-out of cloud-based servers in data centers, with large storage farms for the resulting data.

This is essentially our reality today in 2015, with the shift to mobile devices with their limited processing and storage capabilities further increasing the dependence on the cloud to perform higher-order processing and service creation. As we look toward the future dominated by much simpler, lower power "things" with minimal storage and with the requirement for a 10-year battery life, this phenomenon of shifting computing power to the cloud will be accelerated. The network will again become the bottleneck, in what is in essence a new form of the well-known von Neumann bottleneck in computing systems, in which the processing power of a (cloud) computer is limited by the memory (network) bandwidth. In this case, this will drive the need for the cloud to move into the network to create the CIN with edge cloud facilities, as described previously, and the serving access and aggregation networks to massively scale, as outlined below.

This last requirement, for ultra-high-bandwidth, is one of the most important elements of the new networking era. Because we have reached the limit of almost every physical layer technology – whether it be in the wireless, wireline or optical domain – a shift in network architecture is now required.

The nature of this shift can best be understood by considering the essential characteristics of all electromagnetic transmission systems. As shown in figure 6, there are three dimensions of transmission: the spectrum, spectral efficiency and spatial dimensions. In the first two dimensions, we have nearly reached the fundamental limit of capacity available from the current, relatively centralized architecture. We are close to maximum usage of available spectrum in:

- *Wireless* – within a factor of two of the maximum usage of low-frequency bands
- *Wireline* – within a factor of two of the limit of DSL spectrum usage
- *Optics* – within a factor of two of the number of wavelengths that can be used with current amplifiers

And in all cases, signal processing and interference cancellation techniques have advanced to the point where we are now within a factor of two of the Shannon limit: the information-carrying capacity of each channel that is tied to the spectral efficiency.

Therefore, the only option left is to explore the spatial dimension, by shortening the transmission distances and/or subdividing the transmission medium. The former approach (shortening distance) will allow usage of wider spectral bandwidths (as much as 10x) that can only be carried over shorter distances due to stronger signal attenuation in the transmission medium (copper, fiber or air). The latter approach (subdividing the medium) will allow application of this increased spectrum to smaller user groups. In combination, these two effects will offer the possibility to expand capacity by factors of 20 or more.

In summary, the CIN is comprised of two related and essential architectural elements: the edge cloud that is embedded in the network; and a re-partitioned access and aggregation network that delivers ultra-high-bandwidth and ultra-low-latency to and from this edge cloud. In turn, this distributed cloud connects to the centralized cloud that hosts less performance-sensitive applications, over a similarly adaptive core network.

The rise of the machines

Despite the remarkable history of humankind, it is only recently that we have started to move from a physically-dominated existence to a virtual or digital one. Looking at the technological eras outlined above, it was only with the advent of the scientific-technical revolution beginning in the 1940s that the digitization of *anything* was possible.

Since that time, we have almost completely digitized the communication and entertainment industries, including the production and distribution of published information and audio/video media content. Business-to-consumer commerce or e-commerce is much smaller than "bricks-and-mortar" retail, but growing steadily. We have also digitized some industries, most notably the financial industry, but the vast majority of our physical world remains at best, partially digitized. For example, healthcare is still largely an analog-based industry, as is transportation, manufacturing and its associated supply chains, energy production and delivery, public safety, and the vast majority of all enterprise processes, operations and business-to-business transactions.

On first impression, this may seem like a gross exaggeration, as all current robotic or automated systems have sophisticated digital systems with the ability to communicate and be optimized with a remote control center. And, in addition, the commercial shipping industry has digitally tagged, scanned and tracked objects at each waypoint in the delivery chain. However, even in such relatively advanced cases of digitization, there is little ability to connect these processes to all the other physical systems on which they rely; for example, in the latter case, to the road or rail infrastructure, in order to optimize the route followed. Similarly, although environmental (for example, seismic or weather sensing) has been deployed for decades, many devices still require manual reading and reporting, or the extent of coverage of the more sophisticated digital systems is limited. A simple illustration of the change that we see occurring is provided by the detailed mapping of an earthquake in California by

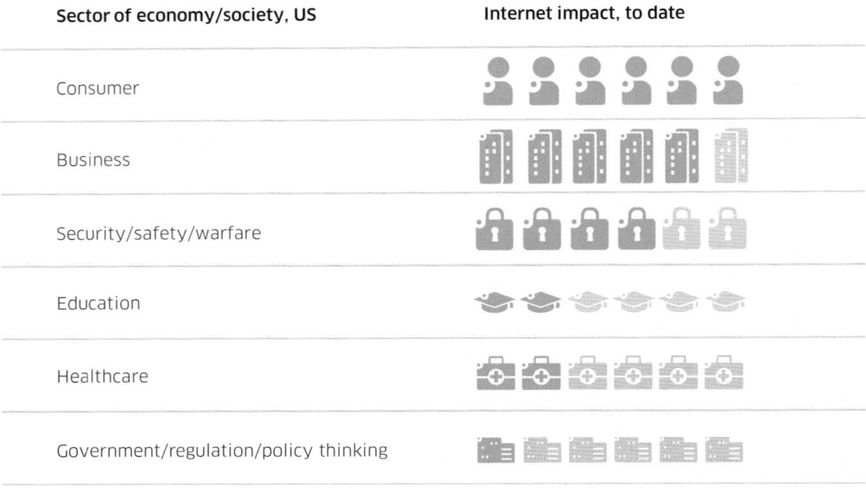

Sector of economy/society, US	Internet impact, to date
Consumer	
Business	
Security/safety/warfare	
Education	
Healthcare	
Government/regulation/policy thinking	

FIGURE 7: Relative impact of the Internet on different industries (Meeker 2015).

"seeing" the motion activity of millions of personal wearable devices.[3] Although this is a somewhat trivial example, it does underscore a massive increase in scale and extent of the mapping of the physical world. The massive digitization that will occur by the addition of billions of such new devices, each with potentially tens of sensors, coupled with the ability to mine the data to convert it to actionable information in the cloud, will make it seem as though, in 2015, we are at the equivalent of the "parchment scroll" stage of digitization of industries, compared to where many consumer services and web businesses are today (figure 7).

What is going to change to accelerate this digital transformation? If one considers the nature of these traditional businesses, they were born and grew massively in the pre-digital era, and so did not, and indeed could not, connect their physical assets to the Internet or the cloud using digital communications interfaces and infrastructure. And so many businesses have remained isolated from the digital world, or have only connected digital assets to isolated (closed) systems, and as a result have remained relatively unevolved and with incomplete intelligence or knowledge.

The key, therefore, is to find a way to connect these legacy physical infrastructures to the digital realm and that is precisely what IoT is intended to do. The Internet "thing" is actually a digital interface that is added to

3 - See https://jawbone.com/blog/napa-earthquake-effect-on-sleep.

a physical object, which allows it to measure and record data and then communicate with, and be controlled by, another system or process across a network (usually the Internet).

Alvin Toffler in *Future Shock* wrote that, "Each new machine or technique, in a sense, changes all existing machines and techniques, by permitting us to put them together into new combinations. The number of possible combinations rises exponentially as the number of new machines or techniques rises arithmetically. Indeed, each new combination may, itself, be regarded as a new super-machine" (Toffler 1970, pp. 28–29). It is this realization that underlies the current momentum around IoT, as there is an increasing awareness that unprecedented scale and efficiency will be made possible, once all the elemental systems are connected and controlled as one. Imagine being able to "see" the state of every raw material, component and finished good in any factory or warehouse and in every container on any truck, ship or plane, and the state of the route it was traversing (the up-to-date condition of the infrastructure and the current weather and congestion conditions) and the up-to-the-minute inventory in every store, as well as the payments received and monies exchanged. Add to this the ability to receive continuous feedback from every part of this chain, and it should be clear that if all this information and data can be turned into knowledge and consequent intelligent automation, we would indeed be living in a very different world – a world where a single (voice or text) command can optimize and automate an entire manufacturing chain, or process or service, and even result in the physical good being produced by a local 3D printing facility and delivered to you within five minutes – the "uber" guarantee, but for everything.

Part of the answer to this quest – turning data into knowledge – lies in the architecture and methods required to collect and exchange this data and will require the following key elements and capabilities:

- Digital applications and services that leverage network information to understand the temporal location of every digital object

- Real-time, streaming data platforms that leverage the edge cloud and allow contextualization of all people and things at scale

- End-to-end, multilayer security systems that leverage the (hyper-local) edge cloud to detect and diffuse the most advanced threats to data theft

- Operations systems and processes that enable rapid creation of services with seamless service assurance and management across global and local networks, and virtual and physical resources

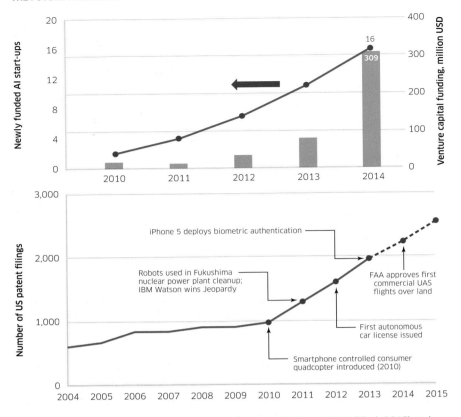

FIGURE 8: Significant increase in AI investments between 2010 and 2014 (Clark 2015) and (Trabulsi 2014), with Bell Labs Consulting analysis.

These capabilities reflect the massive rise of digitally-aware devices and represent the second aspect of the Future X Network architecture. The requirements, once again, demonstrate the need for a scalable, proximal edge cloud architecture, along with the need for ultra-high connectivity with low-latency. The last element of this vision is the ability to turn data into knowledge or intelligence.

The emergence of augmented intelligence

We turn in this section to the third element of our Future X Network architecture: the role of big data and the need for augmented intelligence. There has been much recent attention given to the topic of artificial intelligence (AI) given the significant increase in investments in AI in recent years and the development of "deep learning" algorithms that build on the principles of neural, network-based learning, by adding a layered approach to problem solving (figure 8). The popular narrative around such AI approaches is that

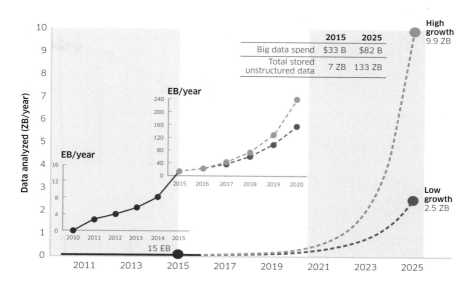

FIGURE 9: Projected increase in data analyzed in the coming decade. The low growth scenario assumes 20% of all new (IoT, UGC) video is analyzed versus 100% in the high growth case (Bell Labs Consulting).

they are intended to replace human beings in various roles, as they will evolve from an initial human-like understanding of the problem at hand, to iteratively solve more and more complex problems by learning from the data and inputs, rather than merely applying the same heuristics, or making subjective or intuitive judgments, as humans are wont to do. Despite the obvious hyperbole around "replacing human beings", it would be foolish to ignore the recent assertion by Stephen Hawking, joining a long tradition of scientists, technology leaders and thinkers, that "artificial intelligence may be humankind's greatest accomplishment, but it may be the final accomplishment" (Hawking 2014).

An alternative approach to AI's active machine training and learning is that of statistical analysis of data sets – the so-called big data approach. With the emergence of massive storage capacity in the cloud, and sophisticated approaches for parallel data analysis (for example, MapReduce, Hadoop, or Cassandra-based approaches), and the large volumes of digital data that have become accessible over the past decade or so, there has also been much attention paid to the ability to parse this data for correlations that can impart information and provide some unique perspective.

As shown in figure 9, there will be a massive increase in the volume of such data analyzed over the coming decade, in large part because of the amount

Augmented intelligence is designed to augment the intelligence and knowledge of the individual and assist in decision-making and further knowledge acquisition, not to replace the person with an artificial system.

of (video-rich) data generated by smart devices and IoT. But such analyses are typically limited in utility, as they reduce the problem to statistical correlations between events and occurrences, and cannot discern causal relationships. In the words of the noted Oxford University Internet Institute professor, Viktor Mayer-Schönberger, "big data tells you *what* happened in the past (or recent present), but not *why*, or what will happen in the future" (Mayer-Schönberger 2013).

Consequently, there is a gap between the "perfect knowledge" approach of AI and the "no knowledge" (only correlation) of big data. We believe that this can and will be filled by a new approach, which we term *augmented intelligence* (AugI).

Augmented intelligence can be thought of as taking big data sets and using inferred similarity between the elements of those data sets to create "imperfect knowledge", meaning that there is a defined probability associated with an answer, based on the computed similarity with all available data, but there is no certainty associated with any answer. Therefore, almost by definition, a human is required to interpret the applicability of the probable answers within the context of the question asked. Such an approach is designed to augment the intelligence and knowledge of the individual and assist in decision-making and further knowledge acquisition, not to replace the person with an artificial system. This new approach and the methods to achieve efficient, scalable *AugI* systems are described in chapter 10, *The future of information*.

In summary, inference-based systems have the potential to facilitate the automation of processes and foster the emergence of new communication and collaboration frameworks. They leverage context and inferred similarity to assist in the automation of interactions. They provide relevant information and aid in the discovery of new content, all of which help to disseminate knowledge and sharing between people, places and things. This is discussed in greater depth in chapter 9, *The future of communications*, and the many other uses of an AugI approach for contextual automation are itemized throughout the book.

In the preceding discussion, we have described what we believe will be the fundamental technological and architectural shifts that define networks and systems in 2020 and beyond. These form two of the essential ingredients of a technological revolution, as we have discussed. But we have yet to address any economic transformation that might be the product of these two elements, which is a requirement for it being a genuine revolution, namely, that the consequences of these technical shifts will significantly impact the social and economic lives of people.

We have alluded to the fact enterprises and industries will be massively transformed by the ability to become automated and to exist independent of physical space and infrastructure — essentially to become virtualized. This is the key economic transformation made possible by the technological shifts: the creation of a new business era characterized by enterprises with global reach and the ability to scale with worldwide demand, but which leverage and require local context for delivery of the service or good. As depicted in figure 10, this era will create a new competitive landscape characterized by:

- The emergence of a new global-local duality of new service providers with either global or local focus:
  - ¬ A set of approximately 10 global service providers that offer global connectivity, cloud and contextualized control and content (C&C) services
  - ¬ A set of approximately 100 local cloud-integrated network providers that offer domestic, hyper-local connectivity, edge cloud and contextual C&C services
  - ¬ The formation of a global-local alliance framework that connects and interworks the local and global service providers, similar to that of the airline industry
- The emergence of two new digital commodities that are key parameters of the new automation paradigm: the ability to save time and create trust by providing protection for data whenever and wherever it is generated, transported and stored

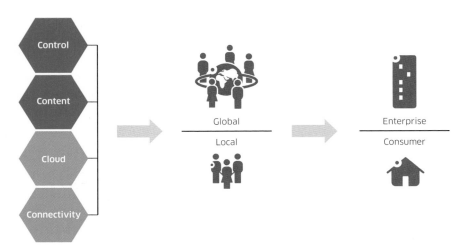

FIGURE 10: The essential competitive dimensions in 2020 and beyond. There will be four key technology dimensions (connectivity, cloud, content and control) and four key market dimensions (global, local, enterprise and consumer).

- The emergence of new virtualized enterprises, which are completely location-independent, and enjoy automated connectivity to and between any process or person or thing, with the ability to scale to meet any demand

- The emergence of new digital commodity exchanges that redefine existing businesses, by leveraging the global-local cloud infrastructure and dynamic connectivity to allow the facile exchange of digital goods or digitally-connected physical goods

- The emergence of a new innovation continuum, which we call continuous disruptive innovation, in which the technological separation between sustaining innovation (by incumbents) and disruptive innovation (by new entrants) disappears

Successful businesses in this new reality will excel in at least one of these areas but, because in many cases they are coupled, leadership in multiple dimensions will likely be required. For example, success as a virtualized enterprise may well depend on owning or better utilizing a new digital commodity exchange platform, which must be able to save time and create trust. The connectivity to this platform could be provided by the enterprise directly (using its own global network facilities) or in partnership with a network operator with global-local reach. Conversely, a network operator could also provide a digital commodity exchange platform, in addition to the connectivity layer, hosting this platform on a high-performance, cloud-

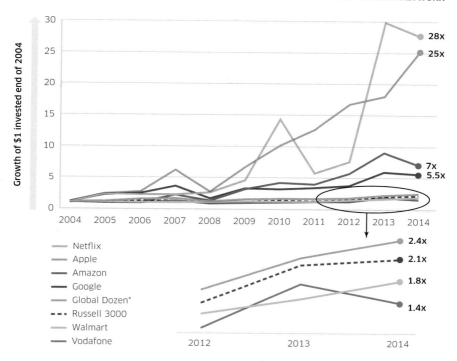

FIGURE 11: Hyper-global (web-scale) company growth significantly outperforms that of traditional multinational companies. (*Global dozen companies in the Interbrand top 20 global brands list from 2005 to 2014 are Coca-Cola, IBM, Microsoft, GE, Toyota, McDonalds, Daimler-Benz, BMW, Intel, Disney, Cisco and HP.)

integrated network infrastructure, and providing new content or control services, for example, an IoT management and contextual analytics service for different industry segments. There are myriad possible combinations that could provide a sustainable business advantage based on continuous disruptive innovation. But the ability to operate on a global scale will be a key aspect of market success, as demonstrated by the growth in value of the hyper-global company valuations over the past decade, compared to even traditional multinational companies (figure 11).

The future begins now

At the end of each chapter of this book, we review elements that are being tested or deployed in 2015 that validate the vision we have outlined for the Future X Network in 2020 and beyond. The technological foundations of NFV and SDN are already clear, but large-scale adoption and deployment have still to occur. Similarly, the edge cloud vision that is also foundational to this future is only in the earliest stages of strategic consideration by network operators, as they look to consolidate real estate assets – central offices, in particular – and complete their migration to all-IP (VoIP) networks. Similarly, webscale providers are moving in the same direction, with Amazon opening 53 distributed data centers closer to end users, of which 20 opened in the last two years alone.

By extension, the required optimization and automation of the IP and Optical metro and core networks using WAN SDN control to respond to the new dynamic edge cloud traffic model is also only now being contemplated for initial trials. Similarly, the massive scaling of the access infrastructure driven by the digitization of everything (IoT) is beginning to commence with evolution of the 3GPP LTE standards to offer improved support for IoT in Release 13. But the larger scale support for IoT will only come with the adoption of new 5G mobile standards, for which research is underway, but first standards are not expected to be ratified before 2018 at the earliest.

So, in short, the essential fabric of the cloud integrated network has yet to be completely defined, standardized and built. Consequently, the new augmented intelligence paradigm and the development of new digital commodity exchange platforms are also currently a vision of what could (and should) be, rather than a reality today. However, the majority of the technological pieces to the puzzle are already available either as advanced research prototypes or early commercial systems; what remains to be done is to put the pieces together to form the complete solution that will enable the next technological revolution.

Summary

A final thought from Marshall McLuhan: "we become what we behold, we shape our tools and our tools shape us."[4] In other words, mankind creates technology and then technology redefines mankind. This process is what underlies the phenomenon of a technological revolution. In this book we argue that we are on the verge of such a technological revolution, and it will reshape our existence around a new global-local paradigm, enabled by, and enabling, the augmented intelligent connection of everything and everyone. This revolution will be well underway in 2020 and will continue throughout the twenty-first century, which will become the century of *augmented digital life*, building on the phenomenal breakthroughs in natural sciences, mathematics and engineering in the twentieth century.

4 - Although widely attributed to McLuhan, this was never published in any work by him, but came from an article written by Father Culkin on McLuhan for the *Saturday Review* (Culkin 1967).

References

Clark, J., 2015. "I'll Be Back: The Return of Artificial Intelligence," Bloomberg Business, February 3 (http://www.bloomberg.com/news/articles/2015-02-03/i-ll-be-back-the-return-of-artificial-intelligence).

Culkin, J. M., 1967. "A schoolman's guide to Marshall McLuhan," *Saturday Review*, pp. 51-53, 71-72.

De Vasconcelos, G., 2015. "The Third Industrial Revolution - Internet, Energy And A New Financial System," interview with David Brown, *Forbes*, March 14 (http://www.forbes.com/sites /goncalodevasconcelos/2015/03/04/the-third-industrial-revolution-internet-energy-and-a-new-financial-system).

Hawking, S., 2014. "Transcendence looks at the implications of artificial intelligence – but are we taking AI seriously enough?" *The Independent*, 1 May (http://www.independent.co.uk/news/science/stephen-hawking-transcendence-looks-at-the-implications-of-artificial-intelligence--but-are-we-taking-ai-seriously-enough-9313474.html).

Mayer-Schönberger, V. and Cukier, K., 2013. *Big Data: A Revolution That Will Transform How We Live, Work, and Think*, John Murray.

McLuhan, E., Zingrone, F., eds, 1997. *Essential McLuhan*, New York: Routledge, p. 283.

Meeker, M., 2015. "Internet Trends 2015," *KPCB web site*, May 27 (http://www.kpcb.com/internet-trends).

O'Brien, S. A., 2015. "Steve Case: We're at a pivotal point in the Internet's history," *CNN Money*, 29 April (http://money.cnn.com/2015/03/14/technology/steve-case-sxsw).

Perez, C., 2009. "Technological revolutions and techno-economic paradigms," *Working Papers in Technology Governance and Economic Dynamics*, No. 20, Talinn University of Technology, Norway.

Šmihula, D., 2009. "The waves of technological innovations of the modern age and the present crisis as the end of the wave of the informational technological revolution," *Studia politica Slovaca* 1/2009, pp. 32–47.

Šmihula, D., 2011. "Long waves of technological innovations," *Studia politica Slovaca*, 2/2011 pp. 50–69.

Toffler, A., 1970. *Future Shock*. Random House.

Trabulsi, A., 2015. "Future of Artificial Intelligence," *Quid Blog*, June (http://quid.com/insights/the-future-of-artificial-intelligence/).

Webster 2015. "Technology," *Merriam-Webster Online Dictionary*, meaning 2 (http://www.merriam-webster.com/dictionary/technology).

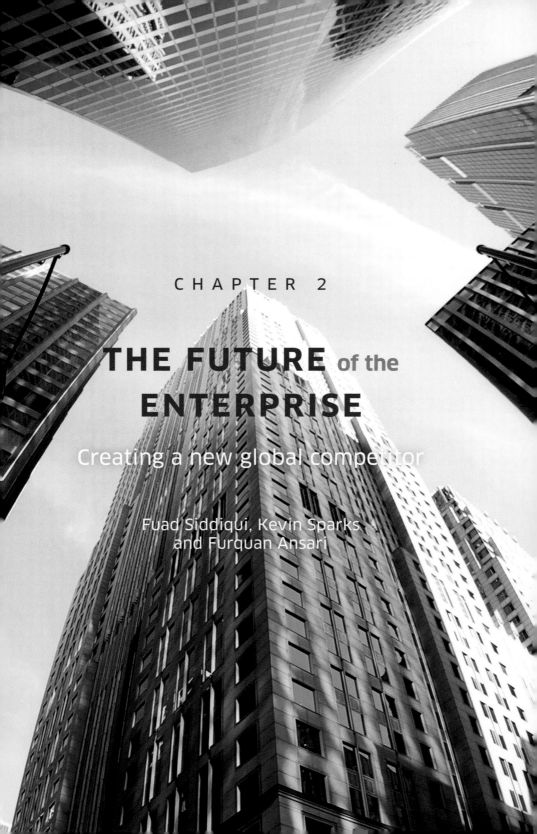

CHAPTER 2

THE FUTURE of the
ENTERPRISE

Creating a new global competitor

Fuad Siddiqui, Kevin Sparks
and Furquan Ansari

The essential vision

Enterprises are in many ways the laggards of the new digital era. Their processes and internal and external services remain rooted in the analog or physical world, and the digital systems they use are frequently disjointed and out-of-date relative to the latest web or consumer services and applications. All that is about to change. Enterprises are about to undergo a massive transformation and will become virtual in the broadest sense. The new virtual enterprise will undergo a comprehensive digital transformation to become adaptive, efficient and location-independent. It will virtualize and digitize its own assets, build on the assets of others and, to the largest extent possible, automate the dynamic utilization of those assets and their connection to its business processes, while continuously optimizing the employee and customer experience.

We see four essential pillars of this enterprise transformation:

1. *The unbounded global-local enterprise* – virtual enterprises will be able to exist beyond their physical infrastructure, or even without any locations of their own, by digitizing their assets and using software to define the connectivity between those assets and their business processes, from any location. Furthermore, using 3D printing, they will be able to recreate

simple physical goods and components wherever they are needed and, thus, optimize the entire delivery value chain.

2. *The systematically secure enterprise* – the virtual enterprise will no longer be able to rely on physical security or static demilitarized zones (DMZs) and firewalls, as every device and user/employee inside the enterprise's virtual global network becomes a potential security threat. New self-learning, adaptive security architectures will be required to police this new reality.

3. *The ubiquitous wireless enterprise* – the age of the LAN cable "tether" is over; fully untethered enterprises will optimize productivity and create a fluid and secure mobile workplace, leveraging seamless and dynamic cellular/ Wi-Fi/Wi-Gig wireless connectivity automatically controlled by a context-aware cloud.

4. *The augmented intelligent enterprise* – the virtual enterprise will be self-learning, intuitive and able to understand how best to organize its employees and connect them to the knowledge and content they need, individually or in groups, in real time, using the optimum set of technologies and systems, made available on demand. This will lead to a new era of *augmented intelligence* in the workplace that is seamlessly interactive, frictionless and personal.

Enterprises that implement this vision will pioneer new innovative "out-of-the-

Enterprises are in many ways the laggards of the new digital era. Their processes and internal and external services remain rooted in the analog or physical world, and the digital systems they use are frequently disjointed and out-of-date.

This profound shift is reflected in the growing realization by enterprises that they must radically transform to survive and thrive in an era where entire industry segments can be transformed almost overnight.

site" business models, elevate employee productivity, engage customers and cut operational costs ahead of their competition. These foundational elements will create new forms of value and redefine global and local competitiveness for the digital era.

The past and present

We are at a unique moment in time. A confluence of critical factors is occurring that will drive a profound shift in how we connect, collaborate and communicate, both at work and at home. This is the product of a true convergence of human needs and behaviors that no longer allows a distinction to be made between work-life and home-life capabilities, because they are both integral parts of a new digital life that has no geographic or temporal boundaries or constraints. All media, content, files, objects and people will always be connected and accessible – with the required resources to ensure the quality of experience. The catalyst for this new digital transformation will not come from the consumer realm, but rather from the enterprise realm, as the need for transformation in how enterprises connect, collaborate and communicate is undeniable, with the enterprise IT and user experience significantly lagging the consumer experience.

This profound shift is reflected in the growing realization by enterprises that they must radically transform to survive and thrive in an era where entire industry segments can be transformed almost overnight. For example, Airbnb, Spotify and Uber have all disrupted their respective sectors – the hotel, music and transportation industries – in a matter of a few years. Their advantage comes from an ability to achieve global scale on demand and to constantly connect to their end customers via applications on their mobile devices. Their success, and the success of an increasing number like them, is putting increased pressure on all traditional businesses to disrupt themselves. It is challenging traditional models and creating an enterprise imperative to take timely action and move with the speed, scale and security needed to win in the new digital playground.

In order to put this new, digital imperative in historical perspective, we will examine relevant past and present trends for several key aspects of enterprise evolution in the next section.

Evolution of enterprise boundaries and reach

Over the past two to three centuries, enterprises have become progressively more geographically dispersed as their products, services and customer bases have all become more global. Prior to the Industrial Revolution, the majority of businesses were local: farmers, bakers, blacksmiths, carpenters, doctors, tailors and sundry local goods purveyors served customers in their immediate area, with limited opportunity to grow.

From the mid-1700s to the mid-1800s, the Industrial Revolution disrupted existing economic structures with a wide range of new technologies, mainly in the form of machines. What was once done by hand was replaced with machines; a new manufacturing-based economy resulted. Much of the distributed and decentralized cottage industry gave way to factories and mills, forcing people to leave their homes for work. The economies of scale enabled by these machines, along with dramatic improvements in technologies for transport across long distances, facilitated the centralization of the production of goods and services.

In the 1840s, telegraphy and Morse code became commercially available giving businesses the ability to communicate quickly over long distances – accelerating decision-making and production. With the advent of the telephone in the 1870s, humans began a long wave of economic transformation, with enterprises gaining a universal, easy way to connect and extend the geographic reach of their operations and products. The decades of the late nineteenth and early twentieth centuries saw the rise of large national industries and highly organized assembly line manufacturing.

Private branch exchanges (PBXs), starting with manual operator-run switchboards and evolving to automated systems, established internal communications systems within enterprises, saving costs and further boosting productivity. Telco-hosted PBX or Centrex services integrating internal and external calling became available in the 1960s, as an early example of outsourcing enterprise communication technology needs. This period also saw the rise of large multinational corporations, on the back of more efficient global communication, as well as the expansion of global (airline) transportation and associated infrastructure.

In 1982, TCP/IP was introduced as the standard protocol for ARPANET – the first IP network in the world. The Internet age then began in the 1990s with the coalescence of personal computers (1970s), the World Wide Web (1989), and the beginnings of IP-based Internet services via modems and phone lines. Ethernet and TCP/IP technologies were initially introduced into enterprise data networks in the late 1980s for internal use, bringing ubiquity of information connectedness between employees, databases and applications. The great success of these technologies led network providers in the 1990s to adopt them in their enterprise connectivity service offerings, bringing enterprises much greater speed and efficiency to link their enterprise LANs across sites and enabling highly effective and distributed operations.

Enterprises also leveraged the new technologies and the emerging ubiquity of the Internet as an efficient means of reaching customers everywhere, in many cases in reaction to the explosion of disruptive "dot-com" Internet start-up companies in the late 1990s, the most successful of which expanded at phenomenal rates to become the web-scale giants of today.

Looking back over these industrial epochs, it is the nature of an enterprise's product – how it is produced, delivered and consumed – essentially the network that it requires, which determines how quickly the enterprise can expand. The Moore's Law-driven trend of products moving from analog to digital, from hardware to software, and from physical to virtual has resulted in products that can be mass produced or simply copied, being shipped from localized automated warehouses or just downloaded or streamed, and used anywhere via apps on mobile devices.

It is instructive to consider just how dramatic such product shifts have been in transforming various industries. The advance of digital goods and services for six different industries are shown in figure 1, where one after another the digital form of the product rapidly displaces the physical form and dominates the industry. It is the companies that lead these revolutions – whether disruptive entrants or established players disrupting from within – that benefit the most from the "lighter weight" digital production/ distribution economics and market expansion agility, taking share as the industry becomes more concentrated and more global. The above examples of upstarts from the "sharing economy" in the taxi, lodging and car rental space are good examples of this phenomenon.

These same digital shifts have transformed the way enterprises operate internally as well. The introduction of virtualization technologies and cloud services (for example, SaaS, PaaS and IaaS[2]) is transforming how and where enterprises implement internal enterprise applications and enterprise resource management (ERM) functions, as well as customer-facing services and customer relationship management (CRM) functions. This is enabling more efficient resource utilization, rapid and flexible scaling, and lower deployment risk with new options to balance operating expenses (OPEX) versus capital expenditures (CAPEX) on new services and capabilities. In fact, enterprises can now more effectively outsource large parts of their operations, retaining just the functions that are essential to sustaining their differentiation and intellectual property; extending what many companies have long done with contract manufacturing, franchising and licensing.

Not unlike the effect of digital goods discussed above, reducing the "weight" of an enterprise's physical asset base can also make it more agile, able to more rapidly expand its operations in both scope and geography. Never before has there been such a rich set of technology tools available for enterprises to leverage and expand their boundaries, nor such an imperative to get ahead in the adoption of these tools in an increasingly "winner-take-all" competitive environment.

Evolution of enterprise security

Businesses have always needed to secure their assets. Prior to the Industrial Revolution, security mechanisms and solutions (for example, fences, bars, locks, defensive weapons, guards and dogs) were needed to protect against theft of material, tools, finished products and documents; similar safeguards were needed against theft of money or currency as well. Because the assets that needed to be protected were physical, the mechanisms to secure them were also physical.

From the mid-1700s to the mid-1800s, the Industrial Revolution transformed the economy with machines, which were also physical assets. Thus, security

2 - Software as a Service, Platform as a Service and Infrastructure as a Service.

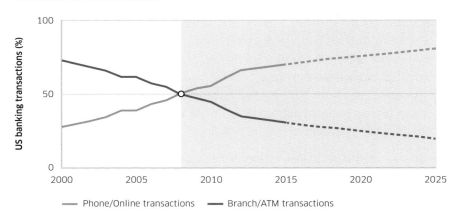

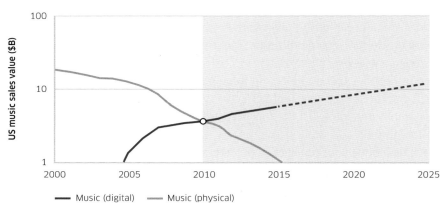

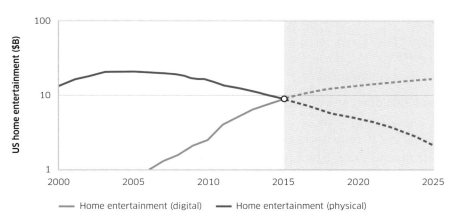

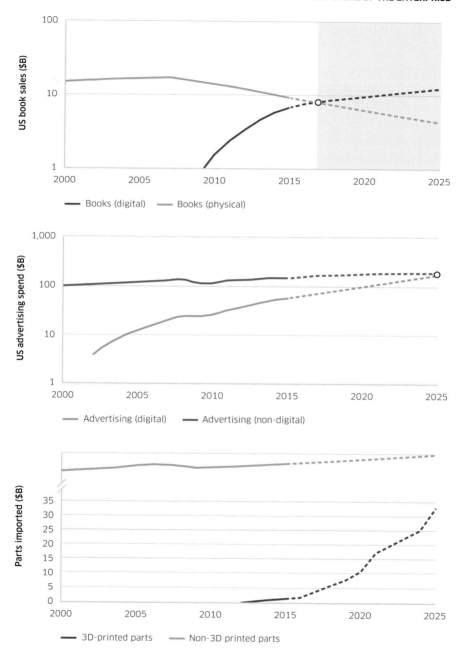

FIGURE 1: Rapid market transformations from digital goods and virtual delivery.[1]

1 - Charts were taken from an analysis done by Bell Labs Consulting based on data models and drawing from multiple data sources (Bell 2009, Cryan 2012, DEG 2011, Dixon 2015, Federal Reserve 2013, Gemalto 2009, Greenfield 2013, IAB 2011, Neely 1977, PNC 2012, Resnikoff 2014, Richter 2013, Statista 2015, Thompson 2014, US Census 2015, Vicari 2015).

mechanisms to protect the new enterprises continued to be physical solutions, just at a larger scale.

With the advent of the telephone in the 1870s, another wave of economic transformation was set in motion based on connection of distant physical locations and assets. However, this also led to an initial loss of privacy and security. Telephony allowed enterprises, suppliers and customers to talk to each other remotely and frequently – compressing business cycle times and accelerating the economy. The telephone, however, was initially highly insecure with no support for privacy. Because switching technology took time to evolve, it was necessary in the beginning to have switchboard operators involved in a call – making it possible for them to not only know who was calling whom but also to listen in.

The introduction of automated PBXs within the enterprise in the 1960s allowed for more secure communication because circuits were carried over private networks that required no operator intervention to route. The introduction of digital telephony in the 1980s allowed enterprise communications to evolve to more feature-rich and efficient digital PBXs. Along with digital telephony, enterprises also embraced PCs, which required data connectivity. However, the circuit-based systems were inefficient and expensive to operate due to the need for reserving dedicated paths and inability to leverage unused capacity – leading to the adoption of packet switching. By the turn of the millennium, the dominant data protocol in the enterprise became IP over Ethernet, supporting PCs and increasing voice over IP-PBXs. This opened up more efficient methods and new ways of doing business. With the growth of IP/Ethernet in the enterprise LAN and the rise of the Internet and associated web services, IP technologies were also adopted and adapted to provide secure communication channels between locations via virtual private network (VPN) services (using IP/MPLS and related L2 and L3 VPNs).

The global IP connectivity and ubiquity that network operators and enterprises adopted in order to enhance the cost/performance and reach of their networks also opened the door for a new type of IP network-based crime – cybercrime. The goals of cybercriminals are many and varied, but theft is still a major goal. With computerization of the enterprise, there are now high-value virtual assets that include intellectual property and customer information, such as credit card numbers, email addresses and social security numbers. Destruction is another major goal. But, now destruction comes in the form of denial of service (DoS) attacks to take down a system or the erasure of company records. Figure 2 shows the recent trends and projected rise in malware and the number of cyberattacks impacting enterprises.

To protect themselves against these attacks, today's enterprises created what is called a DMZ. The DMZ is a physical or logical sub-network that buffers the company's internal network from an external untrusted network, such as the Internet. It is also where the company exposes their external-facing services, including web services, email services and communications services. The DMZ is implemented using firewalls, and in communications systems, firewalls are available as part of session border controllers.

The advent of mobile computing – first the laptop, then smartphones and tablets – enabled employee telecommuting, field work and business travel. It became necessary to provide access to the enterprise network from outside the office, initially through dial-up over relatively secure telephone circuits. But the need for increased capacity, flexibility, and cost efficiency drove the adoption of shared packet networking infrastructure – specifically the Internet in the case of remote user connectivity – where connections could be efficiently isolated "logically" rather than physically. This need to protect remote employee connections, as well as low-cost interconnections for smaller branch locations, gave birth

The global IP connectivity and ubiquity that network operators and enterprises adopted in order to enhance the cost/performance and reach of their networks, also opened the door for a new type of IP network-based crime – cybercrime.

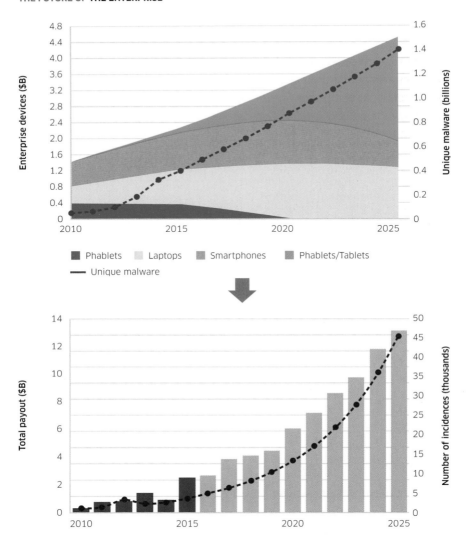

FIGURE 2: (a) Compounding threats, attack points, and (b) economic impact to enterprises[3]

to overlay virtual private network (VPN) connectivity, enabling secure IP connections from the enterprise to any device, anywhere.

These remote-access VPNs "tunnel" traffic over the public network to VPN gateways that typically reside at the enterprise DMZs. These tunnels are terminated at the gateways before the data traffic gets forwarded to

3 - Charts created with analysis done by Bell Labs Consulting using data from multiple sources
(McAfee 2015, Pyramid 2015a-b, Smith 2014, Verizon 2015).

networks within the enterprise. Such VPNs have become commonplace, with IPSec being one of the most popular technologies employed, but other VPN protocols are in use, including SSL/TLS, OpenVPN, DTLS and SSH. These tunneling protocols provide encryption to ensure privacy and protect against theft of information through methods such as deep packet inspection. VPN services include user authentication using some combination of passwords, two-factor authentication, certificates and/or biometrics. VPNs also ensure the integrity of the data from tampering. With workers connecting to the enterprise using cellular-based devices, such as smartphones and tablets, mobile VPNs have become more common. In the future, these mobile VPNs will allow users to roam across cellular and Wi-Fi networks, seamlessly changing IP addresses without service disruption or user authentication requests.

Enterprise architectures will continue to evolve, taking advantage of new technologies such as the cloud and the Internet of Things (IoT). Enterprise business structures will also evolve, moving to more distributed and decentralized workforces with large or even predominantly contractor populations. There will also be a shift back to cottage-industry-style production, as many smaller enterprises leverage digital manufacturing technologies, such as 3D printing. These major changes in technologies and business structures represent a turning point where existing security solutions will not be adequate in capability or scale. This will require innovative approaches and sophisticated data-driven solutions to successfully secure the enterprise of the future against a challenging new threat landscape, as described later in this chapter.

Evolution of the enterprise LAN

The employees, customers and suppliers of enterprises have always needed to exchange information, between themselves and with the enterprise, about products, orders, ideas, projects and schedules, and to maintain up-to-date contracts, budgets, payments/receipts and inventory. Prior to the Industrial Revolution, this was achieved mostly orally and using pen and paper for documentation and letters, delivered by the postal system. Written transactions had to occur between fixed addresses and the process was slow.

The advent of the telephone in the 1870s dramatically reduced the time required for oral communication and, at least partially, eliminated the need to travel for meetings. This accelerated the overall end-to-end business process. Most enterprises were still paper based, and thus relied on the post, although the typewriter quickly replaced the pen, starting in the mid-1800s. With the invention and introduction of the mainframe computer in the 1950s – accelerated by the mini in the 1960–70s and the personal computer in the 1980–90s – the recording and maintenance of contracts, budgets, payments/ receipts and inventory became digital and thus more efficient and accurate. Again, this evolution shortened the overall business process, but enterprises were still highly reliant on paper and post for documented transactions, a reliance that did not begin to wane until the introduction of data networking. Indeed, even today many businesses are still heavily paper based, relying on couriers and the post for critical documentation. Yet, even when fully digitized, enterprise information transmission and access still required employees to use physically tethered terminals – until the invention of wireless technologies.

The specification and adoption of TCP/IP in 1982 facilitated the start of the Internet Age in the 1990s and provided a key building block of the modern, all-wireless enterprise. Widespread, commercial mobile phone service over licensed spectrum also began in the 1980s, making wireless communication and eventually wireless broadband possible for businesses. In parallel, products to support wireless LAN connections to enterprise systems started coming to market in the late 1990s. The first version of the Wi-Fi 802.11 protocol was released in 1997, providing up to 2 Mb/s speeds. Today, Wi-Fi access points can offer up to 1,300 Mb/s speeds using multiple input, multiple output (MIMO) antennas and frequency bands. The primary unlicensed spectrum frequencies used by Wi-Fi are 2.4 GHz and 5 GHz.

The broad adoption and integration of Wi-Fi and wireless broadband technologies into smartphones, tablets, laptops and other custom devices has brought a greater degree of mobility to the enterprise. For enterprise employees, wireless devices along with VPNs have become indispensable for business travel, field work, and even basic connections within the office, as reflected in the growth of wireless device usage (both adoption of ultramobile tablets and small notebooks, as well as in overall wireless data usage) in the enterprise over the past five years (figure 3).

Wireless LAN connectivity saves on in-building wiring installation and maintenance, and allows extension of the workspace and the work environ- ment to many public and private locations. It can also make it easy for office visitors to connect to the enterprise LAN. From figure 4, it is clear that the shift of enterprise traffic to Wi-Fi and cellular is accelerating, and that the transformation of the enterprise from an all-wired system to an all-wireless

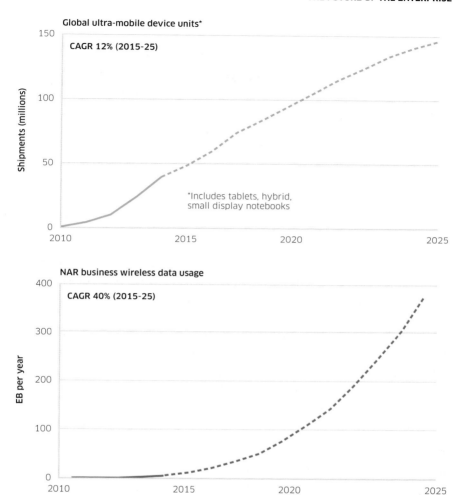

FIGURE 3: Escalating enterprise use of mobile

one is underway – although, as argued in chapter 8, *The future of the enterprise LAN*, there is still a long way to go, with needed shifts in spectrum use, wireless architectures and service provider business models yet to happen. Mobile enterprise device adoption is expected to grow linearly, with corresponding use of wireless data to grow exponentially. By 2025, virtually all access will be via wireless – using Wi-Fi, Wi-Gig and 4G/5G – as discussed in depth in chapter 8.

Mobility is also beginning to change enterprise customer behavior in terms of the applications they use. A recent Internet trends report (Meeker 2015) shows

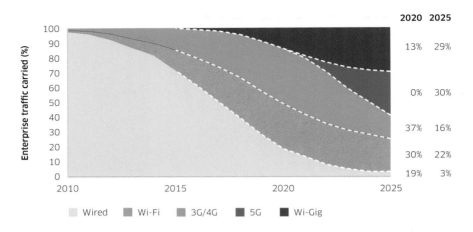

FIGURE 4: Shift of enterprise traffic from wired to wireless access[4]

that in China the very popular OTT messaging app, WeChat, is not used just for pure communications. It is also pioneering social commerce and conversational commerce by partnering with enterprises to make it easy for its subscribers to integrate images/links of products into their chat sessions and to purchase the product by enabling the subscriber to enter credit card information inside the app.

Given the expectation for mobile commerce and mobile business to rise significantly, it is becoming critical for customer relationship management (CRM) to evolve accordingly. As shown in figure 5, businesses are increasingly interacting with their customers through their mobile devices, going beyond traditional means of voice calls and emails to using apps, text messaging, and other emerging communications capabilities. This requires tighter integration between the mobile communications services used by the consumer and the enterprise communications system.

Evolution of enterprise business processes

Enterprises have slowly but surely evolved their technology stacks and policies in order to improve efficiency and productivity, as well as enable security, beginning with mainframes and terminals in the 1960s, corporate controlled PCs with standard IT-managed application builds in the 1990s, to the bring-your-own-device or BYOD revolt by employees bringing their devices and social networking habits into the workplace in the 2000s. We are now at a point where many of the advances are being driven by services

4 - Graph created with analysis done by Bell Labs Consulting using data from multiple sources including Korotky 2013, Dell'Oro 2015, Cisco 2008-2013

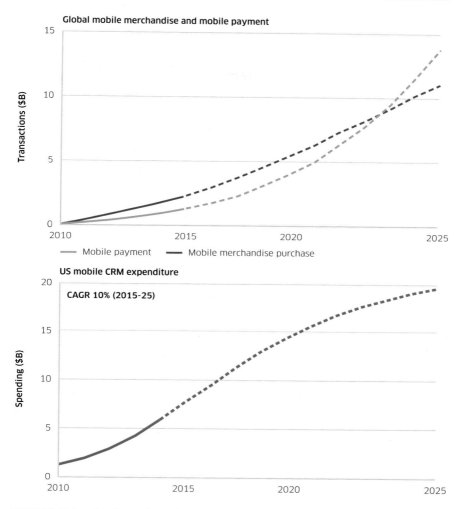

FIGURE 5: Enterprises increasingly dependent on mobile to connect with customers[5]

and applications in the consumer marketplace and a desire on behalf of the consumer/employee to replicate similar experiences within the enterprise, as schematically illustrated in figure 6. Historically, technology adoption has been led by commercial, governmental or academic needs; for example, telephony, email, and Internet access were all first adopted by one or more of these entities in order to drive greater productivity in collaboration and information exchange (telephony for governments and enterprises; the WWW and email for academia). Adoption in the consumer space occurred as employees brought

5 - Charts created using data from multiple sources including Johnson and Plummer 2013 and Gartner 2013.

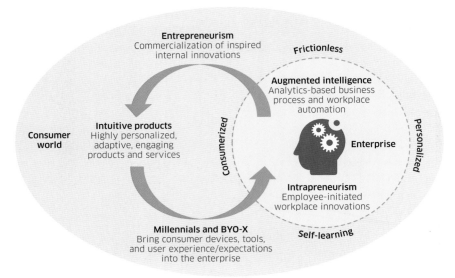

FIGURE 6: Intuitive enterprises dynamically interact and intelligently adapt with the consumer world

these with them, wanting to have similar convenience and capabilities at home. However, over the past decade or more, the rise of the mobile device ecosystem has effectively reversed this trend. As discussed in chapter 8, *The future of the enterprise LAN*, enterprises have, for a number of reasons, been slow to fully integrate mobile devices into their operations or fully exploit the potential of wireless access within the enterprise, with the result that consumer behavior is now driving the evolution of the enterprise.

The unprecedented global adoption of smart mobile devices and the rise of the cloud and big data analytics have already had a dramatic impact on the enterprise landscape and will continue to increase the pace of change. In just the past few years, many enterprises have achieved gains in operational efficiencies and dramatically improved productivity, reduced costs and service deployment timeframes. Web-scale enterprises have leveraged the new cloud technologies, achieving productivity gains of 10x–100x (Puppet 2015), reducing operational costs by 5x–20x and shortening the time to deploy applications and services from months to hours or even minutes. These gains are not just due to technological innovation, but also organizational and cultural changes within the enterprise. Key to this evolution has been the metamorphosis of IT from a rigid and siloed organization into a dynamic, responsive, cross-organizational entity that is strategically aligned with the business it supports.

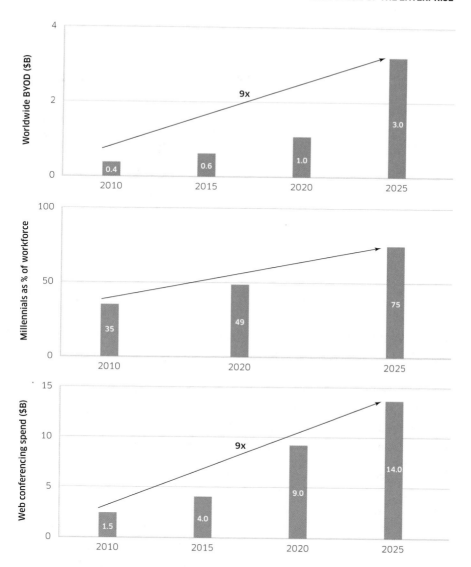

FIGURE 7: Shifts impacting the consumerization and personalization of the enterprise[6]

The pace of change is also increasingly driven by a new generation – the millennials. Having grown up in a digital world, their role will continue to be one of bringing new tools, technologies and expectations into the enterprise, as depicted in figure 6. It has been the increasing numbers of millennials joining the workforce, which has driven the widespread adoption of BYOD and Internet-based social networking and collaboration (figure 7). They are

6 - Charts created using data from multiple sources (Frost & Sullivan 2008, Trotter 2014, U.S. Dept. of Labor 2013).

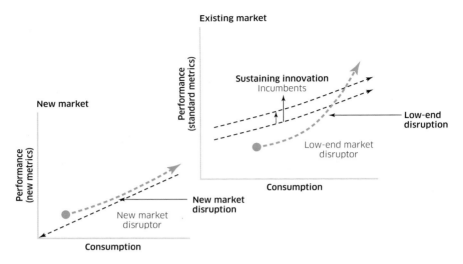

FIGURE 8: Illustration of the difference between sustaining and disruptive innovation (Christensen 2003).

changing expectations and introducing changes, which are leading to the rapid consumerization and personalization of the enterprise.

The future

Clayton Christensen's theory of *disruptive innovation* describes the process by which a product or service initially takes root in simpler, "good-enough" applications at the bottom of a market and then relentlessly moves up-market, eventually displacing established players, who are focused on sustaining innovation for their increasingly high-end customers (figure 8). The end result is disruptive innovation, as the entire market changes over and new players replace old. Essentially, a new value paradigm evolves from the lower end of the market, which was typically underserved by the incumbents, due to their different focus.

This is an increasingly accepted view of our present reality, with multiple confirming cases recently, including the disruption of the hotel and rental business by Airbnb, the taxi industry by Uber, the customer relationship industry by Salesforce and many more. Looking to the future, disruptive innovation will further accelerate the widespread adoption of cloud, virtualization technologies, mobile smart devices and IoT. It is therefore imperative for incumbent enterprises to understand the implications of this profound digital market shift, and be prepared to harness these innovations and create similar technological and business model disruptive innovations of their own.

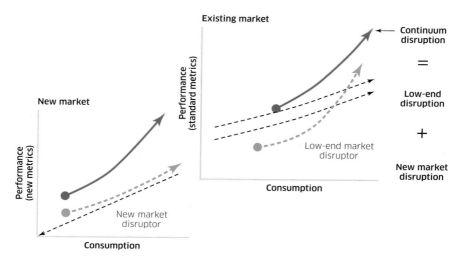

FIGURE 9: Disruptive Innovation reimagined for the new cloud and mobile era.

We believe that the future *virtual enterprise* will become part of a "continuum of disruptive innovation", where both incumbents and new entrants are both capable of similar levels of disruptive innovation by leveraging the same tools and technologies and exploring new business models, as depicted in figure 9. This has previously been unlikely due to the manifest difference in systems and capabilities employed by incumbents and new entrants, but that difference has, in effect, been removed by the adoption of a common set of technologies and enabling devices and elements. The only remaining barrier is the enterprise's traditional, organizational "mindset" and whether or not it can fully embrace this new mode of operation.

In the following section we describe the key operational, technological and architectural shifts and essential capabilities that will define the successful future virtual enterprise.

The unbounded global-local enterprise

Public cloud services offered by companies, such as Amazon Web Services (AWS), have freed companies, such as Netflix, Airbnb and Uber, from the need to own servers and networks. Not only do they avoid the capital costs associated with deploying the equipment and housing them in facilities, they are also able to rapidly scale up or scale down to meet changing consumer demand, wherever it comes from in the world. Similarly, enterprises that embrace continuous, disruptive innovation will be freed in how/when/where they locate their employees, run their operations and provide their goods and services to their end customers, by leveraging and adapting the same set of technologies:

Virtualization and containerization – As key foundational technologies, virtualization and containerization of enterprise applications and IT software functions allow for flexible placement and scaling anywhere within the enterprise's own, or third-party, data centers. This allows services to be delivered when and where needed, instantiating them on common pools of general purpose commodity servers and shared storage – a well-proven economic approach. The same benefits apply to customer-facing service and support functions. The ability to quickly scale up capacity and scale out geographically will ensure a quality user experience during rapid expansion.

Software-defined networking (SDN) – A key architectural concept that physically decouples the network control function from the network forwarding function and enables policy-driven, dynamic programmability of the abstracted, underlying infrastructure (for example, routers or Ethernet switches). By programmatically defining and controlling the connectivity, quality of service (QoS) and associated security rules (see following section) between all of the enterprise's virtualized functions and endpoints, it is possible to create enterprise processes as dynamically orchestrated service chains of such virtualized functions – independent of whether these virtual functions are located inside or outside the enterprise premises. Further, it provides the ability to instantaneously adapt to changing demands along with automatic application and/or modification of requisite policy controls. Thus, SDN works hand-in-hand with software virtualization to make the integrated enterprise infrastructure elastic, dynamic and secure.

Data centers are at the heart of enterprise IT, hosting all critical data, applications and intelligent operational and business "logic" functions. It is, therefore, vital that they are optimized in order for the enterprise to function smoothly, efficiently and cost effectively. SDN will enable data center resource utilization to be optimized by appropriate placement and interconnectivity of workloads as they are dynamically created by the SDN orchestration layer. An internal study by Bell Labs shows that optimal traffic forwarding between virtual machines leads to substantial improvement – in the order of 150% – in the access link and server capacity utilization (figure 10). It also reduces operational costs by more than 50% compared to legacy technologies. In addition, SDN allows bursting on non-mission-critical workloads, transferring compute and store processes to low-cost, data center

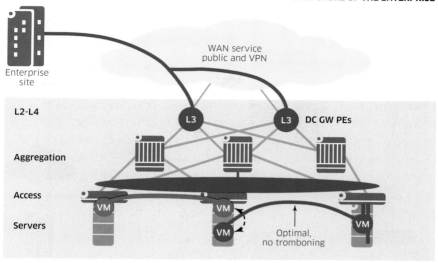

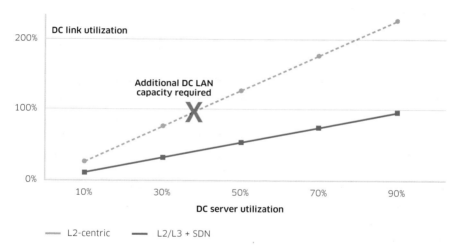

FIGURE 10: Benefits of SDN within data centers

facilities and services, such as those offered by public cloud providers. This allows a further degree of enterprise process optimization.

Software-defined VPNs – Use of SDN within the enterprise will extend well beyond the data center and into the WAN and LAN. As described in chapter 8, *The future of the enterprise LAN*, software-defined VPNs (SD-VPNs) represent a new breed of VPNs. They use SDN policy controls to create secure, software-based tunnels overlaid on any IP network to provide connectivity between locations. SD-VPNs can connect all enterprise sites and locations of any size to create truly secure, massively connected

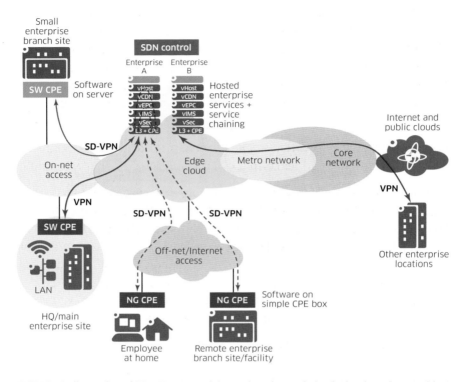

FIGURE 11: Illustration of SD-VPN connectivity options (example is of a local service provider hosting in edge cloud)

digital enterprises. Not only will this allow any enterprise site to be connected – headquarters, regional branch offices or local/home office – but also any warehouse, retail location, transportation or service fleet, or individual employees, wherever they are. As we enter the era of IoT, any asset can be connected as well, independent of location. As illustrated in figure 11, the SD-VPN CPE functionality can be instantiated on any server located on the enterprise premises. It allows any site to be connected and provides the requisite network attachment, traffic management and secure connectivity. Alternatively, if a server is not available, the software can be installed on a "server in a box" – a next-generation CPE (NG CPE) and shipped to the enterprise where it is self-installed and automatically provisioned and managed.

It is also possible to deploy SD-VPN software on local data center infrastructure for larger enterprise sites (for example, HQs, DCs) that have dedicated connectivity to the cloud (for example, over a service-provider-owned and managed access network). This configuration only requires a simple layer two VPN (L2 VPN) at the enterprise site, with layer three and higher CPE functionality (L3+CPE) hosted on a server in the cloud. This approach further simplifies the solution by removing the need to deploy software within the enterprise IT domain, and, thus, any need for local server availability and accessibility at a site. There is also no need to ship and manage NG CPE.

In all three cases, the SD-VPN instance can connect to additional services hosted in the service provider cloud and service-chained under SDN control.

This approach was previously impractical because the creation of VPNs required enterprise VPN routers, WAN accelerators and firewalls at each site. This not only made the solution CAPEX intensive, but each router had to be installed, tracked, configured and managed, resulting in additional operational expense. Furthermore, the hardware had to be periodically upgraded, as the enterprise service needed changes. The resulting excess CAPEX and OPEX costs effectively limited the number of locations that could be economically served, as well as the speed at which enterprises could adapt and expand.

With the shift from dedicated hardware to virtual software instances, it is now possible to create a VPN instance on any server at any site, or in any local cloud or data center location, with vastly improved economics (figure 12). This will make possible "virtual enterprises" that do not have any formal, owned locations or facilities but are defined by the set of SD-VPNs that connect their virtual offices and facilities.

Virtual Services Outsourcing – Just as enterprises utilize hosted enterprise applications, they will increasingly look to third parties (web service providers, IT system providers or communications service providers) to provide their essential communications services, content hosting, ERM and security systems and services. Increasingly, these services will be offered as cloud-based services, either as a platform (PaaS) on which enterprises can build their own services, or as a complete software system (SaaS). These services can either be hosted within the enterprise cloud, or on a third-party cloud, from a virtualized multi-tenant cloud infrastructure. Network operators, content distribution providers or cloud providers with local presence will provide localized edge clouds. They will bring performance-optimized, virtualized compute-and-store capabilities very close to enterprise users. This development will enable even latency-sensitive and high-throughput applications to be efficiently hosted in the cloud. Whole

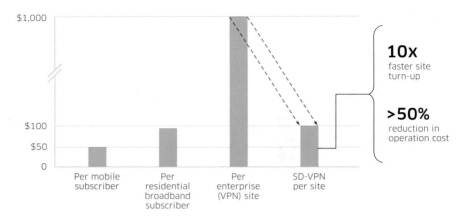

FIGURE 12: Relative economics of SD-VPNs compared to traditional hardware-based VPNs (Bell Labs Consulting)

new classes of cloud applications will emerge to bring highly interactive capabilities, for instance, using augmented or virtual reality to transform customer information and entertainment services or employee productivity and collaboration. See chapter 5, *The future of the cloud*, and chapter 9, *The future of communications*, for a more in-depth perspective on the evolution of edge clouds.

The high-performance edge cloud will also allow enterprises to realize additional value by leveraging innovative network function virtualization (NFV) and SDN-enabled services, such as domain-specific, virtual network slices, on-demand consumption of network bandwidth, dynamic, network service chains, policy-driven traffic differentiation and enhanced usage-based control. As an example, consider an automobile enterprise that uses connected fleet services and IoT technology to maximize delivery or repair efficiency. To optimize its cost structure, the enterprise leverages a network slice of a mobile packet core (virtual EPC function), which is instantiated in an edge cloud provided by a network service provider (figure 13). An internal Bell Labs study of this example showed a 30% revenue increase due to the enhanced control and differentiated services.

This delivery of new services from an edge cloud represents a new synergistic business model between network service providers or edge cloud providers and enterprises. In figure 14, we estimate the revenue potential associated with providing these edge cloud hosted services. For the edge cloud provider it represents an opportunity to nearly double the revenues associated with

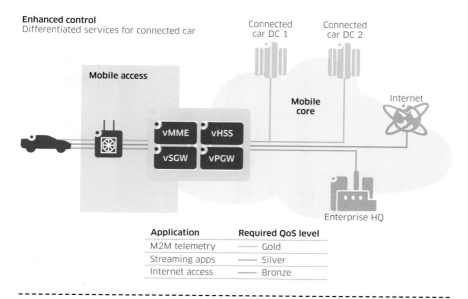

Application	Required QoS level
M2M telemetry	Gold
Streaming apps	Silver
Internet access	Bronze

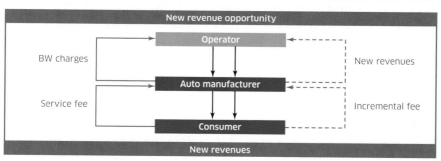

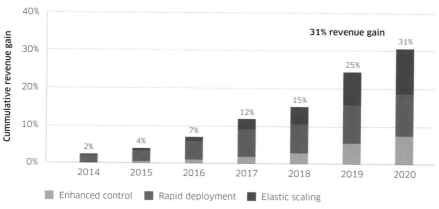

FIGURE 13: Illustration of how network slicing of a mobile packet core can generate new services revenue

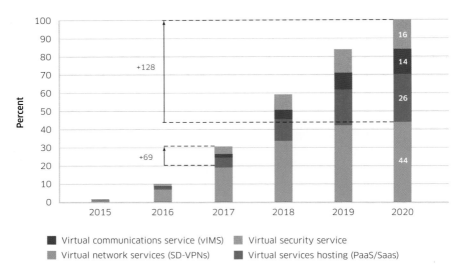

FIGURE 14: Value-added service revenue for an edge cloud provider

the new SD-VPN services, while for an ICT services enterprise it depicts a potential opportunity to add a new revenue stream to its portfolio by using over-the-top SD-VPNs and deliver these value-added services to others.

For a non-ICT enterprise that simply consumes these services, there is also tremendous gain from operational cost savings and the speed and scale of growth that can be realized by utilizing such edge cloud services. Outsourcing these virtual functions frees up capital, increases flexibility, scalability and reach – and SD-VPN connectivity makes them easy, automated and secure. In this way enterprises, especially those with only a virtual presence, will be able to trial new products and operating models, develop those that succeed, and only in-source the most cost-sensitive and differentiated elements. This will allow rapid expansion into global markets with new services, just as web-scale companies have done over the past two decades.

Digital Commodity Exchanges – Imagining and developing new, lighter-weight forms of digital products and services, with innovative ways of creating value and monetization, is an entirely different (but complementary) dimension for enterprises to exploit. There are many examples of Internet-based businesses gaining tremendous value through software assets that are used to create a new marketplace for goods, which are owned or supplied by third parties or have zero marginal cost to produce. Uber is an example of a software platform that creates a new digital marketplace for the exchange of a good (a ride), with a consumer (the rider), and with greater efficiency (automated billing, 5-minute service guarantee) than the existing physical service, and leveraging third-party assets (cars, drivers and credit facilities). The same is true for Airbnb and many other new web services. We term this the rise of digital commodity exchanges.

Taking these principles further, enterprises will begin to drive an ecosystem of shared success, extending physical goods service delivery with 3D printing, where goods transform from digital to physical only at the point of sale or use, opening up new business models that leverage the strengths and assets of partners. As an example, the automobile industry is adopting this technology to build prototypes and miniature models of cars, as well as complex parts for assessment and testing, before going into full-scale production. In some cases, such as Bentley and Lamborghini, they are already beginning to integrate select 3D-printed parts, such as dashboards and consoles, into production models (Kessler 2015). Such applications extend the ability of enterprises to scale and compete with revamped cost structures and investments.

In order to be a leading innovator and market orchestrator in the world of digital goods, an enterprise must be able to move and adapt more rapidly than its tech-savvy rivals, create digital commodity exchanges using an internal or outsourced enterprise infrastructure and leverage a highly dynamic and adaptive operations model. However, its heavy reliance on software and virtualization opens its business to new potential threats, so a more systematic approach to security will be critical to its long-term viability.

The systematically secure enterprise

The threat trend projections shown earlier, in figure 2, clearly demonstrate the compounding threats and the associated potential economic impact on the new virtual enterprise. Security for the future enterprise will need to leverage many of the same virtualization and software-defined technologies that it is seeking to protect. In order for its business to be automated, adaptive and scalable, it will need a security approach that has the same essential characteristics, simply to deal effectively with the constantly changing constellation of mobile devices and location endpoints. This is discussed in detail in chapter 3, *The future of security*, so we will only summarize the essential considerations for the enterprise here.

The wide-ranging array of potential devices, work habits and locations obsoletes the current perimeter-based approach to security protection, as it is practically impossible to define where the organizational perimeters are and

Our vision of future enterprise security is a multi-layered, heuristics-based approach that holistically encompasses people, processes and technology.

protect the many places in which data and processes could be located. This is further complicated by the continued rise in BYOD and other such millennial-driven trends, where barriers to simplicity, functionality and agility will not be endured in the name of security. Therefore, any successful, future approach to security will not be about just controlling endpoints, but rather, protecting and enabling secure access to the requested data across both managed and unmanaged endpoints, while maintaining a consistent and unified experience across them all.

Our vision of future enterprise security is a multi-layered, heuristics-based approach that holistically encompasses people, processes and technology. Comprehensive end-to-end security will be built on "security autonomic" platforms that will be composed of a full range of analytic tools including, endpoint analytics, IT systems analytics and network security analytics, with the data from all these tools systematically processed by big data analytics in order to construct a complete, near-real-time threat and vulnerabilities picture. SDN will be the key enabling technology that provides dynamic flexibility and automatic network adaptability for rapid security defense, mitigation and remediation purposes. It is essential at this scale to secure the endpoints and their dynamic network connections with policy-driven, multi-layer protection that learns to react rapidly and effectively to each new threat as it emerges.

As highlighted in chapter 3, *The future of security*, encryption will also be a key part of an enterprise security solution. Security solutions that recognize malware based on known signatures, or using deep packet inspection (DPI), will neither scale nor be effective against the new advanced persistent threats (APTs), with their diffuse and changing attack vectors. In the post-Snowden world, enterprises will adopt end-to-end encryption and encapsulation to safeguard against unauthorized snooping. Unfortunately, cybercriminals will follow suit, also using encryption and thus appearing like a legitimate and undetectable enterprise flow or process. The solution will be more robust security detection and remediation mechanisms that rely on big data analytics and heuristics.

They will generate models of legitimate versus illicit behavior gained from analyzing large volumes of data from many different sources. This will make it harder for the next generation of threats to evade detection, because the attackers would need awareness of how the models are created from the data sets in order to create attack mechanisms that can effectively bypass security.

Even with the above protections in place, it is essential that future enterprise security will also leverage the collective learning from large pools of community data, gathered and analyzed across hundreds and/or thousands of enterprises. This is a significant departure from today's much more isolated enterprise islands that rely on aggregation of security threat knowledge as part of open source Snort-based communities or antivirus software companies.

Security will remain a continuous cat-and-mouse game. As soon as a security solution is deployed, attackers will immediately look for ways to subvert or evade it, whether it is for the purposes of hactivism, cybercrime, espionage or any other agenda they may have. It is and will remain a constant battle. It will be impossible to fully eliminate cyber-risk without also completely compromising the agility and innovation of the enterprise. Therefore, the goal of the game will be to bend the cost-curve, making the cost of launching and managing an attack more expensive for the attackers than the potential rewards.

The ubiquitous wireless enterprise

There will be a strong trend toward an increasingly virtual, all-wireless enterprise in order to support mobile-centric users, whether consumers or employees, who progressively conduct business in a location-independent way. By analogy with the rapid rise in the mobile banking transactions at the cost of physical transactions shown in figure 1, mobile apps will be increasingly common tools for the enterprise to engage and interact in a personalized contextual way both with the consumer and with employees.

The future enterprise will be built around a fully untethered infrastructure that utilizes licensed, unlicensed, and shared spectrum assets and next-generation cellular, Wi-Fi and Wi-Gig technologies. As discussed in chapter 8, *The future of the enterprise LAN*, the enterprise wireless architecture will extend beyond buildings and even campuses so that the distinction between LAN and WAN disappears. End devices will move seamlessly through cells and Wi-Fi zones coordinated by context-aware, cloud-based SDN control. The enterprise user will be unaware of the wireless access technology currently meeting their connectivity needs, as it will vary by service needs and location. Key elements of this solution include:

- The low-latency of the Long Term Evolution (LTE) air interface (<50 ms) and very low-latency promised by 5G (<5 ms) provides a high-performance channel for highly interactive, real-time applications, which, in combination with very high-capacity unlicensed and shared spectrum, will enable the enterprise to utilize a wireless LAN for all applications

- Coordinated blending of LTE physical layer and medium access control (MAC) on licensed spectrum with the analogous Wi-Fi layers operating in unlicensed spectrum will achieve a throughput that is at least ~30% better than the sum of the two separate interfaces, on average

- Coordinated integration of Wi-Gig for ultra-capacity, very short reach connections opens up many opportunities for improving the quality of experience (QoE) of wireless access through capacity optimization

With regard to the latter two points, there is mounting evidence that Wi-Fi alone cannot provide the requisite QoE required for mission-critical applications, such as communications and cloud services. Consequently, there will be a need to leverage in-building cellular technologies and cellular spectrum as a complement to Wi-Fi. One example of this is the current work on standardization of LTE Licensed Assisted Access (LAA) combining licensed LTE carriers with LTE Unlicensed (LTE-U) carriers – using LTE MAC and physical layers on the unlicensed Wi-Fi bands to achieve two to three times the performance of the Wi-Fi MAC on the same spectrum (internal Bell Labs study). Another example are initiatives to use spectrum sharing techniques to allow better utilization of spectrum previously dedicated for specific government or commercial use. Looking ahead to 5G, other bands, including mm-wave bands (above 20 GHz), are expected to be leveraged in combination to meet increasingly diverse and escalating traffic demands.

As described in chapter 8, *The future of the enterprise LAN,* today's multi-technology (LTE + Wi-Fi) and Wi-Gig access nodes will likely be deployed inside the enterprise by a new operator type – the "in-building operator" (IBO) – a new business model for the outsourcing of the LAN. IBOs will be the single point-of-contact for managing enterprise subscribers, and installing and managing in-building access. The IBO will amortize its application investments across multiple enterprise clients using SD-VPNs to virtually separate client operations from one another. It will coordinate among multiple cellular operators and IT providers to manage in-enterprise deployment, as well as roaming agreements, public Wi-Fi access for mobile employees and support for BYOD.

The augmented intelligent enterprise

Looking at the preceding discussion of the creation of the new wireless, secure and unbounded enterprise, there is a high degree of automation invoked throughout, and also per-user and per-service optimization. This requires a contextualization of the future enterprise at the individual user/employee level, which we call the creation of the augmented enterprise. The *augmented enterprise* will leverage IoT and smart devices to automatically learn, understand and adapt to constant changes in enterprise operations, supply chain, product performance in the field and the activities and interactions of employees, suppliers and customers. Contextual knowledge will be derived through continuous analytics, using information from all connected endpoints, machine or human.

This *augmented intelligence* will be used to continuously adapt the physical and digital enterprise environment to seamlessly connect employees, suppliers and consumers with the right intelligence at the right time and at the right place. Users interested in a specific topic will automatically have the context-specific background provided, including the past interactions, participants and materials – all the relevant information organized and available, ready for decision-making and action. This new era of augmented intelligence is discussed more extensively in chapter 10, *The future of information,* but in essence it takes big data and uses inference-based methods to reduce information to the smallest possible data set that contains the essential insight. If this "small" data set is presented to the user in an intuitive way that allows human intelligence to interpret it, then new knowledge and action can result.

The augmented enterprise will need to leverage the high-performance edge cloud to enable processing of massive volumes of data from devices and systems.

The augmented enterprise will need to leverage the high-performance edge cloud to enable processing of massive volumes of data from devices and systems.

This edge cloud processing will also enable new services, such as the correlation of information from large arrays of sensors in real time to guide a technician to the source of a problem – a sort of enterprise mapping function – automatically suggesting possible corrective actions. Such a service could also be used to enable users to visually identify objects and parts, access repair and maintenance information and history, and give guidance around structural building layout. Similarly, inventory could be maintained and updated by continuous analysis of the warehouse stock through video recognition of objects, barcode scanning or embedded wireless IDs and sensors. The enterprise infrastructure, including that of suppliers, could be similarly mapped as vehicles, rolling stock and containers transit along their routes. These are critical additional levels of the automation that will augment enterprise operation and efficiency.

The unbounded enterprise revisited: the unbounded business

Earlier in the chapter, we introduced the concept of the *unbounded enterprise*, where limitations imposed by physical assets and goods production are removed, facilitating rapid growth. We now return to the "unbounded" theme, but this time to consider a broader goal of unbounded business success that could result from the combination of the preceding dimensions.

Figure 15 illustrates how the concepts discussed build on one another, to create a new virtual enterprise with unbounded business scope and reach:

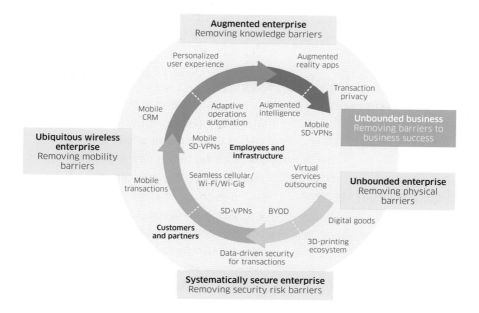

FIGURE 15: The unbounded business, virtuous cycle

The unbounded enterprise – Enterprises will remove physical barriers by limiting and virtualizing the majority of physical assets (offices, factories and warehouses), connectivity, processes, goods production and delivery – anything which limits the rate and scope of geographic expansion and new business creation

The systematically secure enterprise – Enterprises will remove security risk barriers – but without compromising security – that would otherwise limit their ability to automate and become more efficient, by implementing an adaptive multi-layered (endpoint, network and perimeter), heuristic security autonomics approach encompassing people, processes and technology

The ubiquitous wireless enterprise – Enterprises will remove all fixed connectivity barriers that would otherwise constrain their freedom to seamlessly interact and transact from any location (own facilities or any public or private space), by leveraging an all-wireless infrastructure provided by an in-building operator, using all available wireless resources and optimized for productivity

The augmented enterprise – Enterprises will remove knowledge barriers that currently prevent optimized decision-making, learning and information exchange by leveraging edge cloud infrastructure and augmented intelligence to create a new personalized, contextualized information paradigm

The future begins now

Many leading businesses are already leveraging some of the key technologies and architectural shifts in order to transform themselves toward the unbounded enterprise. Large financial institutions, industrial manufacturers and especially web-scale providers are driving operational efficiencies through automation and seamless multi-channel, multi-device user experiences. For example:

- Facebook has deployed an automation and remediation engine called "FBAR" that monitors, detects and reports system failures while automatically correcting and self-healing in order to ensure high service reliability. FBAR is reportedly managed by just two full-time administrators and monitors and manages 50% of the entire Facebook infrastructure doing the equivalent job of 200 full-time system administrators.

- Many European and North American financial institutions are now using hybrid clouds with seamless SDN-based orchestration between their private and public cloud infrastructure. Many of these institutions are developing individualized and personalized offers that can be managed and delivered at massive scale but with efficient and economical use of infrastructure.

- Boeing is using augmented-reality glasses so that assembly workers on the 747 aircraft have a reduced need to consult manuals to identify drilling locations and proper wiring connections.

Leading network equipment vendors are innovating across the entire spectrum of networking elements and functions. This includes SDN control and cloud orchestration solutions for data centers, and recent announcements of carrier-level SDN control plane solutions designed to compute optimal network paths across the WAN, based on innovative algorithms. There are many proofs of concepts and/or initial commercial offerings for virtual network functions, starting with those most amenable to virtualization and offering the greatest benefits:

- Virtual CPE are being offered in Latin American and European markets to lower the cost of deployment of new services (Telefónica 2014)

- Virtual IMS and 3GPP evolved packet core solutions are being evaluated for voice over LTE and machine-to-machine/IoT platform deployments in North America and Asia-Pacific (Alcatel-Lucent 2014, NEC 2013)

- There are multiple commercial deployments and trials of SDN within data centers and for enterprise branch network connectivity using SD-VPNs (Nuage 2014)

- There is significant momentum around the use of open-source technologies to accelerate development of common assets for SDN control, as well as cloud management and orchestration functionality (Nelson 2015, OpenDaylight 2015)

Network operators are acquiring cloud and converged networks assets, as well as new IT skills and capabilities to accelerate their transformation strategy. It is clear that the networking industry as a whole has begun a wholesale transformation toward cloud and virtualization of the network leveraging NFV and SDN, as evidenced by the increasing number of proofs-of-concept, field trials and commercial offerings beginning to emerge in this space. Some operators are also cloud-enabling their existing data centers and central offices to allow them to deliver high-performance, unified IT and virtualized network services:

- AT&T is consolidating and redistributing its internal IT and network data centers into large integrated cloud centers with a simplified architecture that unifies infrastructure platforms and network services (AT&T 2013). Such services can be provisioned, orchestrated and fulfilled in a completely automated and self-service fashion typical of cloud services. This will be followed by a consolidation of network services, capabilities and business policies at the network edge, creating a rich set of APIs that can be leveraged to seamlessly manage, manipulate and consume network services on-demand and in near-real-time.

- Deutsche Telekom (DT) is offering a cloud VPN service, an SD-VPN-like service that fully automates service provisioning and service delivery of highly secure and scalable business Internet services, such as site-to-site branch VPNs, firewall, web security and other similar applications (T-Systems 2015). The cloud service is self-serviced through an online portal where customers can select, subscribe and activate various business services. NTT is offering similar self-serviced, self-managed virtual network services to enterprise customers (Matsumoto 2014).

Summary

We are still in the infancy of this massive global shift – reimagining the enterprise in response to the new digital imperative – nonetheless, the journey has begun. By 2020, this shift will be well underway, with a complete redefinition of the boundaries of the enterprise, enjoying global reach, at speed and with the ability to offer and scale new services like leading web service companies today. Leveraging the optimized performance and global ubiquity of edge cloud facilities and localized production facilities, this *unbound enterprise* will serve its employees, suppliers and customers with securely and wirelessly-delivered augmented intelligence, thus initiating a new period in global commerce where, freed from physical limitations, business will know no bounds.

References

Alcatel-Lucent 2014. "Alcatel-Lucent delivers suite of virtualized network functions, ushering in the next phase of mobile ultra-broadband for service providers," *Alcatel-Lucent Press Release*, February 19 (https://www.alcatel-lucent.com/press/2014/alcatel-lucent-delivers-suite-virtualized-network-functions-ushering-next-phase-mobile).

AT&T 2013. "Domain 2.0 Vision White Paper," *AT&T Vision Alignment Challenge Technology Survey*, November 13 (http://www.att.com/Common/about_us/pdf/AT&T%20Domain 2.0%20Vision%20White%20Paper.pdf).

Bell, C. J., Hogarth, J. M. and Robbins, E., 2009. *U.S. Households' Access to and Use of Electronic Banking, 1989-2007*. The Federal Reserve Board.

Christensen, C. M. and Raynor, M. E., 2003. *The Innovator's Solution*, Harvard Business School Press.

Cryan, D., 2012. "US Audiences to Pay for More Online Movies in 2012 than for Physical Videos," *IHS Technology Press Release*, March 22 (https://technology.ihs.com/389509/us-audiences-to-pay-for-more-online-movies-in-2012-than-for-physical-videos).

DEG 2011. *DEG Year-End 2010 Home Entertainment Report*, January 6 (http://degonline.org/wp-content/uploads/2014/02/f_Q410.pdf).

Dell'Oro Group 2015. *Ethernet Switch Layer 2+3 Report Five-Year Forecast 2015–2019*.

Dixon, C., 2015. "SVOD extends dominance of digital video spending," January 27 (http://www.nscreenmedia.com/svod-extends-dominance-digital-video-spending).

Federal Reserve 2013. "The 2013 Federal Reserve Payments Study," *U.S. Federal Reserve Bank Services web site*, December 19, updated July 24, 2014 (https://www.frbservices.org/communications/payment_system_research.html).

Frost & Sullivan 2008. *World Web Conferencing Services Forecast*.

Gartner 2013. *Forecast: Mobile Payment, Worldwide*, 2013 Update.

Gartner 2015. *Forecast: PCs, Ultramobiles and Mobile Phones*, Worldwide, 2012-2019, 2Q15 Update.

Gemalto 2009. "Mobile Financial Services," *Slideshare* (http://www.slideshare.net/daniel_b4e/evolucin-de-servicios-financieros-mviles-en-latinoamrica-2487277).

Greenfield, J., 2013. "E-Retailers Now Accounting for Nearly Half of Book Purchases by Volume, Overtake Physical Retail," *DBW web site*, March 18 (http://www.digitalbookworld.com/2013/e-retailers-now-accounting-for-nearly-half-of-book-purchases-by-volume).

IAB 2011. *"TV Ad Spending Largely Unaffected by Growth Online,"* May (http://www.iab.net/research/industry_data_and_landscape/1675/1707493).

Johnson, P. A. and Plummer, J., 2012. "Mobile Marketing Economic Impact Study," *MMA Study*, May 9 (http://www.mmaglobal.com/files/whitepapers/MMA%20MEIS%20Report%20 FINAL%20May%209%202013.pdf).

Kessler, A., 2015. "A 3D-Printed Car, Ready for the Road," *New York Times,* January 15 (http://www.nytimes.com/2015/01/16/business/a-3-d-printed-car-ready-for-the-road.html).

Matsumoto, C., 2014. "NTT uses NFV to Power Services-on-Demand," *SDX Central*, May 29 (https://www.sdxcentral.com/articles/news/ntt-uses-nfv-power-services-demand/2014/05).

McAfee 2015. Threats Report. McAfee Labs.

Meeker, M., 2015. "Internet Trends 2015." *KPCB web site*, May 27 (http://www.kpcb.com/internet-trends).

NEC 2013. "NEC Launches World's First Virtualization Mobile Core Network Solution," *NEC Press Release*, October 22 (http://www.nec.com/en/press/201310/global_20131022_03.html).

Neely, M. C., 1997. "What Price Convenience? The ATM Surcharge Debate," *The Federal Reserve Bank of St. Louis web site*, July (https://www.stlouisfed.org/publications/regional-economist/july-1997/what-price-convenience-the-atm-surcharge-debate).

Nelson, L. E., 2015. "OpenStack Is Ready — Are You?" *Forrester web site* May 18 (https://www.openstack.org/assets/pdf-downloads/OpenStack-Is-Ready-Are-You.pdf).

Nuage Networks 2014. "Nuage Networks Brings Data Centers and Enterprise Branch Offices Closer with SDN-Powered Virtualized Network Services," *Nuage Networks Press Release*, November 12 (http://www.nuagenetworks. net/news/nuage-networks-brings-data-centers-enterprise-branch-offices-closer-sdn-powered-virtualized-network-services).

OpenDaylight 2015. "OpenDaylight's Third Open SDN Release Broadens Programmability of Intelligent Networks and Support for Virtualized and Cloud Environments," *OpenDaylight Press Release*, June 29 (https://www. opendaylight.org/news/platform-news/2015/06/opendaylight%E2%80%99s-third-open-sdn-release-broadens-programmability).

PNC 2012. "The PNC Financial Services Group Inc.," *Morgan Stanley Financials Conference*, June 12 (http://www.sec.gov/Archives/edgar/data/713676/000119312512267657/d365749dex991.htm).

Puppet Labs 2015. "The Business Case for IT Automation." *Infoworld* (http://resources.infoworld.com/ccd/assets/74320/detail).

Pyramid Research 2015a. *Global Mobile Data Forecast Pack*, June 2015, Pyramid Research Inc.

Pyramid Research 2015b. *Global Fixed Forecast Pack*, June 2015, Pyramid Research Inc.

Resnikoff, P., "The Music Industry: 1973-2013," *Digital Music News*, August 26 (http://www.digitalmusicnews.com/2014/08/26/music-industry-1973-2013).

Richter, F., "US eBook Sales to Surpass Printed Book Sales in 2017," *Statista*, June 6 (http://www.statista.com/chart/1159/ebook-sales-to-surpass-printed-book-sales-in-2017).

Smith, E., 2014. "Global Mobile PC Forecasts by Form Factor Touchscreen and Operating System 2009-2018," Strategy Analytics, October 3 (https://www. strategyanalytics.com/access-services/devices/connected-home/consumer-electronics/market-data/report-detail/global-mobile-pc-forecasts-by-form-factor-touchscreen-and-operating-system-2009-2018?Related#.VdIOrFwk_ww).

Statista 2015. "Digital advertising spending in the United States from 2012 to 2018," *Statista* (http://www.statista.com/statistics/270985/online-advertising-expenditure-in-the-united-states).

Telefónica 2014. "Telefónica and NEC showcase first use case of automated deployment over NFV Platform on Intel-based Technologies," *Telefónica Press Release* October 13 (http://pressoffice.telefonica.com/jsp/base.jsp?contenido=/jsp/notasdeprensa/notadetalle sp&selectNumReg=5&pagina=11&id=66&origen= notapres&idm=eng&pais=1&elem=21047).

Thompson, D., 2014. "Facebook and Google Own the Future of Advertising– in 2 Charts," *The Atlantic Monthly*, March 25 (http://www.theatlantic. com/business/archive/2014/03/facebook-and-google-own-the-future-of-advertising-in-2-charts/359568).

Trotter, P., 2014. "Employee-Liable Mobile (BYOD) Connections Forecast: 2014–19," *Ovum web site*, January 15 (http://www.ovum.com/research/employee-liable-mobile-byod-connections-forecast-2013-18).

T-Systems 2015. "Deutsche Telekom Launches Cloud VPN Service," *T-Systems Press Release*, March 2 (http://www.t-systems.hu/news-and-media/news/deutsche-telekom-launches-cloud-vpn-service).

US Census 2015. "Monthly Retail Trade Report," *U.S. Census Bureau web site* (http://www.census.gov/retail/index.html).

US Department of Labor 2013. *Employment Projections - 2012-2022*. Bureau of Labor Statistics, 2013.

Verizon 2015. 2015 *Data Breach Investigations Report* (http://www. verizonenterprise.com/DBIR/2015).

Vicari, A., 2015. "The 3D Printed Part Market Will Reach $7 Billion in 2025," *Lux Populi,* April (http://blog.luxresearchinc.com/blog/2014/04/the-3d-printed-part-market-will-reach-7-billion-in-2025).

CHAPTER 3

THE FUTURE
of SECURITY

Mark Clougherty
and Brian Pratt

The essential vision

Human behavioral evolution is increasingly dependent on the information that defines us as individuals, communities or enterprises. The volume of stored data per capita has been growing and will continue to grow at a staggering rate, as humans effectively evolve into "digital hoarders". Consequently, the richness of this data makes it a valuable target for those who want to exploit it for financial or political purposes, and, conversely, the loss of control of this information can have devastating financial and/or social consequences to its owner.

As a result of the value of this information, bulk, botnet-driven, distributed denial of service (DDoS) attacks, which can bring down specific web servers for political or economic reasons, are no longer the dominant security concern. Indiscriminate phishing[1] attacks aimed at stealing arbitrary sets of credentials are no longer as threatening on their own, as people become more cautious about accessing sites or opening attachments based on an email message. However, a more targeted phishing attack to gain credentials that are used as part of a subsequent targeted attack — a so-called advanced persistent threat

1 - Phishing is the attempt to acquire sensitive information such as usernames, passwords and credit card details (and sometimes, indirectly, money), often for malicious reasons, by masquerading as a trustworthy entity in an electronic communication (Wikipedia, "Phishing").

(APT) attack – is potentially devastating. These attacks target specific high-value information and, unlike the bulk attacks of the past, the perpetrators of these attacks are ever-more sophisticated, well-financed criminals, hacktivists, competitors and even governments. They will take advantage of the massively increased threat surface that is created by the ultra-connectivity of users and devices, and the future prevalence of simple, poorly-secured Internet of Things (IoT) devices connected to the network and cloud services.

Fortunately, there is hope in countering this threat through technology and cooperation. In an "us versus them" world, there are simply more of "us" (those trying to protect data) than there are of "them" (those trying to steal it). We believe that this numerical advantage will form the basis of the solution, by employing massively scalable stream analytics in the edge cloud to detect the presence of attackers, and by intensifying widespread sharing of this information among cloud and network providers, shifting the balance away from the attackers.

The increasing use of end-to-end encryption is an attempt to protect the privacy of user data in transit between two points. Perversely, this also makes it difficult for security tools to effectively monitor networks for evidence of infiltration and compromised devices. Thus, a great irony of our age is that the very mechanisms consumers and web services are using to protect our data and privacy may actually put our information at greater risk. In the future, we will need approaches that allow users to provide multiple levels of encryption and permit limited access of information by explicitly agreed-upon, trusted parties. This will form one component of a comprehensive end-to-end security architecture that encompasses endpoint (device and server) and network-based cloud security functions, which we will discuss further in this chapter.

Finally, we will consider the recent advances in quantum computing devices and consider the impact on current encryption techniques in a post-quantum computing world.

The past and present

Humans have come to depend on the ease of access to information. Everything in our lives is a digital file, or a session where information is created, exchanged, or stored. As indicated in figure 1, we are generating massive amounts of data, including images and videos, which constantly accumulate and are turning us into unintentional digital hoarders. This effect is currently most pronounced in

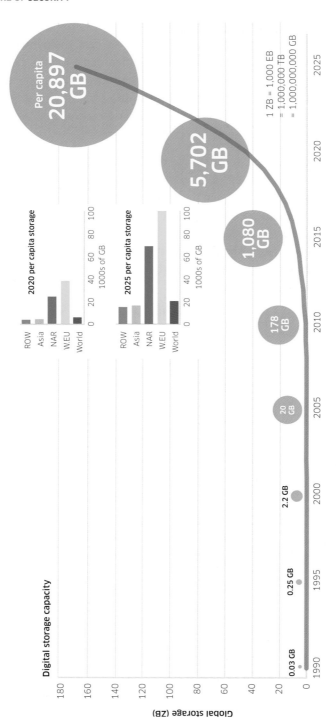

FIGURE 1: Humanity's increasing dependence on data and our tendency toward digital hoarding (EMC 2014)

Europe and North America, but per capita storage growth in emerging markets is outpacing that of mature markets and will eventually catch up.

The rise of the World Wide Web over the past 20 years, together with the ability to search it, has made the collective digital data of all humanity available to everyone with an inexpensive device and access to the Internet. Memorizing information about any subject is becoming obsolete; with the vast majority of the known information about any subject a few keystrokes, taps and gestures away. This phenomenon, while seeming to render human memory less relevant, offers the possibility to allow our brains to focus on synthesizing and analyzing the vast array of data at our fingertips and creating useful knowledge.

It is this convenience of data storage and transmission to remote storage systems, rather than locally in physical assets (human memories and physical texts), which presents the opportunity for exploitation. A very insightful observation on the motivation of cybercrime is attributed to Kevin Poulsen, also known as "Dark Dante," who was a hacking pioneer in the 1980s and 1990s and was sentenced to five years in a US prison in 1995 for cybercrime activities: "Information is secure when it costs more to get it than it's worth." With the advent of ubiquitous IP connectivity to the Internet, and an increasingly rich set of freely available tools and methods, the cost to access the data is tending toward zero, making most data insecure by this definition.

"The likely annual cost to the global economy from cybercrime is more than $400 billion. A conservative estimate would be $375 billion in losses, while the maximum could be as much as $575 billion."

(CSIS 2014, p. 2)

There continues to be cybercriminals who assemble devices into botnets for spam and DDoS attacks, and groups who focus on *ransomware* that seize control of your device and data, only relinquishing it when you pay their "fee". The FBI reported 992 consumer and enterprise victims of CryptoWall ransomware from April 2014 to June 2015 to the tune of $18 million: some as low as $200 per incident but averaging $18,145 per incident.

The most sophisticated cybercriminals have largely abandoned efforts to randomly attack average individuals, turning instead to more lucrative targets, where the biggest prizes are high net-worth individuals, and corporate and government data. Today's cybercriminals are well researched and buy services from each other to further the goals of their particular endeavor. Consequently, hacking has gone well beyond the pranks and petty crimes of the twentieth century. In 2014, a report from the McAfee sponsored Center for Strategic and International Studies (CSIS) estimated that, in 2013, "the likely annual cost to the global economy from cybercrime is more than $400 billion. A conservative estimate would be $375 billion in losses, while the maximum could be as much as $575 billion" (CSIS 2014, p. 2). Cybercrime is now a well-financed global and vertically integrated set of businesses that encompasses everything from criminals looking purely for economic gain, to "hacktivism" of political movements, to state-sponsored warfare and terrorism. This new industrial and economic reality is a function of the incredible value and potential of the ever-growing collection of data and the dependence of our lives and economies on it.

The dramatic growth in cybercrime is easy to understand by comparing statistics for a traditional crime, such as bank robbery, to that of a common cybercrime, such as identity theft, as shown in figure 2. According to data reported by the FBI, the total amount of money stolen in 2014 by identity theft amounted to 25 times that of bank robbery, with identity thieves having 11 times less risk of being apprehended. Furthermore, bank robbery is a time- and manpower-intensive activity averaging less than one crime per perpetrator per year, while identity theft is easily performed by individuals averaging more than two crimes per year. The data is clear: the risk-reward proposition of cybercrime is far better than virtually any other form of criminal activity.

The security industry has always been a continuous cycle of "cat and mouse", where a smart attacker finds a way to penetrate a security system once thought to be impenetrable. The breach is eventually discovered, the vulnerability addressed, and the attacker looks for a new way in. In

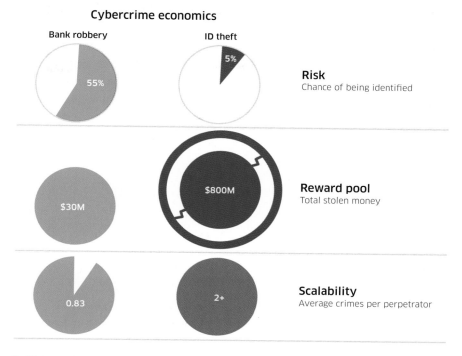

Cybercrime economics

Bank robbery

55%

ID theft

5%

Risk
Chance of being identified

$30M

$800M

Reward pool
Total stolen money

0.83

2+

Scalability
Average crimes per perpetrator

FIGURE 2: Cybercrime is much more lucrative and much less risky than robbing banks
(FBI 2014a-b, Finklea 2014, Grimes 2012, PWC 2014, US Attorneys 2014)

cybersecurity this cycle is no different, but the timescale is orders of
magnitude faster. Attacks come from a multitude of sources (sometimes
simultaneously) and with a more diverse set of tools, and more organization.
Over the years, cybercriminals have developed a structure of sorts, with
different groups and individuals focusing on different aspects of cybercrime.
There are entities that specialize in reconnaissance to identify a specific
target. Once identified, the attack is planned by identifying vulnerabilities,
not just of their network and IT infrastructure, but of all the people that
can unknowingly "help" to penetrate the target. Increasingly, for example,
cybercriminals start their reconnaissance with social media, such as LinkedIn
and Facebook, to identify employees whose permissions and credentials can
be the starting point of their cyberattack, and who may be vulnerable to an
attack that appears to be a request from a trusted colleague or organization.

Other specialized groups focus on the hacking technology itself. They sell
this capability to other groups, or offer it as freeware, either to further
their political cause or elevate their status in the *black hat* community. The

creativity of this segment of the cybercrime industry is impressive. Their general attack vector attempts to persuade targets to click on a link or install software on their device that installs a server application that connects to their command and control (C&C) server over the Internet. They need not exploit a specific vulnerability in the operating system, if they can trick users into installing something that looks legitimate and asks for administrative privileges for the installation, which is often innocently given without thinking about it. Attacks typically take the form of emails with infected links or attachments, or links in Facebook messages or LinkedIn messages. Attacks have also been known to start with planting USB memory sticks in the parking lot outside of the targeted enterprise's offices, relying on employees to pick them up and insert them into their corporate laptops at some point. In 2011, the US Department of Homeland Security discovered that 60% of those who found flash drives planted outside of government and contractor buildings plugged them right into their networked computers, and even worse, when the drives were labeled with an official corporate logo, this number jumped to 90%.

Historically, we have depended on antivirus software to protect us from such attacks. Antivirus software mainly protects devices from known attacks using only known infected files and links. However, cybercriminals rapidly evolve their attack signatures, protocols and C&C locations, as well as using *polymorphism* (malware that is massively reproduced with subtle changes in each copy) in order to circumvent antivirus protection. In 2014, an executive from Symantec (a leading supplier of antivirus software and other cybersecurity software products) told the *Wall Street Journal* that "antivirus is dead" as it was now only catching about 45% of cyberattacks. On their own, antivirus software and device-based firewalls are simply not sophisticated enough to protect users from installing malware on their device, or to monitor their Internet connection for anomalous traffic among all of the different connections that are made by the user, their browser and the software applications they run. One approach is for devices to self-monitor, but sophisticated self-monitoring is resource intensive (on both processor and memory) and is also an incomplete if not flawed security strategy because hackers have repeatedly demonstrated creative ways to defeat self-monitoring security, including antivirus software. A new approach is needed.

The industry term for this new form of stealthy, well-concealed attacks is *advanced persistent threats* (APTs). To infiltrate an enterprise, the attacker just needs to be successful at establishing a connection to one employee device, often by collaborating with an insider. The APT C&C connection is from the infected device to an IP address not known to be a risk. The installed agent then waits, often not exchanging any data for days or weeks before communicating, and when it does, the communication is sometimes encrypted, running over HTTPS, mimicking an increasing proportion of web and social network traffic, including Google, Facebook and LinkedIn. For an enterprise IT team trying to protect the network and its thousands of devices from data leakage and DDoS attacks, it is an almost impossible task to identify the intrusion, as they don't know what to look for and often are not able to verify if the data transfer is illegitimate. To make matters worse, employees increasingly want the convenience of bringing their own devices (BYOD) into the enterprise's secure network. The BYOD trend has increased the attack surface and complexity for IT security teams who cannot control the security of these devices, and with employees expecting BYOD access to the same sensitive and valuable enterprise data as enterprise-issued devices.

In 2014, an executive from Symantec... told the *Wall Street Journal* that "antivirus is dead" as it was now only catching about 45% of cyberattacks.

With the APT connection established to a compromised device, the cybercriminal waits until the device is inside the enterprise and connected to the corporate network and then begins *discovery* – scanning for interesting data and for login and password credentials to corporate systems, both in local files, as well as those potentially left in specific areas by the IT team itself. The attacker then searches for other devices to connect to from the compromised device and scans those devices, navigating from one device to another in a process which the industry terms *moving laterally*, constantly looking for valuable data at each step. It is even possible for the attacker to remotely activate the web cam on a laptop or the camera on a smartphone to observe the surroundings of the device, or activate the microphone of these devices to eavesdrop on conversations. All of this can be accomplished on devices with up-to-date antivirus software, and without the user's knowledge.

The cybersecurity industry estimates that APTs often go undetected in enterprise networks for seven months or more. More often than not, it is an outside entity that makes the enterprise aware of the dangerous connection, rather than it being detected by the enterprise IT security team, who are ill-equipped to deal with the more sophisticated cyberattacker. There is also an inherent imbalance: the cybercriminal needs to be successful just once, whereas IT security needs to be successful in defending thousands or tens of thousands of users and devices, every day.

The ultimate goal of the APT depends on the funding source. The attack may be in search of specific data, such as customer credit cards, personal information of employees, trade secrets or the movements of assets. The goal may simply be to disrupt the enterprise's operation by compromising the activities of employees or the delivery of services or goods for days or weeks by crashing servers, wiping out databases, etc. In an industrial setting, the APT can even be used to tamper with production processes; for example, make a turbine spin faster than it should, causing failure or damage resulting in inoperability, or affecting the quality of the output, rendering the enterprise's end product either useless or of far lower value.

Industry estimates put the cost of such data breaches caused by APTs at about $5 million per breach. The highly publicized breach at Target in 2013, caused by an APT delivered through the device of a Target supplier, is estimated to have cost the business over $160 million. The breach at Sony, in late 2014, is estimated to have cost $35 million. Most notably, in 2010, an APT called Stuxnet, which initially targeted Windows devices, had the ultimate goal of

80% store personal information

70% communicate in the clear

80% allow weak passwords

60% without secure interface

60% without secure software update

FIGURE 3: IoT is a new and very large attack surface, often with insufficient attention to security (HP 2014)

attacking Siemens industrial devices, and reportedly destroyed close to a fifth of Iran's nuclear centrifuges by causing them to spin out of control. And, it remains to be seen, whether extortion will be used against the 37 million registered users of ashleymadison.com, which was breached in 2015.

The attack surface has grown substantially in recent years, with cybercriminals expanding to smartphones and tablets, especially Android devices. Apple devices are not immune to malware, as there are many examples of attacks that target them. However, the more open security model of Android (for example, the ability to easily download an app from any source), has led to these devices being targeted at a rate approximately 10 times higher than iOS devices. As part of the global effort to combat cybercrime, the Alcatel-Lucent Motive Security Labs collect samples of malware from a variety of different sources, and test them to characterize their command-and-control protocols. In 2014, Android malware samples grew by over 160% and passed the threshold of one million by mid-year.

As discussed throughout this book, the next wave of Internet growth is from the IoT. The industry estimates that there will be some 50–60 billion devices connected to the Internet by 2020, and as many as 500 IoT devices per household. The attack surface is massive, with everything becoming connected, from cars, airplanes, traffic lights and parking meters, to the locks on doors, thermostats, ovens, refrigerators and lights in the house. Many of these devices are simple and inexpensive, with limited processing power and memory. A 2014 HP study of IoT devices (figure 3) found a majority of IoT devices to be designed without sufficient safeguards to protect them from APTs, leaving them vulnerable to compromise. Today, botnets of compromised IoTs are mainly used to launch DDoS attacks[2] or as platforms for spam.[3] However, as both the number of IoT devices and the amount of confidential data collected, stored or accessible by these devices increases, the threat grows. The possibilities for attacks are nearly endless: eavesdropping or watching people in their homes, taking control of a car to injure its occupants, opening the locks on a house or building, turning on an oven to start a fire; our imaginations may be the only limit. In 2015, there have been widely publicized incidents of commercial flights being hacked and controlled, cars on highways having their engines shut down remotely and on it goes.

With our dependence on digital data increasing, the frequency and severity of cyberattacks will continue to grow. Cyberspace has, in effect, become the new battlefield among corporations and nation states. As indicated in figure 4, reported incidents of cybercrime doubled roughly every 16 months between 2009 and 2014, a growth rate even higher than Moore's Law.

This observed growth rate is not surprising, as all of the driving factors are themselves growing exponentially. First, CPU power is growing at Moore's Law rates, doubling every 18 to 24 months, making it easier to launch more attacks. Second, as discussed earlier and shown in figure 1, the sheer volume of information stored online is growing exponentially, providing a much larger target (and potential payout) for attackers. Third, the rapid growth in the number of connected devices (especially poorly secured IoT devices) significantly increases the attack surface, resulting in a much larger opportunity for attackers to infiltrate the network. These factors combine to create a cybercrime growth rate that is likely to exceed Moore's Law for the foreseeable future.

The key questions of the chapter are naturally raised at this point: will the security risks of increasingly sophisticated APTs in a digital-everything, IoT world quench the Internet's evolution and end human digital development and convenience? And, given that antivirus software is now akin to a mounted cavalry trying to defend against stealth bombers, how will we combat these APTs in a way that stays ahead of the black hats? We consider these daunting questions in the following section.

2 - In 2014, a Spike botnet launched a DDoS attack from IoT devices that reached 215 Gb/s (PLXsert 2014).
3 - In January 2014, Gigaom reported that a botnet army that included fridges sent 750,000 spam emails (Gigaom 2014).

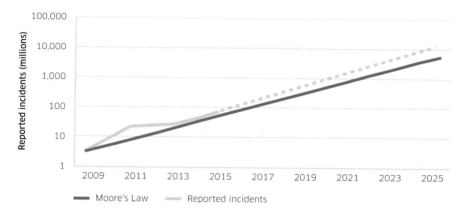

FIGURE 4: Cybercrime is growing tremendously, even faster than Moore's Law (PWC 2015)

The future

The ultimate goal of cybersecurity is not to make it impossible for criminals to access critical information, as this is a task that cannot be accomplished with any degree of certainty. Rather, it is to address Poulsen's observation: the fundamental goal of cybersecurity is to make the cost of illegally accessing sensitive information much higher than the value of the information to the attacker. Only in this way will the cyber war be effectively fought.

In the endless, cat-and-mouse game between cybercriminals and cybersecurity, there are reasons for optimism. The cybersecurity industry should be able to leverage a variety of innovations in the new global-local cloud infrastructure to keep digital lives and businesses secure.

Security everywhere

Until recently, IT security teams believed that adequate firewalls, along with intrusion detection and prevention systems, could defend their internal networks from attack and keep safe everything behind this protected perimeter. Over the last decade, the industry has gone through cycles of relying on *perimeter-based* security versus device-based or *endpoint* security. But APTs have demonstrated that we can never assume that anything behind the perimeter is safe. Like a prison, it is not enough to have high walls and a barbed wire fence: there must be checkpoints and locks within its walls and constant patrolling of the hallways and conduits. The combined use of endpoint, perimeter, and *network-based* security will be required for protecting data and applications.

Looking first at endpoint security, current antivirus software has become insufficient on the endpoint, but a new class of endpoint detection and response (EDR) solutions has been created to meet the new needs. A variety of companies have found innovative ways to put lightweight agents on any device (server, PC/laptop, tablet or smartphone). EDRs monitor key activities of the operating system, memory, processes and network interfaces, and report data to a centralized security analytics engine running in the cloud. This combination of local monitoring and centralized processing is able to optimize the detection of APTs, and reduce the attacker's available time to discover, navigate, and exfiltrate sensitive data. Without necessarily knowing what to look for, these solutions analyze activities across many devices and sift through massive amounts of data to focus on the most suspicious events. Endpoint solutions also include innovative traps and deceptions that are only triggered by attackers, and put them in a confusing environment that neutralizes their potential lateral movements within the enterprise network.

As we migrate all of our data and applications to the cloud, perimeter and other network-based security appliances will be virtualized using SDN control to be automatically associated with application virtual machines and workloads that are dynamically spun up, moved and shut down – increasingly without the need of human intervention. By 2020, nearly all security appliances and algorithms will run as virtual applications in the cloud, both in centralized clouds (for lower throughput analysis) and distributed edge clouds (for high throughput analysis). Due to the scalability of the distributed cloud, perimeter and network security solutions will be able to process massive amounts of data to search for and prevent APTs. They will be able to replicate and store all of the network activity, acting like a network digital event recorder, offering the ability for a massively scalable security analytics engine to replay the network activity back and forth from past to present, to identify anomalous activities.

By 2020, the emergence of the new global-local cloud paradigm will provide security professionals with a powerful new tool set for combating cybercrime. The flexibility and scalability offered by the distributed cloud will allow the network to rapidly and automatically adapt to threats in order to confuse, redirect, block and contain the attacker before any damage is done. Importantly, localized edge cloud resources will allow security functions to be applied at the network edge, where they can best protect the network and scale most effectively. As shown in figure 5, the network throughput of a typical edge cloud is estimated to be approximately 8,600 Gb/s by 2025; it is much more feasible to perform analytics in a distributed fashion on this volume of traffic than it would be to attempt to perform analytics at the regional level where traffic will reach over 280,000 Gb/s.

When the analytics algorithms determine that an attack may be underway, security *autonomics* may take one or more different actions. These security autonomics will take input from security appliances and analytics throughout the cloud infrastructure, determine the best course of action and use cloud and SDN controllers to rapidly adapt the network to counter the threat. Flows that are clearly malicious will be blocked immediately, while SDN may be used to transparently redirect suspicious flows to more sophisticated security functions for deeper analysis before taking action. If a particular application instance appears to be under attack, autonomics may choose to automatically relocate the instance to a quarantine zone where it is prevented from accessing sensitive information, but allowed to remain accessible by the attacker in order to facilitate study of the attack. Other actions that security autonomics may take

The distributed cloud will allow the network to rapidly and automatically adapt to threats in order to confuse, redirect, block and contain the attacker before any damage is done.

in response to attacks include scaling up a security appliance instance that is experiencing a heavy volume attack or initiating the destruction and recreation of infected application instances. Autonomics may also choose to create ephemeral decoy targets to confuse the attacker and draw attention away from real information while an appropriate defense is mounted. The global-local cloud will provide network operators with agility and scalability that is not present in today's centralized and disjointed network security infrastructure, allowing them to meet a highly dynamic threat with an equally dynamic response.

Security organization

The security teams that face the daunting task of protecting an enterprise network and its data assets in the face of rapidly accelerating attacks will have to adapt both organizationally and functionally. Most enterprises have developed a separation between their security strategy/planning and security operations organizations. Traditionally, strategy/planning is more business-aware (for example, focusing on business cases for new security tools), whereas operations manages day-to-day security focusing on technology (for example, managing firewall rules and investigating security incidents). This separation renders traditional risks assessment methods too slow and often obsolete for effectively securing highly dynamic networks. A new security approach is needed that is consistent with the virtualized, SDN-controlled networks of the very near future.

By 2020, security assessment and enforcement will have completely shifted from a predominantly manual activity into *dynamic risk management* (DRM), or continuous situational security awareness complemented by proactive and reactive capabilities. There are two key elements to DRM. The first will be new models, systems, and methods capable of capturing and measuring the:

- Dynamic state of the underlying IT and networking systems
- Threat information gathered from both internal and external sources
- Overall organization mission and business objectives

Each of these areas will provide input to a new kind of dynamic and more thorough risk assessment as a first and crucial step toward comprehensive DRM.

The second element will be dynamic risk management actions; the DRM output will be used by SDN and NFV technologies to implement advanced proactive capabilities to address and mitigate prominent and acceptable risks. Similarly, risk-aware reactive capabilities will be designed to respond to ongoing, coordinated, and simultaneous attacks in the network and IT systems. The lines between enterprise security strategy/planning and security operations will be blurred to the point of being almost indistinguishable.

In the same way that the cybercriminals have organized themselves vertically, the cybersecurity industry is also becoming more effective in specializing, researching and sharing threat intelligence. By 2020, broad-based security information sharing will be commonplace, allowing service providers, enterprises, and security specialists to share information about attacks in near-real-time. As depicted in figure 5, threat intelligence gained by one entity experiencing an attack will be quickly disseminated across the industry, allowing detection and mitigation techniques that were found to be effective by one company to be applied by the security analytics and autonomics engines used by other companies. Effective crowdsourcing of security will force cybercriminals to become even more elusive, incurring higher and higher costs as they try to outwit the collective intelligence of the global IT, network and applications security communities. This focus on threat intelligence sharing has become a cornerstone of the cybersecurity framework across industries, including, for example, the National Institute of Standards and Technology (NIST) serving critical infrastructure in the US. In 2015, we have seen the emergence of industry-specific information sharing and analysis centers (ISACs), which are the beginnings of this trend. These ISACs have been created for financial services, oil and gas, and even the automotive industry, in anticipation of the risks to connected cars as they become more mainstream.

Security of things

Smartphones and tablets offer an attack surface that has yet to be fully exploited by cybercriminals. Their potential to be harnessed into massive botnets has not yet been realized in any significant way, beyond DNS amplification DDoS attacks that have brought down DNS servers, for instance, in Spark, New Zealand, in 2014. Motive Security Labs is a leader in infection detection in mobile networks: it has observed that, although the rate of infection among smartphones is growing quickly, it remains relatively small compared to Windows-based devices (PCs and laptops). Cybercriminals continue to optimize the return of their activity, focusing on the devices that they know best and that typically contain the most valuable data. However, by 2020, we will experience large-scale smartphone botnet attacks that inflict DDoS attacks on targets such as enterprise call centers (for example, their toll-free, 1-800 customer service numbers), emergency systems (for example, 911 numbers),

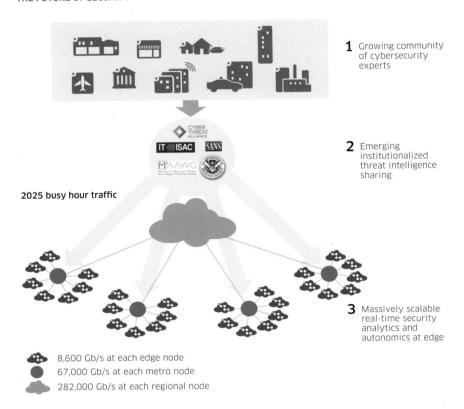

2025 busy hour traffic

1 Growing community of cybersecurity experts

2 Emerging institutionalized threat intelligence sharing

3 Massively scalable real-time security analytics and autonomics at edge

- 8,600 Gb/s at each edge node
- 67,000 Gb/s at each metro node
- 282,000 Gb/s at each regional node

FIGURE 5: Increasingly sophisticated sharing of threat intelligence will drive security analytics at the edge (M3AAWG 2015, Cyber 2015, IT-IASC 2015, Homeland 2015)

messaging systems (for example, SMS) and voice spam. Smartphones are also ideal spying devices that can be leveraged by APTs, with *already* many examples of cybercriminals infecting them with "trojanized apps" and controlling them to access their cameras, microphones, location, etc. The same approach of using security analytics and endpoint security will be used to defend against smartphone attacks.

In addition to smartphones, the attack surface will be expanded even further by the rise of IoT. IoT covers a vast array of devices with a huge range of capabilities from very simple low-cost tracking devices, to relatively sophisticated control systems, such as may be found in vehicles. As a result, it is difficult to make general statements about security in the IoT domain. There will likely be a lag in the targeting of IoT devices by cyberattackers, as they maximize their return targeting devices that they know well and where they already have large committed investments in their criminal activities. However, there is no doubt that IoT will be a heavily attacked domain by 2020. Sophisticated IoT devices may have processing power equivalent to simple personal computers, making it perfectly reasonable to expect to protect them with the same EDR mechanisms described above. But at the opposite end of the spectrum, cost and energy considerations may make it impractical for simple monitoring and reporting devices to support EDR software. The positive side of this is that the simplicity of these devices also makes them poorly suited as jump-points for cyberattackers. For these devices, security will rely heavily on the use of hardened and regularly updated platforms built to present a minimized attack surface exposing only the functionality absolutely necessary to perform their limited tasks. Perimeter security and analysis of traffic to and from these devices is also essential to ensuring their security. The local resources of the global-local cloud will provide an ideal location to implement localized security analytics for clusters of simple IoT devices. This will be the security architecture, regardless of whether IoT communication is transported over 5G wireless networks or emerging LPWA networks using LoRa, SigFox, or some other new protocol, because IP will continue to be the underlying unifying protocol and all traffic will be transported over the same cloud integrated network.

Security of identity

The evolution of identity management with the combination of location and biometrics will play a key role in authenticating users and identifying valid authorized transactions. The authentication industry relies on three mechanisms to confirm someone's identity: either using something they know that no one else knows (for example, a password), something they have (for example, a token or key) or a physical characteristic (biometrics). However, passwords are widely known to be prone to compromise, and the use of tokens can be costly and complicated to manage. This leaves biometrics as the best approach assuming that it also can reach reasonable cost points. Recent studies show that the combined use of multiple biometric factors or *multi-modal biometrics* (voice; physical characteristic such as fingerprint, iris, even DNA) and biomechanical attributes (such as handwriting, patterns of keyboard and mouse activity) can give 99.97% accuracy with very low false positives and false negatives. The use of biometrics has become more common in recent years and, by 2020, will be commonplace. There are those who express concern on the privacy implications of biometric data, and believe that any form of biometric data can be spoofed: they argue that

Privacy of Internet traffic is a hot topic in 2015, but the great irony is that the techniques being used to protect privacy may actually put our information at greater risk.

there is always the possibility to use high resolution images of a person to create fake biometric data. To address these risks, the biometrics industry is adding liveness detection capabilities, which, for example, can distinguish a fake fingerprint from an actual finger, or a photo from a live human being. The telecom industry can also add elements of location and habits to support biometric data. Combined with *liveness detection*, the adoption of industry de facto standards such as fast identity online (FIDO) 2.0 and the introduction of wearable IoT gadgets will accelerate the use of biometric authentication. By 2020, a liveness-verified heartbeat rhythm on a user's wearable device may be sufficient authentication for them to access their home, place of employment or financial assets.

Security and privacy

Privacy of Internet traffic is a hot topic in 2015, but the great irony is that the techniques being used to protect privacy may actually put our information at greater risk. What has become known as the "Snowden Effect" dramatically increased the public's awareness of the extent of government surveillance both in the US and globally, and network security and privacy is a genuine concern among Internet users. According to a recently published survey by the Centre for International Governance Innovation (CIGI),

- 62% of users are concerned about government agencies from other countries secretly monitoring their online activities
- 61% of users are concerned about police or other government agencies from their own country secretly monitoring their online activities
- Of those aware of Edward Snowden (60%), 39% have taken steps to protect their online privacy and security as a result of his revelations (CIGI 2014)

An important means of protecting online privacy is through the use of cryptographic protection of data that traverses networks. Strong encryption protects the data even if a third party were to somehow intercept it, because without access to the encryption key, it is virtually impossible for the eavesdropper to read the data. For this reason, encryption has long been used to secure exchanges of critical data, such as financial transactions, authentication credentials or sensitive enterprise or government data. However, recently public awareness of privacy and security issues has resulted in a significant increase in the use of encryption for less sensitive data as well.

While encrypted traffic by no means dominates the Internet today, it soon will. As indicated in figure 6, the fraction of North American fixed network service provider traffic being encrypted has risen from less than 3% in 2013 (at the time of the Snowden NSA surveillance revelation) to 29.1% in April 2015. Most of the top web domains are now accessed via HTTPS (encrypted HTTP), including Google (and YouTube), Facebook, BitTorrent, Twitter and Wikipedia, with more expected to follow suit as major search engines now rank encrypted sites higher than non-encrypted sites. In its April 2015 quarterly earnings statement, Netflix announced that it plans to use HTTPS to deliver streaming video to "protect member privacy" while viewing content, which will result in a significant increase in such traffic. The fraction of traffic using encryption is forecast to reach over 64.7% by 2016. The trend is expected to continue until virtually all network traffic is encrypted. Because malware seeks to hide its C&C traffic in plain sight by making it look like normal traffic, encryption of malware traffic is expected to roughly track that of general Internet traffic, with Gartner forecasting that 50% of malware will be encrypted by 2017.

While network encryption prevents eavesdroppers from snooping on end-user data, it also makes it difficult for the network monitoring techniques described above to be used effectively and can reduce the effectiveness

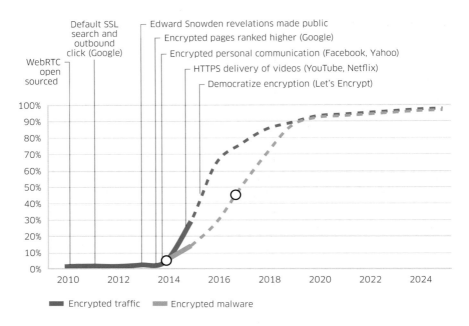

FIGURE 6: Internet traffic is increasingly encrypted; malware is expected to follow suit[4]

of security analytics solutions in the detection of APTs and other malware. Paradoxically, the actions taken to protect data from eavesdroppers may actually make it somewhat easier for criminals (cyber, civil and terror) to go undetected. This is somewhat analogous to disabling the security cameras and sensors in your home; privacy is indeed increased as no one can monitor your activities, but security is also reduced, as the entity tasked with ensuring the security of your home is blind to any activity in it. Network security analytics are not completely compromised by the use of encryption; it is still possible to observe patterns such as the number, duration and diversity of connections, all of which can be used as signatures of malware C&C traffic. Nonetheless, they are certainly rendered less effective.

The problem with current encryption solutions is that they offer an all-or-nothing approach to encryption. Traffic is either sent in the clear and available for anyone to see, or it is encrypted and opaque to the network. However, in many use cases, it may well be acceptable for a trusted third party (such as a connectivity service provider) to have limited access to the traffic headers or even the payload. This would enable the service provider to

---

4 - The timeline is based on analysis by Bell Labs Consulting and references: Kao 2011, Let's Encrypt 2015, Sandvine 2014-2015, WebRTC 2015, Winkler 2014.

provide the security monitoring and analytics functions necessary to identify and counter cyberattacks. A new type of end-to-end security model is being considered in the industry that does just this: allowing communicating endpoints to negotiate a level of security that is appropriate for the application, while allowing trusted third parties to gain some visibility into specific portions of the packets. In the future, we believe, this type of fine-grained security model will be the de facto architecture, to the benefit of all.

In this model, the client and server agree on a security context (or contexts) for the session in which a trusted third party may have access to some specific fields of the data (for example, HTTP headers). The third party is not involved in the end-to-end security association negotiation, nor can it alter the association in any way. A new security proxy function is introduced as a trusted third-party element and is provided with the third-party key(s), allowing it to access specific portions of the payload in order to perform performance optimization and security management functions, without interfering with the end-to-end nature of the encryption. When such a capability is available in the network, a simple policy that blocks all traffic that does not allow some level of access to the encrypted data (even if only to HTTP headers) could provide a significant step in countering the detrimental effects of encryption on security tools; ensuring that at least some visibility is available for all flows through the network.

Some groups object strongly to any technology that provides greater visibility to Internet traffic, even for public safety purposes. The debate and balance between privacy versus security predates the Internet by many centuries, and will continue in 2020 and beyond.

Quantum insecurity?

Finally, we look at an interesting development with impact on future security architecture: quantum computing. Today's public-key cryptosystems are only as secure as their underlying algebraic problems. For example, the widely used RSA cryptosystem uses large prime numbers as encryption keys and is immediately broken should the attacker be able to factor large numbers. As it stands today, these are hard algebraic and computational problems to solve. Algorithms to crack RSA encryption require so much processing and time (on the order of thousands of years) that it is viewed as being impractical to attempt.

With advances of algorithms and computer hardware, however, the industry rightly chose to be conservative and use large parameters for cryptosystems, despite increasing their operational cost. For example, since 2013, NIST requires use of minimum 2048-bit RSA encryption. Systems with smaller parameters were previously thought to be safe because of theoretically requiring thousands of years to be compromised, but have since shown to be more vulnerable. In 2010, a 768-bit RSA cryptosystem was cracked by a team led by former Bell Labs researcher, Arjen Lenstra. Despite hypothetical estimates that six months of processing on 100,000 workstations would be required to accomplish this task, the two-year effort only required the equivalent of almost 2,000 years of computing on a single 2.2 GHz CPU (Kleinjung 2010). In 2012, a team of researchers at Fujitsu and Japan's National Institute of Information and Communications Technology (NICT) cracked a 923-bit pairing-based cryptogram. Researchers who worked with pair-based cryptography had in the past expressed confidence that 900+ bit cryptograms would take hundreds of thousands of years to crack. However, Fujitsu and NICT achieved the feat in a mere 148 days, running on a 21-computer cluster with 252 cores (Fujitsu 2012).

However, quantum computing presents a direct threat to RSA and other cryptosystems. It makes direct use of quantum-mechanical phenomena, such as superposition and entanglement, to perform operations on data, through the use of quantum bits or *qubits*. In theory, a quantum computer would be able to perform some computational functions far faster than conventional digital computers, namely sorting using Grover's algorithm (Grover 1996) and factorization using Shor's algorithm (Shor 1994). In particular, based on Shor's algorithm, a quantum computer's ability to factor in polynomial time (much faster) gives it the potential to crack any encryption scheme based on prime numbers in a fraction of the time of conventional computers. However, it depends on the ability to scale a quantum computer with multiple qubits and combine multiple qubits into a large-scale quantum computer. This continues to be elusive despite significant ongoing research on the topic. Bell Labs has made significant progress toward the creation of a new type of topological qubit device (figure 7). It promises to be immune to the decoherence phenomena – and the unacceptably short lifetimes of quantum states that result – which have plagued quantum computing for many decades.

If, by 2020 or 2025, a quantum computer with multiple qubits becomes a practical reality, communication protected with prime-number-based encryption may become vulnerable. The cryptographic community, as well as government organizations such as NIST, closely monitor developments in quantum and classical computing and adjust the recommended/required

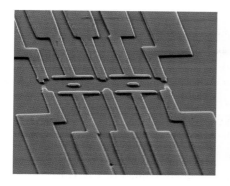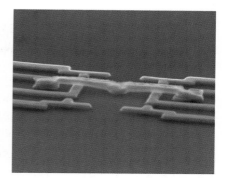

FIGURE 7: Left: a prototype topological qubit. Right: a top gate device used to test non-Abelian statistics (used in a topological qubit) in a two-dimensional electron system (from Bell Labs)

key sizes accordingly. In a 2014 *Washington Post* article, Daniel Lidar, a professor of electrical engineering and the director of the Center for Quantum Information Science and Technology at the University of Southern California, stated: "The irony of quantum computing is that if you can imagine someone building a quantum computer that can break encryption a few decades into the future, then you need to be worried right now" (Rich 2014). However, as the cat and mouse continue to play, other algorithms suitable for quantum computing or harnessing of other physical phenomena may also provide new methods of securing data that researchers have not yet considered. Elliptic-curve cryptography (ECC) emerged as a compelling alternative to prime-number-based RSA encryption; although research showed that it too was breakable (in theory) using a quantum computer with a sufficiently large number of qubits. Then, a twist on ECC was discovered that shows promise against quantum computing: specifically, the Supersingular Isogeny Diffie–Hellman Key Exchange of De Feo, Jao, and Plût, which is an elliptic-curve-based alternative to Elliptic Curve Diffie-Hellman and is not susceptible to Shor attack (Kao 2011). In 2009, Craig Gentry proposed a fully homomorphic lattice-based cryptosystem that is also believed to be safe from quantum decryption (Craig 2009).

Even without the emergence of practical quantum computing, other new computing paradigms may emerge whose power and ability to parallelize processing may affect the practical parameters of prime-number-based encryption. Innovations in optical computing (using optical phenomena to create a highly parallel analog computer), spintronics (using the spin of an atom to represent digital state, instead of a transistor), memristors, and memcomputing (UC-San Diego concept) all offer potential in this area.

End-to-end security of everything

If we summarize the preceding discussion, we believe that the approach to cybersecurity in 2020 and beyond will focus on several key technologies, as depicted in figure 8:

- *Endpoint introspection* – This will be deployed on all devices with sufficient embedded intelligence, from airplanes, cars and parking meters, to door locks, refrigerators and ovens. Software introspection will also be applied to all applications running in the cloud.

- *Permission-based access* – Sophisticated biometric authentication will be used to improve authentication of users, and security proxies will be deployed that support protocols offering flexible levels of encryption for visibility into some payload traffic, for both quality of service purposes, as well as security, law enforcement and civil defense.

- *Security analytics* – This will be deployed on a scale capable of recording and storing virtually all network activity among all devices (including the IoT domain), allowing both real-time APT detection and mitigation, as well as forensic analysis. It will be deployed as a virtualized function in enterprise data centers, as well as in central and local edge clouds.

- *Security information and event management systems (SIEMs)* – These will evolve into security autonomics platforms based on the latest advances in big data analytics and will rapidly activate or modify security policies automatically based on the perceived threat.

- *SDN* – This will be a key technology that provides dynamic flexibility to allow the network to rapidly and automatically adapt to security threats. The combination of SDN and SIEMs evolving together with security analytics and security autonomics will enable the overall orchestration of cybersecurity defenses.

- *A tighter network of threat intelligence and security expertise* – This will involve network and cloud service providers and enterprises working closely with government agencies to ensure the protection of data and the infrastructure that depends on it.

This model will be common across the infrastructure of large and small enterprises, as well as web service providers and the internal networks of communications service providers (CSPs). Both mobile and fixed service providers, however, will face the additional issue of security of consumer devices, including consumer IoT. The value of most consumer data is higher

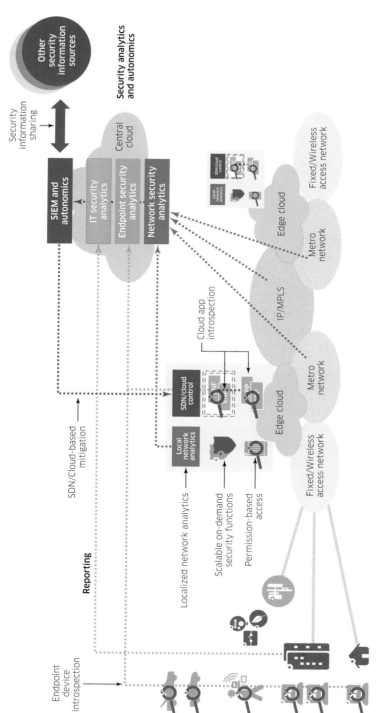

FIGURE 8: 2020 end-to-end security architecture

to the consumer than it is to cybercriminals, with the exception of identity theft and financial data for high net worth individuals. Consumer-focused security services will be appealing to a wide cross-section of the consumer market, and will provide opportunities for new revenue-generating, value-added security services. With the widespread adoption of IoT in the home on the 2020 horizon, there is a clear opportunity for service providers to offer value-added services based on security analytics running in the edge cloud.

The future begins now

The security vision described above is an extension of the direction already being taken by forward-looking players in the industry today. The technologies and techniques necessary to realize this vision are already beginning to emerge, and significant new research to advance the state of the art is underway. The challenge over the next decade will be as much about implementation as about innovation. However, by 2020, there will be far fewer security companies and solutions than there are today; companies with good ideas and strong business cases will merge and separate themselves from companies unable to scale on their own or ineffective in staying ahead of cybercriminals.

Security has become a key focus for network equipment vendors, in order to ensure that the network elements are hardened from a security standpoint. Access equipment, routers, optical transport and applications increasingly embed critical security features, such as DDoS protection, role-based access control, strong encryption support and secure key-management solutions. Vendors also conduct regular and thorough security audits of their products to ensure that security best practices are followed and security vulnerabilities are quickly identified and resolved.

In the area of security specific products, the first generation of network security analytics solutions are becoming available in the market and are beginning to get traction. While many of these solutions have tended to rely on known traffic patterns, or signatures, to identify the presence of malware, the latest generations are shifting to a more behavioral analysis approach which, when combined with big data analytics, allows them to identify malware that they have not seen before. Companies are engaged in ongoing research into improving these analytics algorithms, which continue to improve in their ability to detect *zero-day* attacks.

Endpoint security has begun evolving well beyond the basic antivirus solutions of the past. In the last couple of years, sophisticated solutions that combine detailed endpoint monitoring with big data analytics have emerged. They can detect suspicious and coordinated activity within an organization's endpoint devices that are telltale signs of an APT attack. More recently, a class of endpoint solutions has appeared that are designed to misdirect and confuse attackers in order to delay them while a defense is being mounted. These endpoint solutions are applying many of the same concealment and adaptation techniques used by attackers in order to make it difficult for malware to detect their presence.

Security and network equipment vendors are beginning to explore the integration of these sophisticated analytics solutions with autonomics. This approach will allow the solution to apply what has been learned about the attacker to automate the mitigation of the threat and adapt the network and applications. This is still in the exploration phase, but we expect solutions to emerge before 2020.

Virtualization, cloud and SDN are also key technologies that will enable effective mitigation of attacks by providing the tools necessary to move and scale virtual security appliances and network applications, as well as to automatically reroute, throttle and/or block suspicious traffic. The first generation of cloud and SDN control solutions is available today, and vendors are actively working to implement their security appliances and network applications as virtual solutions in the cloud, providing the flexibility that is essential to this automatic mitigation.

In the service provider space, several large providers have already begun building a distributed edge cloud infrastructure that will ultimately support the flexibility and automation necessary to achieve the vision described above. The ETSI Industry Specification Group for Network Functions Virtualization (ETSI ISG NFV) has a work item on security monitoring covering security analytics and autonomics. It is closely aligned with the mechanisms described in this chapter, with the goal of applying the flexibility and automation provided by NFV and SDN to improving the security of Telco networks and their customers.

At the other end of the spectrum, industrial enterprises and critical infrastructure players are in the early stages on confronting the risks of APTs to their *operational* technology (OT) versus their *information* technology (IT). At one time, they believed that an "air gap" protected their OT from Internet-based threats. OT remains a domain with a variety of proprietary protocols, many of which are serial and non-IP-based. As such, security technologies that address the specific needs of OT are being developed and deployed that are very different from traditional IT security. OT and IT are becoming less distinguishable, and air gaps are being shown to be effectively non-existent. A new class of security for industrial control systems and OT will also emerge.

Summary

We are increasingly dependent on digital data stores and augmented intelligence assistants to help us synthesize and act on the ever-growing amounts of data at our fingertips. The risk to our society is growing; whether through financial or political disruption by acts of cybercrime, cyberterrorism or cyberwarfare. As Marc Goodman stated in his book *Future Crimes*, "The Internet has lost its innocence. Our interconnected world is becoming an increasingly dangerous place, and the more we incorporate assailable technologies into our lives, the more vulnerable we become" (Goodman 2015). The perpetrators are sophisticated and well-financed attackers and specifically target our high-value data.

Having come of age in the open source era, these attackers crowdsource their tools and techniques, allowing them to apply the collective intelligence of the black hat community to achieving their goals. As in the traditional open source community, this approach enables them to learn and adapt their software very rapidly, making it very difficult for individual security vendors to keep up. The combination of the potential for high payoff (financial or political, depending on the motivation of the attackers), low barriers to entry, low risk of apprehension, rapid growth of the attack surface (due to poorly secured IoT devices), and the rapidly increasing volume of target data makes for a cybersecurity "perfect storm". As a result, the rate of attack is continuing to rise, roughly doubling every 16 months – faster than Moore's Law.

The goal of cybersecurity is not to make successful attacks impossible, but to make it (on average) more costly for cybercriminals to access critical data than the value of that data to them. Their costs can be measured in terms of people (the cost to develop, launch and manage attacks), time to achieve goals and risk of apprehension. Tools are emerging that use some of the same adaptive and concealment techniques as attackers to surreptitiously monitor endpoints and networks. In addition, security researchers are developing sophisticated analytics algorithms capable of identifying behaviors consistent with cyberattacks and automating responses designed to misdirect, confuse and block the attacker.

As we move toward 2020, these analytics algorithms will operate on increasingly large sets of monitoring data, including data from endpoints, networks and security appliances in order to provide a true end-to-end view of the network, including visibility of coordinated activity across the network. Extending beyond the boundaries of an individual service provider or enterprise, institutionalized sharing of threat information and mitigation techniques and tools will allow for the crowdsourcing of security responses among a larger community than that of the attackers, leveraging the fact that there are more of *us* than there are of *them*.

References

CIGI 2014. "CIGI-Ipsos Global Survey on Internet Security and Trust," *CIGI web site*, November 24 (https://www.cigionline.org/internet-survey).

CSIS 2014. "Net Losses: Estimating the Global Cost of Cybercrime, Economic impact of cybercrime II," Center for Strategic and International Studies and McAfee, *McAfee web site*, June (http://www.mcafee.com/us/resources/reports/rp-economic-impact-cybercrime2.pdf).

Craig, G., 2009. "A Fully Homomorphic Encryption Scheme," Ph.D. *Dissertation, Stanford University*, September (https://crypto.stanford.edu/craig/craig-thesis.pdf).

Cyber Threat Alliance 2015. *Cyber Threat Alliance web site* (http://cyberthreatalliance.org).

EMC and IDC 2014. "The digital universe of opportunities," *Infobrief*, April (http://www.emc.com/collateral/analyst-reports/idc-digital-universe-2014.pdf).

FBI 2014a. "Violations of the Federal Bank Robbery and Incidental Crimes Statute," *U.S. Dept. of Justice and Federal Bureau of Investigation, Bank Crime Statistics*, December (https://www.fbi.gov/stats-services/publications/bank-crime-statistics-2014/bank-crime-statistics-2014).

FBI 2014b. "2014 Internet Crime Report," *2014 IC3 Annual Report* (https://www.fbi.gov/news/news_blog/2014-ic3-annual-report/at_download/file).

Finklea, K., 2014. "Identity Theft: Trends and Issues," *U.S. Congressional Research Service Report R40599*, January 16 (https://www.fas.org/sgp/crs/misc/R40599.pdf).

Fujitsu 2012. "Fujitsu Laboratories, NICT and Kyushu University Achieve World Record Cryptanalysis of Next-Generation Cryptography," Fujitsu Press Release, June 18 (http://www.fujitsu.com/global/about/resources/news/press-releases/2012/0618-01.html).

Gigaom 2014. "The Internet of Things Needs a New Security Model; Which One Will Win?" *Gigaom web site*, January 22 (https://gigaom.com/2014/01/22/the-internet-of-things-needs-a-new-security-model-which-one-will-win).

Goodman, Marc, 2015. *Future Crimes*, Knopf Doubleday Publishing Group.

Grimes, R., 2012. "Why Internet crime goes unpunished," *InfoWorld*, Jan 10 (http://www.infoworld.com/article/2618598/cybercrime/why-internet-crime-goes-unpunished.html).

Grover, L. K., 1996. "A fast quantum mechanical algorithm for database search," *Proceedings, 28th Annual ACM Symposium on the Theory of Computing*, May, p. 212.

Homeland Security 2015. *The U.S. Dept. of Homeland Security web site* (http://www.dhs.gov).

HP 2014. "Internet of Things HP Security Research Study," *HP Fortify web site*, June (http://h20195.www2.hp.com/V2/GetDocument.aspx?docname=4AA5-4759ENW&cc=us&lc=en).

IT-ISAC 2015. *Information Technology – Information Sharing Analysis Center web site* (http://www.it-isac.org).

Jao, D., De Feo, L., Plût, J., 2011."Toward quantum-resistant cryptosystems from supersingular elliptic curve isogenies." *Post-Quantum Cryptography*. Springer Berlin Heidelberg, pp. 19-34.

Kao, E., 2011. "Making Search More Secure," *Google Official Blog*, October 18 (http://googleblog.blogspot.com/2011/10/making-search-more-secure.html).

Kleinjung, T., et al, 2010. "Factorization of a 768-bit RSA modulus." *Advances in Cryptology–CRYPTO 2010*, Springer Berlin Heidelberg, pp. 333-350.

Let's Encrypt 2015. *Let's Encrypt web site* (https://letsencrypt org/2015/06/16/lets-encrypt-launch-schedule.html).

M3AAWG 2015. *The Messaging, Malware and Mobile Anti-Abuse Working Group web site* (https://www.m3aawg.org).

PLXsert 2014. "Spike DDOS Toolkit," PLXsert Threat Advisory (http://www.prolexic.com/kcresources/prolexic-threat-advisories/prolexic-threat-advisory-spike-ddos-toolkit-botnet/index.html).

PwC 2014. "The 2014 Global Economic Crime Survey," *PwC web site* (http://www.pwc.com/gx/en/economic-crime-survey/cybercrime.jhtml).

PwC 2015. "The 2015 Global State of Information Security Survey," *PwC web site*, 2009–2014 reported incidents (http://www.pwc.com/gx/en/consulting-services/information-security-survey/index.jhtml#).

Rich, S., Gellman, B., 2014. "NSA seeks to build quantum computer that could crack most types of encryption," *Washington Post*, January 2 (http://www.washingtonpost.com/world/national-security/nsa-seeks-to-build-quantum-computer-that-could-crack-most-types-of-encryption/2014/01/02/8fff297e-7195-11e3-8def-a33011492df2_story.html).

Sandvine, 2015a. "Global Internet Phenomena Report 2010-2014," *Sandvine web site* (https://www.sandvine.com/trends/global-internet-phenomena).

Sandvine, 2015b. "Global Internet Phenomena Spotlight: Encrypted Internet Traffic," *Sandvine web site* (https://www.sandvine.com/trends/encryption.html).

Shor, P. W., 1994. "Algorithms for quantum computation: discrete logarithms and factoring," *Proceedings of the 35th Annual Symposium on Foundations of Computer Science*, pp. 124–134.

United States Attorneys 2014. "Annual Statistical Report, Fiscal Year 2014," U.S. Dept. of Justice Document (http://www.justice.gov/sites/default/files/usao/pages/attachments/2015/03/23/14statrpt.pdf).

WebRTC 2015. "FAQ: What other components are included in the WebRTC package?" *WebRTC web site* (http://www.webrtc.org/faq).

Wikipedia 2015, "Phishing", (https://en.wikipedia.org/wiki/Phishing).

Winkler, R., 2014. "Google Weighs Boosting Encrypted Sites in Its Search Algorithm," *WSJ Dijits web site,* April 14 (http://blogs.wsj.com/digits/2014/04/14/google-may-push-sites-to-use-encryption).

CHAPTER 4

THE FUTURE of
WIDE AREA NETWORKS

Creating the perfect network for a new era

T. V. Lakshman, Kevin Sparks
and Marina Thottan

The essential vision

We are entering a new technological era with unprecedented demand to digitize everything, and with new value created from the automation of lives and systems. The network will be required to deliver information and entertainment content on a massive scale but with an economic outlay that is not vastly different than what we expend today. This requirement comes at a time when we have reached the physical limits of today's network performance; so a top-to-bottom restructuring of networks is required, with a wholesale shift to highly distributed, highly intelligent optimized networking. Today's self-contained, standalone networking constructs – whether it be an operator, network, network layer or element – will be disaggregated and then realigned and reintegrated at different levels using a set of new software platforms and interfaces. The shift will also require a new automated control paradigm, in which networks are truly programmable and become the essential building blocks of a new digital connected reality. Three essential elements are required to realize this new digital networking era.

1. *A network OS* – Truly automated networks will require a network "brain" or operating system (OS) that will control the functions and fabric of the

network. The OS needs to be context-aware and adaptive to dynamically changing services and user demand through simple intent-based interfaces. A set of distributed autonomous controllers will give it a comprehensive end-to-end view of the network across all network domains and technologies.

2. *Tunable network fabric* – Achieving the "seemingly infinite" scale required for this new "digital everything" network affordably will require a new automated network fabric that fuses a distributed data center fabric, a scalable routing fabric and a massively scalable and flexible optical interconnection fabric into a single adaptive continuum under network OS control.

3. *Dynamic connectivity marketplace* – New technological eras bring with them new business models and wholesale economic transformation. A key part of this new networking reality will be the creation of a new global capacity marketplace where capacity can be bought and sold in near-real-time, which will enable the dynamic interconnection of web-scale data center providers, global network service providers and multinational enterprises.

As this vision is realized, the distinction between networks of different types (core, metro, data center, local area) will disappear as they become one automated network fabric, built from arrays of distributed cloud infrastructure that are fused together into cohesive dynamic service platforms.

Today's self-contained, standalone networking constructs – whether it be an operator, network, network layer or element – will be disaggregated and then realigned and reintegrated at different levels using a set of new software platforms and interfaces.

The past and the present

Network automation

Control and management in communications networks have evolved considerably since the telephone was invented in 1876. Early telephony systems had limited reach and operators were used for manual switchboard operation to provide connections among local callers. The operators were located at exchange points, which were introduced to avoid the problem of running wires from each telephone to every other (a fully meshed topology). As telephone services grew in popularity and frequency, so did the need for connection automation. With the invention in 1889 of the Strowger exchange, an electromechanical device that enabled automatic switching, the telephone became faster, easier to use and more private, without the intervention of a manual operator. This required automation of signaling (for example, the exchange of information between network components), mainly for setting up and tearing down voice circuits, which led to the development of automatic exchanges by 1940. Calls between subscribers served by the same local switch could then be automated and a manual operator was only required to perform signaling for long distance calls.

In traditional telephony, once a circuit was set up, no other signaling was performed apart from terminating the call and releasing related resources; therefore, all calls were simple, basic telephone service calls. In 1980, as the demand for richer telecommunication services grew, the ITU-T defined Signaling System No. 7, known in North America as SS7 and elsewhere as C7. SS7 was both a network architecture and a series of protocols that provided telecommunications network signaling and allowed the transfer of data between network nodes. SS7 used out-of-band signaling to enable the provisioning of new calling services to support voice calling in cellular networks, as well as other advanced telephony services.

Figure 1 shows the basic architecture for an SS7 telephony network, which is a highly redundant architecture because of the criticality of SS7 for call processing. A user's phone was connected to the local switch (class 5 switch), and a group of local switches were connected to a long distance switch (class 4 or tandem switch), which were typically connected in a full mesh network through signal transfer points (STPs). STPs are the "packet switches" of the SS7 network, routing incoming signaling messages toward the proper destination and performing special routing functions such as signaling between different networks. Another key element in the SS7 network was the signal control point (SCP), which provided the information and logic

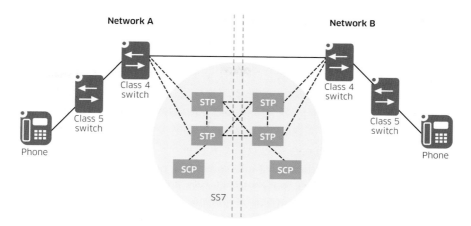

Class 4 switch: Tandem switch (long distance switch)   Class 5 switch: Local switch (central office)
SCP: Signal control point   STP: Signal transfer point

FIGURE 1: The basic architecture for an SS7 telephony network

required for advanced call-processing capabilities such as routing of 800 number time-of-day-based calls.

As the number of subscribers and exchange points grew, a need emerged to develop an operations support system (OSS). This need coincided with the advent of early computing systems, resulting in a wholesale migration to software-based OSS systems, as well as the development of the UNIX operating system and the C programming language at Bell Labs in the 1970s. The OSS systems were designed to automate the interconnection and billing reconciliation of telephony networks. However, the scope of the automation remained at the level of specific network systems or domains, and therefore they continued to evolve separately. Consequently, today we have large networks typically managed by more than 100 independent OSS systems across the different domains and elements of the same network infrastructure. These OSS systems need to be manually integrated in order to provision network services, which leads to the need for OSS integration systems and services.

The big disruption in terms of network automation was the birth of the Internet in the early 1980s, as it was premised on the ability to connect and run systems across a network that is geographically distributed and dynamically changing. The technological revolution that led to the development of the Internet began with research on packet switching in 1961 and the Advanced Research Projects Agency Network (ARPANET) in 1969. The first computer network had four host computers, which were connected together using a low-speed dial-up telephone line. The ARPANET was based on the idea of open architecture networking in which there is no specification of an individual network technology, but the networks were made to interconnect with each other through a meta-level internetworking architecture.

Until the ARPANET, the only method to interconnect or federate networks was traditional circuit switching. The ARPANET introduced the first federated model of networks, with networks acting as peers to each other in offering end-to-end service. The concept of interconnecting networks required a reliable end-to-end protocol that could maintain effective communication in the face of failure. This led to the development of the transmission control protocol (TCP), which provided all of the transport and forwarding services in the Internet. The initial efforts to implement TCP only allowed for virtual circuits, which worked well for file transfer types of applications. However, for some applications it became clear that packet losses should not be corrected by TCP, but dealt with by the application. As a result, the original TCP was broken up into two protocols: the Internet protocol (IP), which provided only for addressing and forwarding of individual packets, and TCP, which was concerned with service features such as recovery from lost packets.

In the ARPANET, two different networks were connected using gateways and routers that contained only network connectivity and reachability information. Routers did not maintain any individual flow-level information. Network connectivity was obtained using a single distributed algorithm, an interior gateway protocol (IGP), which was implemented uniformly by all the routers in the Internet.

Simultaneously with the development of ARPANET, Ethernet was under development at what was then Xerox PARC in the 1970s. At the time, the model for a single interconnected network was interconnecting only a few national-level networks such as those in ARPANET. This model failed to anticipate the rapid proliferation of local area networks (LANs) interconnecting PCs and workstations. The increase in the size of the Internet challenged the capabilities of the routers and the IGP-based routing model had to be replaced by a hierarchical model of routing as shown in figure 2. Some version of an IGP was used inside each region of the Internet (typically each network operator), while operating between them, the border gateway protocol (BGP) was used to tie the regions together. This design permitted different regions to use different IGPs in which different requirements for cost, rapid reconfiguration, robustness and scale could be accommodated. Further, using the BGP, the

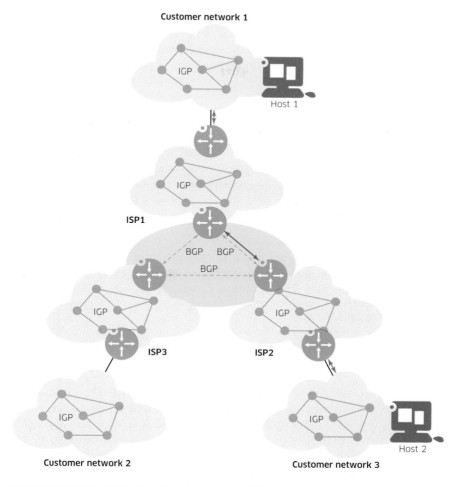

FIGURE 2: Interconnecting operator networks using BGP

independent network regions could limit the information sharing to just reachability without having to expose the entire network topology.

The ARPANET evolved to become the National Science Foundation network (NSFnet) and was subsequently commercialized in 1988. The original network built out of routers by the research community made the transition to a network of networks interconnected by commercial equipment, which evolved to the IP networks of today. To participate in today's Internet, individual network operators now have to manage complex sets of Internet peering relationships to support core services and these peering points account for a significant proportion of network operating costs.

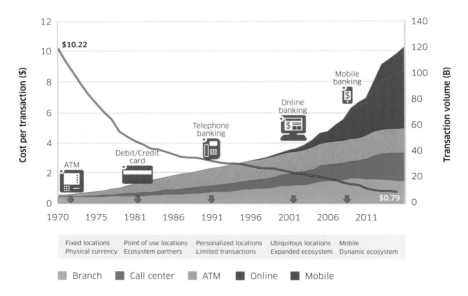

FIGURE 3: Digitization and automation of banking networks has brought huge gains in transaction volume and efficiency[1]

If we look outside the telecommunications and networking industry, analogous evolution scenarios can be found in other sectors with highly distributed autonomous systems that need to be interworked to provide a service with global "access" and scale. For example, the need to provide global consumer and commercial banking services that allow facile transactions anytime, anywhere has led to the large-scale automation of the banking industry. As shown in figure 3, this automation of (consumer) banking services has led to a more than 10-fold decrease in cost per transaction. At the same time, the service was adapted to the instantaneous customer location and device context, moving from the initial requirement for a customer to be present at specific physical banking locations (during limited opening hours), to the flexibility for anytime transactions at more locations (the advent of ATMs), to the use of a global debit and credit system to reduce the need for hard currency transactions, then finally evolving to touch-tone telephonic banking for simple transactions, which over time increased the number and type of services supported, first on the desktop browser, and then, dedicated apps on mobile devices. This evolution has led to a more than six-fold increase in the number of transactions from 20 to 120 billion over the last three decades.

1 - Data analysis and modeling done by Bell Labs Consulting based on various banking sources (Bell 2009, Broeders 2015, Federal Reserve 2013, Martinez 2009, Neely 1997, Phan 2006, Tower Group).

Autonomous financial entities

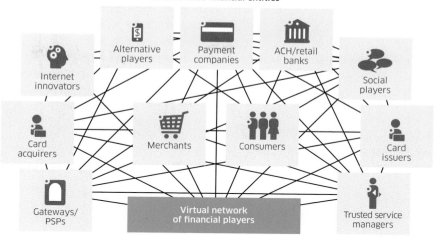

Autonomous network domains

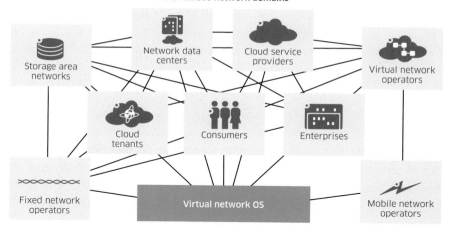

FIGURE 4: Networks of federated autonomous systems providing seamless diverse services, controlling financial transactions (Skinner 2015) or connectivity transactions (Bell Labs Consulting)

It is instructive to consider the key technological enabler of this evolution to a dynamic, adaptive financial network, in order to understand whether there are parallels that apply to the networking and connectivity sector. As shown in figure 4, an ecosystem of partners was created and networked to create and deploy composite services that led to the massive increase in the number of mobile transactions. The complete virtualization of currency was a critical enabler, so that physical currency was no longer needed at the endpoints to initiate and drive the exchanges. This is analogous to the virtualization of physical goods that underlies the Internet of Things (IoT) phenomenon, discussed throughout this book.

The partners in the ecosystem are autonomous entities that federate the information necessary to instantiate composite global services. The increase in service scale and addition of new partners is used to drive new valuation creation. Again, if we consider future networks, network providers could similarly support connectivity transactions to build composite services by leveraging partnerships between traditional network providers and enterprise networks, network data centers, storage area networks and other new market entrants.

One challenge in achieving such globally federated networking services today is that network management systems are usually optimized for specific services and typically coupled with vendor-specific network hardware. In addition, these OSS platforms are tied to the control and management of these specific network elements and are not architected for rapid service creation and adaptation. This ossification of the management system makes the task of automating service creation nearly impossible due to the inability to manage and the complexity of sharing information across network layers and network domains. For example, fulfillment of a new enterprise service provisioning order can be weeks, with considerable time taken to verify the input data and determine the availability of network resources, including the intermediate connectivity needed from each different carrier.

Another significant challenge in network automation is ensuring reliable and secure services delivery. Today's telecommunications networks derive their resiliency through an extensive infrastructure for fast error detection, isolation and recovery in the equipment. Any new network automation paradigm will have to maintain these same levels of service availability to avoid customer dissatisfaction; in the financial sector the analogous philosophy is "high data integrity" (ensures that transaction integrity is guaranteed) at the expense of lower availability of an endpoint service, such as an ATM.

A last element of the banking revolution was the focus on end-customer service in order to provide simple service creation and support; the future network must have a similar focus on technology-agnostic service definition, creation and management.

To simplify the multi-layer, multi-domain network operations and reliably and securely automate the network fabric, the communications industry is undertaking a major transformation with the introduction of software-defined networking (SDN) and network functions virtualization (NFV). This transformation is not only changing the way networks are deployed and operated, but also greatly enhancing the potential for innovative network automation and service creation paradigms. One of the crucial value propositions of the SDN/NFV transformation is to move away from closed platforms and service management systems to more open platforms based on commercial computing hardware and open source software. However, if these open platforms are not designed and optimized to support the required reliability and resiliency, as well as the simplicity of service on-boarding, the potential benefits of the introduction of SDN/NFV could rapidly disappear and time-to-market for new solutions could actually increase, which would effectively prevent or significantly delay the onset of the new digital era.

IP/Optical networking challenges and essentials

In the last four decades the Internet has transformed from a small research network to an essential infrastructure for all communication and a digitized society. Internet applications have evolved from file sharing and email to support many aspects of human and machine-to-machine communication. The evolution of the Internet to support this diverse array of services has required the constant evolution of underlying network infrastructure to support ever-increasing numbers of endpoints at ever-higher bandwidths and with ever-decreasing connection latencies.

The number of endpoints has increased from a few million in the 1990s to billions, which will soon become tens of billions with the rise of the machines and IoT. The first phase in endpoint growth was enabled by the introduction of classless inter-domain routing (CIDR) addresses and route aggregation. This prevented a corresponding explosion in routing table sizes, thereby allowing the Internet to scale three orders of magnitude in endpoint connectivity.

The potential performance impact from the more complex "longest prefix match" route lookups associated with CIDR addresses was offset by the advent of sophisticated, high-performance packet forwarding ASICs and Ternary CAM (T-CAM) memory devices that allowed routers to support the required route lookup speeds. Routers have subsequently ridden the semiconductor technology curve to total forwarding capacities reaching tens of terabits per second. The capacity of optical transport networks has evolved at an even faster rate, due to the additional capacity multiplier provided by the use of dense wavelength division multiplexing (DWDM)

The SDN paradigm enables the required rapid network tuning by giving greater network control and visibility to a small set of federated controllers that facilitate the abstraction of the underlying network.

in the 1990s. More recently we have seen the advancement of long distance optical signal rates to 100G and beyond through the invention of new high-data-rate digital signal processing techniques and the use of additional advanced modulation formats and multiple polarizations with so-called "coherent" signal processing. The fastest currently deployed commercial interfaces employ dual-wavelength optical superchannels at 2x 200G. Routers have also reached 400G speeds, initially following the steep evolution of complementary metal oxide semiconductor (CMOS) and, more gradually, so have high-speed optical interfaces over the past decade. The evolution and convergence of serial interface rates of optical DWDM systems (per wavelength) and routers is shown in figure 5. It is clear from this plot that the need for intermediate multiplexing layers (for example, ATM, SDH/SONET or OTN) to bridge the "rate gap" has diminished over time.

Networking is not the only industry to have to evolve continually to meet capacity demand, so it is instructive to look at a few other industries to see how they have evolved. The airline industry has had to continuously increase the number of global destinations served to meet customer demand and to constantly extract efficiencies in their networks to stay competitive and financially viable. This has led most airlines to adopt a "hub-and-spoke" architecture, which funnels traffic through a few high-throughput hubs,

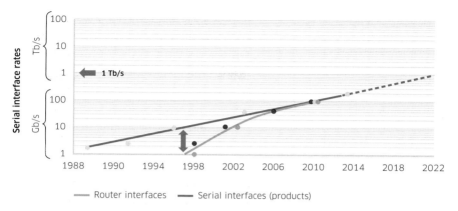

FIGURE 5: Optical system and router serial interface rate evolution (Winzer 2014).

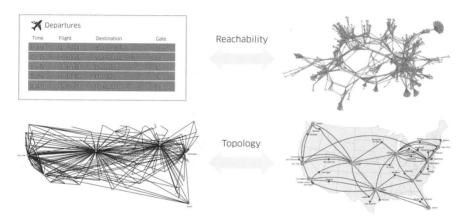

FIGURE 6: The parallel between the air transport and the packet transport industries and networks in terms of overall topology and reachability[2]

allowing airlines to serve more destinations and offer more reachability with a more limited fleet of planes than would be possible with direct point-to-point connectivity. This hub-and-spoke architecture is a classic traffic management approach that offers a flexible and efficient way to aggregate traffic. It is clear that the same paradigm has been at work in the evolution of the Internet to one in which most Internet traffic is funneled over aggregation points to efficient long-haul routes.

The benefits derived by the airline industry from this architecture are shown in figure 7; the number of endpoints served increased almost seven-fold

2 - Departure board graphic from 123RF. United Airline's 1993 domestic routes (West 2011). XO Communications' IP Network (Fierce 2011).

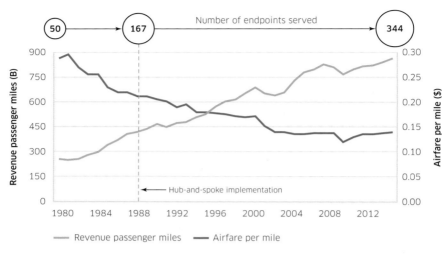

FIGURE 7: Major airlines significantly expanded reach, optimized cost and grew revenue-generating traffic by using a more adaptable hub-and-spoke routing[3]

and the per-passenger miles carried tripled. At the same time, the industry achieved a per-mile airfare reduction of approximately 50% due to the gains in network efficiency achieved by hub-and-spoke adoption and other factors.

With the advent of server virtualization and the concomitant ability to rapidly grow and move compute workloads, the need for rapid tuning of the network to accommodate the dynamic connectivity needs of virtualized workloads becomes paramount. Furthermore, the requirements of multi-tenant cloud computing systems have led to the need for dynamically instantiated "virtual networks".

The SDN paradigm enables the required rapid network tuning by giving greater network control and visibility to a small set of federated controllers that facilitate the abstraction of the underlying network (paths, layers, specific equipment capabilities) and the associated dynamic tuning of the network. SDN achieves this by automating the enforcement of networking policies for each virtualized function to form a new service instance or to combine existing service instances together to form a higher-order service called a *service chain*. The service chain is conceptually the networking equivalent of a web service created using multiple interfaces (APIs) to other web services.

SDN also allows the computation of an optimized path across multiple layers of the network. Traditionally, IP/MPLS (multiprotocol label switching)

3 - Data analysis and modeling done by Bell Labs Consulting based on various sources (Brueckner 1994, Belobaba 2010).

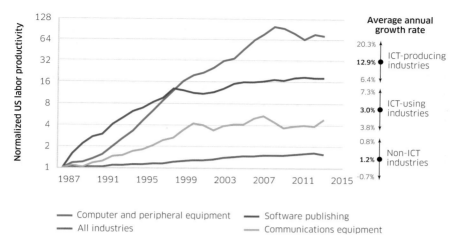

FIGURE 8: Industries implementing automation have realized substantive productivity gains relative to others that have continued to deploy traditional methods (U.S. Bureau of Labor Statistics)[4]

and optical networks have formed separate layers, with the IP routes and MPLS label-switched paths managed together, but separately from the optical wavelengths. SDN provides the mechanism for unified control of all of these layers and the means for multi-layer automation, adaptation and optimization. This unified control, adaptation and optimization spanning the IP, MPLS and optical layers can significantly boost network efficiency and enable a new class of real-time bandwidth-on-demand services and calendaring (scheduled reservation) applications.

To understand the potential benefits of such automation, it is again instructive to look at parallels in other industries. Figure 8 shows the productivity gains stemming from automation in the computer hardware, software and communications equipment industries, ranging from factors of 4 (communications equipment) to factors of approximately 16 (software publishing) to factors of approximately 100 (computing and peripherals).

It is clear that industries most involved in ICT (and especially IT) technologies have far outstripped productivity gains in other industries, as these are the technologies most compatible with digital automation today. All industries that become substantially digitized and leverage this new networking infrastructure will similarly enjoy a steep rise in productivity.

---

4 - Data analysis and modeling done by Bell Labs Consulting based on data available from the U.S. Bureau of Labor Statistics.

Network capacity markets

Domestic telecommunication networks connect together in order to extend services beyond the boundaries of any individual provider's network and ultimately across the globe. Indeed, ubiquitous reachability is a fundamental value proposition for these networks, consistent with Metcalfe's Law.[5]

Mass connectivity services such as voice calling, text messaging and web services/Internet access have been globally connected since their inception, with automated service interconnection provided by well-defined IP interconnect and peering points and associated signaling and control. The core IP/optical infrastructure over which these services flow, as well as managed bandwidth services essential for linking enterprise locations and/or data centers, have also been interconnected between networks for decades, but in a manual or semi-manual manner. As a result, these connections and connection points are relatively long-lived and static in terms of location, capacity and price, and consequently the types of applications and customers they serve. But the same dynamic cloud services creation that is driving individual networks to become more dynamic and elastic will also impact federations of networks and their interconnection.

Network interconnection between many providers is facilitated by common collocation facilities, often referred to as "carrier hotels," and brokers who connect suppliers and consumers of wholesale network capacity. In this way, the global network capacity market resembles other industries such as electrical energy and online ad markets that also deal in valuable but temporal commodities. In considering how the network capacity market will evolve, we again look at dynamics within these parallel markets.

Looking at electrical energy generation, it is obvious that this is a market that is largely premised on dynamic tuning to instantaneously and continuously match demand, with the optimum (lowest cost) source chosen to supply demand at all times. The highest cost sources (for example, natural gas turbines) are only being used to meet the unexpected or uncommon demand peaks. Energy utilities benefit significantly when peaks are reduced and demand is spread out more evenly. This is closely analogous to the information networks, which likewise are most efficient when highly utilized.

---

5 - Metcalfe's Law states that the value of a telecommunications network is proportional to the square of the number of connected users of the system or n^2 (Wikipedia, "Metcalfe's Law").

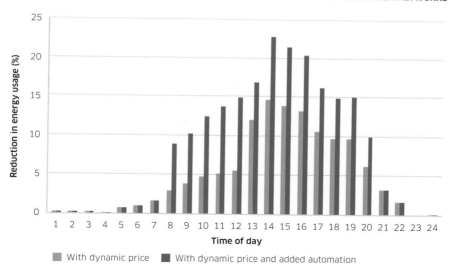

FIGURE 9: Dynamic, adaptive pricing is a key driver of peak usage reduction in energy and other industries (Tholin 2006)

Dynamic real-time pricing, where energy pricing tracks demand level, has been used successfully by energy utilities to reduce peak demand, with significantly greater reductions when automated mechanisms are employed, as demonstrated in figure 9.

The Internet advertising market is another example. This market is an example of real-time bidding (RTB) for another temporal commodity – context-sensitive ad space availability. With RTB, advertisers have just 100 milliseconds to consider the context of a user and bid for each ad space, and the highest bidder's ad is then displayed. Following the deployment of automated demand-side platforms in 2007, RTB has grown steadily to over 40% of the display ad market, with sizable gains claimed as compared with non-RTB ads, as shown in figure 10.

It is clear that such dynamic marketplace mechanisms, finely tailoring offers and usage to personalized needs, can transform markets – driving volume, efficient utilization and enhanced monetization. Again, all of these effects apply equally to the network capacity market, illustrating the great potential of tailoring bandwidth services to the needs of the global cloud ecosystem.

The large-scale compute virtualization trend that swept the ICT industry over the past decade is now driving inflection in the networking industry as well. The dynamic needs of virtualized workloads were instrumental in driving the need

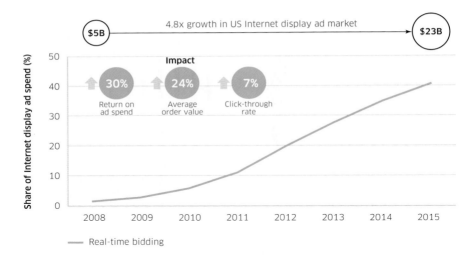

FIGURE 10: Dynamic, real-time bidding has resulted in significant expansion of the online advertisement marketplace (eMarketer 2013, Kenshoo 2013)

for SDN. The same advances in virtualization technologies, multi-core commodity servers and distributed systems software are now making it possible to develop high-performance, scalable network operating systems that interface with and control the underlying infrastructure capabilities based on both virtualized and optimized hardware systems in real time. The confluence of these technology trends is creating the exact set of capabilities or "perfect storm" of technology supply that is coincident with the latent end-user demand and the burgeoning demand of the IoT, and will drive widespread changes in communications and information networking architectures, capabilities and business models.

The future

Future wide area networks will be driven by technology advances in software-defined network control, optimized tunable network fabric and automated federation between networks. In this section, we describe the key innovations that we believe will be instrumental in shaping the future of the WAN.

The network OS

It has long been the dream of network operators to be able to dynamically provision new network services as easily as web services. In computing infrastructure, this automation is offered by the operating system (OS), which manages the sharing and abstraction of the underlying compute, memory,

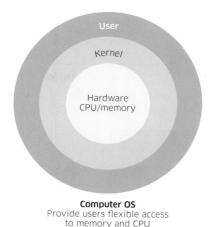
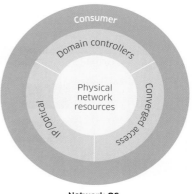

Computer OS
Provide users flexible access
to memory and CPU

Network OS
Provide network consumers flexible
access to network resources

FIGURE 11: Analogy of computer OS and network OS

storage and input/output (I/O) for applications and creates the notion of optimal resources for secure execution for each application. If we extend this analogy, as shown in figure 11, to realize the vision of rapid network service creation will require a network OS.

Just as the PC or smartphone OS platforms provide a programming abstraction that decouples applications from multivendor hardware, the network OS will provide a centralized network abstraction that decouples the underlying specifics of network elements and domains from network apps, which are network services that require real-time interaction with the network. The network OS would be instantiated as a logically centralized platform with an intuitive interface for network consumers that allows the network to be programmed and also provides the perception of seemingly infinite network resources.

A high-level architecture for such a network OS is shown in figure 12. The network OS unifies the fragmented management and control planes of the different network domains and also provides the intelligence needed to rapidly configure end-to-end network services across different technologies. Therefore, the ideal network operating system enables the following:

1. Real-time network control through simple, extensible network interfaces

2. Exposure of service-relevant physical-layer capabilities in the underlying network fabric, extensible to accommodate new innovations

3. Unified operations of disparate network domains and technologies

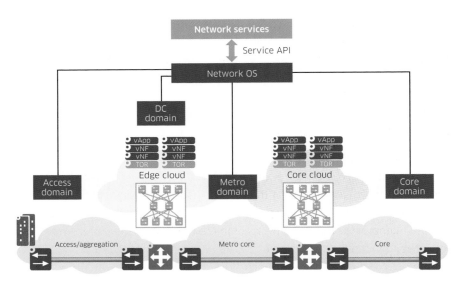

FIGURE 12: Logically centralized network OS

The network OS provides the essential abstraction and control, but it relies on many underlying capabilities to allow optimization of the network resources and the creation of the illusion of seemingly infinite capacity. The key capabilities are the ability to maintain network state, a hybrid control plane, simplified, intent-based service definition, a network description language and federation control. We discuss each of these capabilities in more depth below.

Maintaining network state

Network state includes information about such things as network inventory, connectivity and current resource utilization as well as ongoing and scheduled network outage and maintenance events. The global network state captures the operational view of the complex and fragmented network domains and provides the basis on which services are defined, provisioned and instantiated on the network fabric.

The centralized view of network state supports agile service provisioning and dynamic on-demand network services, and derives real-time contextual information about the network and services to devise appropriate responses to different network events. By leveraging the network state, the network OS scales up or scales down the necessary network resources and functions across network layers and network technologies. In this way, it meets the changing needs of the service and responds to changing network conditions such as failures and underlying network impairments. To support the dynamics of network services and track real-time network events for large network topologies, the network OS maintains an up-to-date view of the network state using scalable performance monitoring tools. An accurate view of the network state is enhanced by a resiliency infrastructure with mechanisms for fast (< 1 second) error or failure detection of logical and physical elements and the subsequent recovery mechanisms that are triggered by these events.

Hybridizing the control plane

Recent studies have shown that using sophisticated path computation on a centralized controller instead of the current distributed constrained shortest path first (CSPF) computation approach can lead to as much as 25% more revenue-generating traffic being accommodated in the network (figure 13). Moreover, by careful spreading of traffic in the network, overall congestion due to highly utilized links on shortest paths is reduced as well.

However, a centralized controller still needs an optimized distributed protocol for reachability information. Distributed control is also needed for near-real-time response for failure recovery. Therefore, with the massive IP reachability and responsive resiliency that is essential for the underlying network connectivity, only a hybrid combination of centralized network control and distributed autonomous domain controllers with distributed routing protocols will provide the combination of scalable, dynamic and application-driven connection setup required for the digital, virtualized world.

Simplifying service creation and fulfillment

Today service creation and fulfillment is accomplished in a piecemeal fashion, separately for each network technology domain, with manual integration of the service components across the different domains, resulting in lengthy service fulfillment times and high costs. Programmability significantly reduces costly delays as illustrated for the recent case of the software-defined VPN service (described in chapter 2, *The future of the enterprise*) where service fulfillment time (including CPE shipping time) is reduced by an order of magnitude from weeks to days and operational cost reduced by more than half (figure 14). The programmability expected of future networks fundamentally requires control interconnectedness, automation and abstraction across equipment from different suppliers and across technology domains.

Intuitive service definition through the service API (figure 12) is a key component of network programmability. This new service definition framework will support customizable service definitions and enable

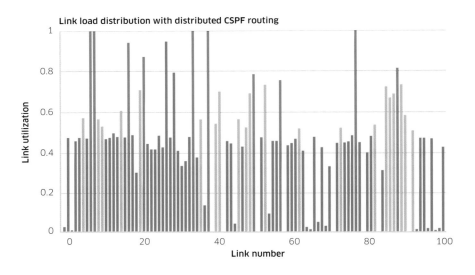

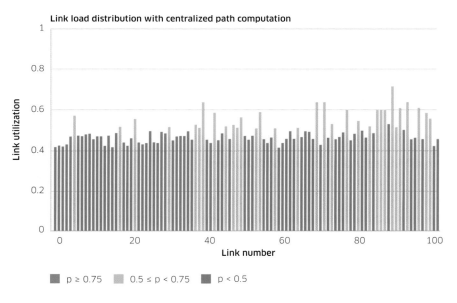

FIGURE 13: Efficiency of distributed versus centralized path computation (Hao 2015)

interaction with the network through an "intent-based", high-level interface that allows the network consumer to intuitively define and consume network services in a technology-agnostic manner and to specify their intended connectivity requirements.

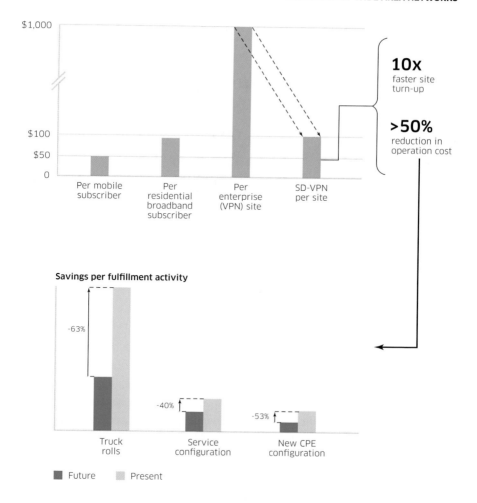

FIGURE 14: Reduction in service fulfillment cost and time through network programmability

As an example, in figure 15 an enterprise connectivity request for data center interconnect (DCI) services can be described to the network OS as a set of endpoint connectivity demands with bandwidth and latency requirements. The intent-based user interface presents the set of possible connectivity options with different quality of service (QoS) parameters based on the available features of the network and allows the user to select the desired connectivity based on a range of different cost/performance options that are consistent with their needs. The network OS will then translate these inputs into actionable commands for configuration and control of network resources.

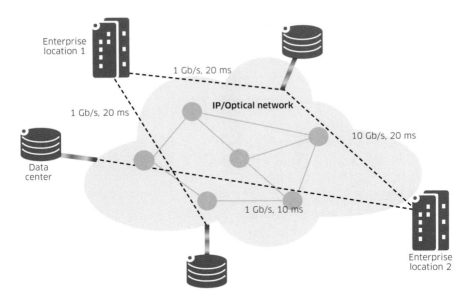

FIGURE 15: Enterprise specification for DCI services

As intent-driven service definition uses centralized network information, any combination of intent-based services can be used concurrently; the intent compiler checks for any conflicting requirements being specified to the network. The simplicity and technology-agnostic specification that is inherent in this intent-based specification approach will make it straightforward to specify network services across network domains and to scale out the service definition framework to support an arbitrary number of network domains, because detailed knowledge of the underlying network is not required. This will enable true end-to-end service definition and creation – something that has eluded communications and data networks since their inception.

Talking to networks
Translating service configuration requests expressed as "intents" into actionable network element configurations will require the equipment-agnostic description of network elements and their functionality. The network OS will address this challenge by employing a network description language that can do two things. It can capture the essential characteristics of network elements and their value-added features, and it exposes the necessary performance metrics and attributes of network elements relevant for service configuration as uniform simple abstractions to the service orchestration function. The mediation provided by network abstractions will enable further physical layer innovation

while allowing network operators to continue to achieve Interoperability for service provisioning. The service orchestrator will be responsible for coordinating the provisioning of multiple intent requests and for generating the sequence of instructions for configuring the network service.

Network configuration protocol (NETCONF) is a new protocol enabling the service orchestrator function for the IP layer using remote procedure calls and notifications to configure network elements described through YANG models (Bjorklund 2010, Enns 2006). YANG is a data modeling language used to describe network elements and their possible configurations and parameter values and states. To enable end-to-end programmability across domains and network layers, the network modeling language must go beyond this element-level configuration and be able to provide description of the entire network.

The network modeling language and abstractions must capture the connectivity of the network and do it in a manner that is independent of specific forwarding technologies and based on models suitable for heterogeneous forwarding technologies. The language should describe network connectivity (topology) and path specification and capture hierarchical aggregations of network connectivity. For example, the modeling language will need to capture the "IP path" as an aggregate construct comprised of a concatenation of optical spans in the physical network. The model must be simple, applicable to both connection-oriented (for example, VPNs, optical wavelengths) and connectionless (for example, IP) networks, and independent of any particular technology.

A network is often modeled as a graph – a simple representation of nodes and their interconnecting links; however, a graph is inadequate as it does not describe how higher level links (for example, router-router connections) are formed from other links in the lower layers (for example, optical DWDM spans). For a complete network model the language must capture the multi-layer topology model for networks and provide a snapshot of an existing or potential network. The model will include links, adaptations, connections, communication paths and bindings.

- Links have different types that characterize the layers of the transport network (for example, IP or wavelength).
- Adaptation includes both encapsulation – the changing of a data stream from one format to another – and multiplexing, which describes the combination of several streams of data into one transport path.

- Connections tie links together to form end-to-end communication paths, using adaptation as needed.

- Bindings allow links at a given layer to be bound to a specific network path in the underlying layers (for example, an IP path may be bound to a set of optical link spans). This ability to capture hierarchical aggregations through bindings makes the model abstractions useful for multi-layer control.

Coordination of network resources across multiple network layers with multiple degrees of freedom allows for optimal service fulfillment. For example, optical networks are typically manually configured with a maximum utilization per wavelength of approximately 60% to allow for unanticipated traffic demands and optical link degradation over time and to provide support for occasional traffic bursts. But without multi-layer control it is difficult to utilize the approximately 40% of available capacity to support dynamic bandwidth requests.

In addition to single (optical) domain control optimization, the network OS will support nesting of IP and optical domain controllers along with the necessary interfaces between the respective path computation engines, allowing real-time mapping of service requests to actions at the appropriate network layer(s). In today's networks the IP layer has a fixed capacity view of the underlying optical network; without multi-layer control the goal is simply to maximize the utilization of this fixed network capacity and the joint IP/optical network is only optimized at much larger time scales (quarterly or yearly intervals). However, the need to support high-bandwidth dynamic network services driven by the multitude of distributed data center traffic sources will now make it necessary to jointly optimize the IP/optical network at least on a daily basis for certain segments of the network. Such network optimization will help network operators drive high network utilization, maximize return on investment and manifestly improve network planning.

To realize the vision of real-time reoptimization of network services and service-aware resiliency mechanisms, novel algorithms will need to be implemented for rapid and scalable multi-layer control that jointly optimizes and ensures efficient utilization of the physical network fabric within a constrained time window. The algorithms must be able to arrive at quasi-optimal solutions in short time scales (for example, a few seconds). In the network OS, this will be achieved by both the autonomous distributed and central controllers implementing feedback loops that take coordinated action within and between network domains.

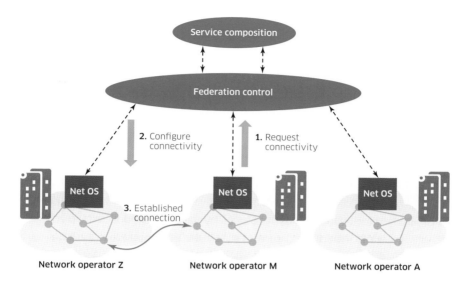

FIGURE 16: Federation of multiple network providers.

Federating control

The concept of federation of network domains is intended to allow the creation of a large-scale and diverse infrastructure for connectivity technologies, services and applications by interconnection of two or more independently managed network domains to create a virtually unified network. The independent participating domains will mutually benefit from participating in the federation. Today end-to-end service provisioning requires configuring distinct network domains that have very little visibility across them especially when involving different network technologies. In the IP domain network federation mechanisms such as BGP are already used to achieve the high level of automation in the Internet today, but the current BGP lacks QoS mechanisms to support guaranteed services. To support such guaranteed end-to-end automated services, federation mechanisms will need to extend to all network technology domains. For instance, referring back to figure 12, the illustrated end-to-end enterprise service traverses multiple network domains – access, metro and core networks. Even within the access domains the service can involve different access technologies, both fixed access and multiple radio access technologies.

In the network federation architecture shown in figure 16, the domains are geographically dispersed and owned by different organizations.

In the network OS such domains are considered as part of a single virtual environment, and the centralized federation coordination function within the network OS imposes only a minimum overhead to the owners of network domains and their customers. Network domains will publish their services, features and capabilities, and a balance must be found between service creation speed and efficiency and fine-grained management and multi-domain complexity to enable practical, scalable, robust end-to-end network control.

The operation and management of the federated network infrastructure will use new (common) mechanisms and tools to do the following:

1. Describe the network in terms of a set of common network abstractions

2. Locate network connectivity resources by using a common federated network state, including all current blocked paths and failure events

3. Orchestrate services and infrastructure based on intent-based networking and with multi-layer control that is common across the federated set of networks

The orchestration mechanism uses the network abstractions and graph-based description of network services to enable the composition and invocation of services. The network OS will also support a brokering mechanism to generate the service request and to check infrastructure availability before provisioning the service. Such federation must clearly be dynamic and evolve over time based on network and service requirements. The network OS will support these dynamic federations through the interworking of service-level compliance checkers and policy management systems that monitor service performance of ongoing federations and, based on previously configured policies and network state updates, chooses the optimal federated connectivity.

Finally, the same federation mechanism that is described above for different network domains can, and will, apply to the federation of different operator networks, to create a global-local "cloud integrated network" fabric for the new digital everything, anywhere era.

The preceding section describes the concept of a future network OS and the capabilities it will possess to optimize the control of the underlying network fabric and all future innovations in the networking layers. In the next section, we describe the essential innovations we foresee in the key networking domains, from the optical layer up to the routing layer.

The tunable network fabric

The disruptive shift of computing infrastructure to an "instantly adaptable and elastic" capacity paradigm, initiated by the advent of Amazon Web Services

(AWS), is causing networks to evolve to a similar paradigm of the "instantly adaptable and elastic" network. This will be made possible by the concept of the "tunable network fabric", a highly flexible network-wide architecture optimized for automation under network OS control. Together, the tunable network fabric and network OS will enable fully top-to-bottom programmable networks that can piece together the right network components at any scale, creating the perfect virtual network for an application in near-real-time and with the same ease as "ordering from Amazon."

Creation of the perfect virtual network will be enabled by virtualization of networking elements and functions leading to the ability to automate the "assembly" of virtual network components to form these perfect virtual networks. To support and adapt these dynamically instantiated virtual network elements, the underlying physical infrastructure will become disaggregated into more elemental flexible systems, which will be coordinated, optimized and programmed by end-to-end SDN control.

In addition, today's meshed hub-and-spoke architecture will evolve to encompass many more distributed data centers hosting virtual network components and applications. These localized edge clouds, centralized core clouds, and large web-scale clouds will become the new main hubs of the network, where a significant proportion of the traffic terminates or is aggregated, switched and service-chained through the multiple virtual components of a virtual network.

> Creation of the perfect virtual network will be enabled by virtualization of networking elements and functions leading to the ability to automate the "assembly" of virtual network components to form these perfect virtual networks.

In the subsections that follow, we elaborate on the key technologies that are integral components of a tunable network fabric, starting from the data center, where increasingly all flows will originate and terminate, and extending out to examine the evolution of all key networking elements in the wide area network (WAN) as well.

Software switches as the first hop in virtual networks

In data centers today, the number of virtual machines and containers exceeds the number of physical servers, so that the majority of network ports are virtual and not physical. Virtual machines connect to virtual ports and virtual software switches in the hypervisor are then the "first hop" switches that provide network services to virtual machines (Open vSwitch). Traffic leaving the physical machines is essentially tunneled across physical switches in a virtual network that is decoupled in configuration from the underlying physical network. This decoupling enables the programmatic control and instantiation of the virtual network needed for the "instantly adaptable and elastic" paradigm, and allows these virtual network overlays to be built on top of any physical layer infrastructure.

In the future, it will be critical for software switches to be able to achieve forwarding performance that can drive the 40 Gb network interface cards (NICs) common today and the 100 Gb and beyond interfaces of tomorrow. Just as commodity hardware has evolved to include specialized instructions for floating point, multimedia and vector instructions, operating systems and NICs will increasingly incorporate more and more specialized optimizations for acceleration of packet processing.

Virtual appliances as the building blocks for virtual networks

Today networks support many enhanced services beyond basic packet forwarding. Virtual networks will need similar services on a per-virtual network basis. Examples include firewalling, deep packet inspection (DPI), intrusion detection system (IDS), load balancing and WAN acceleration. The dedicated hardware appliances that have traditionally provided these network services are not designed for the "instantly adaptable and elastic" paradigm and are therefore unsuited for the future tunable network fabric. This deficiency is increasingly being addressed by implementing network services as virtual appliances that can be instantiated and managed in much the same manner as applications running inside virtual machines or containers. Then service-chained virtual appliances become the primary building blocks used by the tunable network fabric to automatically assemble virtual network components to form an on-demand "perfect" virtual network.

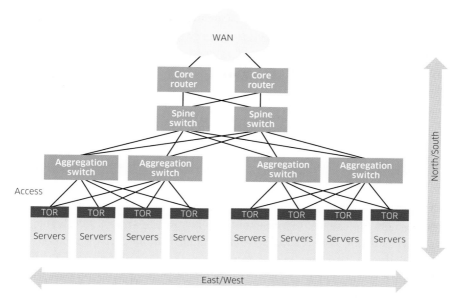

FIGURE 17: Data center network fabric

High-volume switching adds power to the new network hubs

Because traffic aggregated into hub data centers typically needs to be chained through multiple network services before egress to the destination, traffic entering and leaving the data center (north-south traffic) is effectively multiplied within the data center by generating additional east-west traffic. Indeed, server-to-server east-west traffic within the hub data center can be as much as five or more times higher than traffic entering and leaving the hub data center, based on Bell Labs models. In addition, with the high multiplexing of workloads on multi-core, multi-processor server cards, driven by containerization and virtualization of the OS, there will be increasing demand for high-speed network I/O, with server network-interfaces continuing to evolve from 10 Gigabit Ethernet (GigE) to 40 GigE and 100 GigE and beyond. Consequently, server interconnection within these networking hubs will require a new, highly scalable data center switching infrastructure to handle this high east-west traffic. The trend to avoiding east-west bottlenecks by eliminating oversubscription will further drive up (non-blocking) switching capacity needs in the data center.

Because of the multiplier effect from service chaining, switches carrying the east-west traffic (figure 17) perform most of the switching in the network and need to be low cost (capital and operational expense).

Similarly, the rise of distributed service architectures and algorithms based on MapReduce-type algorithms and Hadoop or Cassandra clusters, server disaggregation to improve server utilization, and web-service "mash-ups" and the large amount of east-west intra-data center traffic they generate have also led to adoption of large numbers of simple large-scale commodity switches in data centers by web-scale data providers (Jain 2013). While this has resulted in lower capital costs per unit switching capacity, lowering operational expenses requires modifying the switch management software to make switch management similar to server management.

Web-scale providers have typically developed multiple utilities to manage these switches as Linux-based appliances and so integrate switch management with server management (Open Compute Project 2011). This trend will simplify network management in data centers, lower operations costs for the massive data center switching infrastructure and facilitate automated operations by lowering the boundaries between cloud, server and (electrical) switch management.

An alternative approach to data center switching that is gaining momentum is the use of more photonic switching, as a complement to or replacement for some electrical switches. A couple of issues arise with electrical switching; each electrical switch interface requires an optical interface (to connect to the fiber infrastructure that connects these switches to each other and to the top of rack switches that connect to the servers) and higher latency can occur when a packet gets queued in a switch as it is forwarded across multiple switches, one or more of which may experience instantaneous congestion (as the data center architecture is not non-blocking for all traffic scenarios). These two factors result in excess capital cost as the optics are increasingly the most expensive element of the switch, and the useful computation that can be done across the set of servers in the data center is reduced by the latency, resulting in the need to expand data center capacity to compensate. In addition, there is an operational cost associated with the management of the fiber infrastructure. We therefore expect to see much lower-cost optical techniques being applied in the data centers, based on shared optical sources (lower cost) and multiple wavelengths (less fiber), and the addition of a photonic switching layer to optimize the dynamic reconfiguration of low-latency data center network paths.

In addition to intra-data center photonic switching, we expect to see high-scale (>1000 ports) SDN-controlled photonic cross-connects appearing, to

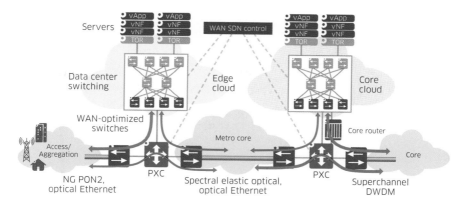

FIGURE 18: Programmable edge cloud and core cloud hubs

enable flexible reconfiguration of interconnections between WAN domains as shown in figure 18. However, in order for such photonic switching to become prevalent, new photonic cross-connect (PXC) designs will be required to achieve significant improvements in scale, cost and footprint.

Though north-south WAN-facing traffic is considerably smaller than the east-west traffic at hub locations, the fundamental cost profile of WAN infrastructure requires that these links be highly utilized without adversely affecting the QoS for the huge number of virtual connections carried on these WAN links. The optimization of WAN utilization has been a primary motivation for web-scale providers' use of SDN in their wide area DCI networks. The future network OS we envision will again play a critical role in the evolution of DCI networks, leveraging the centralized view of network resources, tracking and coordinating traffic demands and optimizing capacity usage for flows and aggregations of flows across the network. These controls rely on traffic management functions (for example, queuing, policing, shaping) in the data path in order to ensure QoS for each managed aggregated flow class and VPN. Doing this at scale at the data center switching/WAN boundary requires much greater capabilities and wide area path controls than available on generic data center switches. Consequently, a new set of high-throughput, low-latency, WAN-optimized switches will emerge to complement the generic data center switches for WAN-facing traffic interfaces.

Using the complement of capabilities described above, the new set of network "edge cloud" data centers will become the intelligent hubs around which the rest of the network is built.

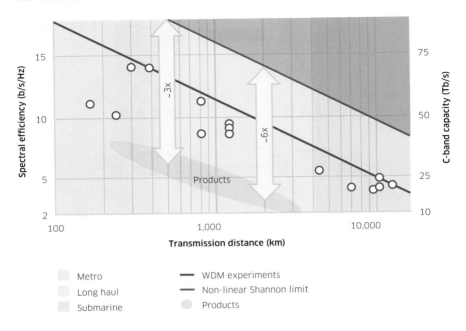

FIGURE 19: Optical systems approaching the non-linear Shannon limit

Scalable, flexible and adaptive optics

Interconnecting all of these large data center hubs, as well as the centralized web-scale data centers, will require orders of magnitude scaling of optical networks with corresponding economies of scale. Enabling the level of network programmability needed to accommodate dynamic demands will also require the optical transport to be dynamically flexible to a far greater extent. The following essential innovations will be key to realizing these needs in the foundational layer of the tunable network fabric of tomorrow.

Ultra-scalable optical networks

Current fiber optical transport systems are already exploiting the amplitude, phase and polarization of the light to encode and transport data. And using these approaches, the maximum information density per unit of frequency, or spectral efficiency, of fiber – the so-called "nonlinear Shannon Limit" – is being reached as shown in figure 19 (Winzer 2014).

Current DWDM with around 100 laser frequencies and data modulation rates of around 30 Gbaud can achieve capacities exceeding 10 Tb/s over long-haul distances. Achieving further gains of greater than 10 times that capacity will require leveraging the two remaining physical dimensions not yet exhausted – frequency and space. Cost-effective expansion of the frequency (wavelength) space will be enabled by ultra-wideband optical amplifiers, transceivers and wavelength steering components, allowing utilization of DWDM operation in the more challenging longer and shorter wavelength "L" and "S" bands of the low-fiber-loss window, adjacent to the conventional "C" band (figure 20 [Winzer 2014]). The resulting system capacity scales directly with this increase in (optical) spectrum usage, resulting in a three to five times capacity growth potential with larger gains possible in unamplified spans using the full transmission window.

However, for each of the available wavelengths, it is still important to maximize the transmission rate in order to maximize the total throughput of the optical transport system. Nevertheless, we are almost at the limit of the digital signal processing speeds required for sophisticated higher-order modulation schemes, so alternative approaches will need to be employed, using multiple parallel carriers with lower-order modulation and more efficient usage of spectrum to achieve 1 Tb/s rates. Alternatively, we could use multiple spatial paths in parallel. Both approaches are forms of what are called "superchannels".

There are two primary types of superchannels, as shown in figure 21.

Spectral superchannels – leverage smaller carrier frequency spacing than the conventional 50 GHz (typically a multiple of the "FlexGrid" 6.25 GHz spacing, such as 12.5 GHz), with each subcarrier being modulated at a lower rate to achieve the target overall rate with the desired economics. Figure 22 shows three Tb/s examples, illustrating different trade-off points between symbol rate and number of subcarriers.

Spatial superchannels – a form of space division multiplexing (SDM) in which transmission occurs over multiple parallel physical fibers (multiple distinct light paths or optical modes inside a multimode fiber), allowing data to be transmitted over multiple parallel fiber paths to achieve the desired higher overall rate between two points. Large-scale integration of the optical transceivers, amplifiers and other components will be required to drive economies of scale with this SDM approach, as shown in figure 23, compared with the conventional wave-division multiplexing (WDM) plus spectral superchannel approach.

Demand-adaptive metro

While core optical networking demands overall scalability and relatively coarse levels of flexibility, metro optical networks connecting edge clouds

Attenuation of various fiber types

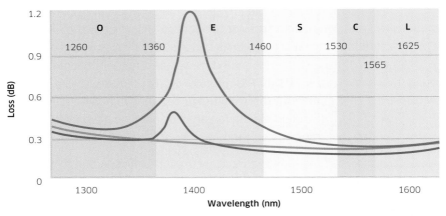

FIGURE 20: Extending DWDM system bandwidth from C to C+S+L wavelength bands

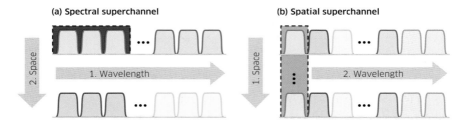

FIGURE 21: Spectral and spatial superchannels for ultra-high-speed interfaces (Winzer 2014)

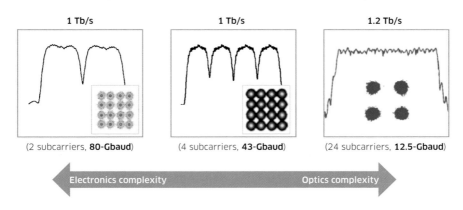

FIGURE 22: Tb/s spectral superchannels (Raybon 2012, Renaudier 2012, Chandrasekhar 2009)

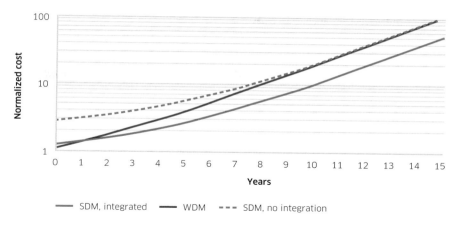

FIGURE 23: Cumulative cost gain from SDM with integrated components, with capacity growth at 40% CAGR (Winzer 2014)

and enterprises must be optimized to handle much larger-scale variations in traffic at different times. In order to address this need, new elastic optical networking technologies will emerge, combining the dynamic bandwidth flexibility of packet technologies with the cost effectiveness of optical layer transport. Two approaches will likely be employed in future:

1. Using new "optical-Ethernet" switching approaches to efficiently inspect and forward flows based on (simple forwarding) information that resides between the optical and the Ethernet layers

2. Elastically varying the amount (width) of spectrum used for each optical path based on demand by using intelligent, auto-tuned "coherent optical modems" that autonomously adjust signal modulation and therefore the rate of each wavelength to achieve the highest capacity operation possible over the fiber span. Such dynamic rate-reach adaption will also be important in core DWDM optical systems, allowing capacity to be optimized as paths dynamically change due to restoration and optimization actions.

It should be apparent from the above that a wide variety of different technologies and techniques will be employed to create a tunable network fabric across the optical and IP domains. However, there is another dimension of potential optimization that will likely occur above the IP layer, as described in the next section.

Content routing

While the tunable network fabric can provide elastic bandwidth on demand from any endpoint to virtually any endpoint, the efficiency of the tunable network fabric can potentially be improved by judiciously caching information and content closer to the user. This is the principle behind today's content distribution networks (CDNs), which are architected as an overlay service on top of the IP layer, using techniques such as DNS and HTTP redirection to direct the request to the optimal local cache.

Over the past decade there has been much research into a concept known as *content-centric networking* (CCN) or, equivalently, *information-centric networking* (ICN), for which the idea was to replace the IP layer (which is host address based) with a new I/CCN layer that is content name based. The fundamental thinking has been that the location of an object (file) was less important than the name of the file, because the requested file was stored and could be found in multiple locations, so routing based on the name was more efficient than routing based on the location. However, this logic becomes questionable when the number of objects becomes very large and the interest in each object more limited, as will be the case with the advent of more user-generated content and with the rise of IoT. Therefore, we believe that a hybrid approach, which combines the current IP-based data plane with edge SDN control and cloud-based *interest brokers* and translates user-expressed interests to eventual content-server addresses, will be adopted in the future, rather than the clean-slate routing paradigm envisaged by I/CCN.

This hybrid approach is illustrated in figure 24. Here, a user interested in a type of video content sends an interest packet expressing that interest using a well-defined namespace. This interest packet is forwarded at the IP layer to an edge SDN switch, which sends the packet to an interest broker. The interest broker translates the request to a specific content item using metadata about content items and users' interests, and returns the content name to a local stream broker, which then returns an appropriate local server address for this content item to the user. The user then contacts the content server directly to obtain the desired video stream. In this architecture, the global interest brokers and the local stream brokers have the necessary metadata about user interests and content locations for translation of user interests to the appropriate stream server address. The IP forwarding plane together with edge SDN control is then just responsible for forwarding requests between the users and interest brokers and for setting up the paths for stream delivery.

This interest-based model can be applied to all forms of content, including static, dynamic and streaming data as well as generic client-server applications.

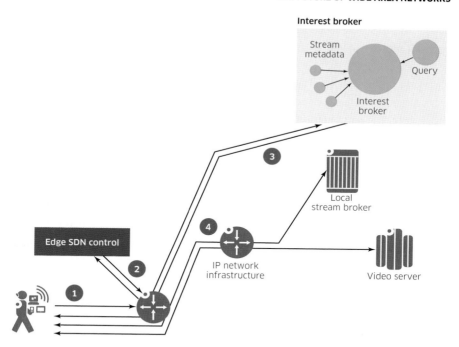

FIGURE 24: Interest-based content retrieval architecture showing how clients are connected to content via interest brokers over an IP network

Furthermore, it leverages the current IP data plane and does not require the tunable network fabric to support a completely new forwarding plane, as is the case with conventional I/CCN approaches. In addition, any new form of content-based routing can be supported by simply modifying the interest broker function, rather than modifying the underlying IP network fabric.

Dynamic connectivity marketplace

As discussed earlier, as networks become programmable, and connectivity customers of all types adapt their operations to take advantage of new elastic service flexibility, dynamic cloud networking will become the fastest growing segment of the overall network connectivity market. Large multinational enterprises, global service providers and CDNs, and web-scale cloud data center companies have the scale and sophistication to be the first to take advantage of this new networking paradigm. Because these types of companies are typically global in scope, new forms of dynamic multi-provider programmable networks will be essential to satisfy the predominant upfront demand. Multiple new business models will likely emerge, with two primary options being what we term "alliances" and "brokered marketplaces".

Similar to the Star Alliance and OneWorld alliances in the airline industry, global network provider alliances or consortia will likely form to establish automatically federated *cloudnets* that provide global reach with ubiquitous local access in each geographic market. These alliances will likely each be anchored on one or more large global network providers. These global-local federations will be made possible by the combination of the tunable network fabric and programmable SDN control within each provider network, and the broader network OS, which federates control over this consortium of autonomous networks.

Ultimately, an open brokered connectivity marketplace will develop. As we have seen across a variety of industries, any market with fast and easy access to a valuable but ephemeral commodity (seats on a flight, ads on a web page, electrical energy on a transmission line, cloud capacity, or gigabits per second on a fiber) will naturally move toward a spot market with dynamic or even real-time pricing. Spot markets more efficiently balance supply and demand, raise the utilization (and thus efficiency) levels of expensive infrastructure, and ultimately maximize economic value within an industry.

Indeed, this development will bring about many changes and new opportunities:

- As in other spot markets, dynamic brokers and aggregators will emerge (or evolve) to efficiently match up large numbers of consumers and suppliers of on-demand capacity in real time

- Large cloud provider and network provider data centers will become the natural exchange hubs of this market, as they represent the source and destination for much of the demand

- Real-time access will open up a much greater opportunity to mix and match different suppliers, types of connection technologies (fiber, microwave, satellite, free-space optics) and reliability levels, as diverse alternate paths become plentiful and readily accessible – all of which will enhance the creation of new economics and business models

- Like other dynamic marketplaces, the supply and demand, as well as the pricing and provisioning of this on-demand connectivity exchange, will be primarily transaction-driven, instead of relying on long- or even medium-term contracts

- New multi-tiered pricing and auction models, making extensive use of analytics-enabled optimization, will be designed to address the dynamic, context-dependent nature of capacity transactions and facilitate real-time bidding to ensure maximum monetization of service opportunities

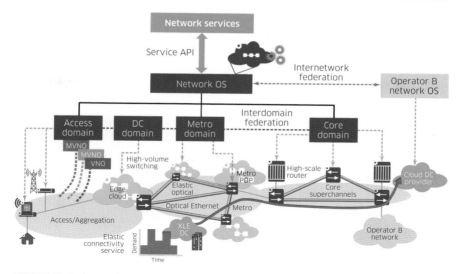

FIGURE 25: End-to-end perfect network architecture

- Ultimately, as on-demand programmable network functions and capacity become the norm, the legacy static capacity remaining for lease will become relegated to discount stock, in some cases absorbed as part of higher-value on-demand offerings by the real-time aggregators

In such a dynamic world, those entities (cloud providers, enterprises, service providers, network providers or brokers) with the most automated operations and best alliances with global-local reach will stand to benefit the most, and will be able to leverage their dynamic agility – as supplier or consumer of bandwidth – to win in the marketplace.

End-to-end perfect future network

In the coming "everything digital" era, the networks of the future must be built around fundamentally different operating principles that allow optimization from top-to-bottom and end-to-end, with massive scale and multinetwork automated service federation, all accessed by programmatic interfaces. The overall architecture of this network is depicted in figure 25.

The essential building blocks of this network with seemingly infinite capacity are:

- A highly scalable, tunable network fabric built around data center switching hubs embedded in the network as edge clouds, disaggregated SDN-controllable network functions and components, adaptive impairment-aware multi-layer IP/optical optimization, elastic optical metro transport and ultra-scalable core routers and transport

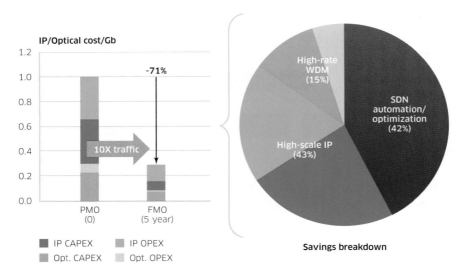

FIGURE 26: Estimated cost/GB savings of high-scale tunable IP/optical network, with 10x traffic growth over five years

- A network OS that fully automates and exposes the tunable network fabric and optimizes the federation of control across multiple network domains and network operator boundaries

Bell Labs estimates (figure 26) that the TCO savings for such a network, built to support a 10x increase in traffic over five years, will be approximately 70% on a per-unit capacity basis. Over half of this benefit is attributable to the high-scale, high-density routing and optical transport advances, while over 40% of the benefit arises from SDN automation and network optimization.

This near order of magnitude change in the economics of supply and demand confirms that a technological revolution is indeed upon us, driven by the dynamic networking of new digital goods (new information, data, content streams from people and things), and resulting in an economic transformation.

The future begins now

Although the preceding discussion sounds rather futuristic, the start of this ambitious transformation of networks is already underway. Here we present a few examples of the ongoing transformation of networks toward the "instantly adaptable and elastic" paradigm, spanning standardization, open source initiatives, collaborative industry efforts and product trials and releases:

- Network equipment vendors are implementing cloud orchestration and automation systems and SDN controllers that together can create virtual networks on demand to connect virtual machines within data centers and via secure tunnels to virtual machines in other data centers and in the enterprise. This network automation is managed as part of the virtual machine life-cycle automation, allowing a virtual machine's network connectivity to follow it even if it is migrated from one server to another.

- The IETF's NVO3 group is enhancing data plane and control plane protocols to enable IP-based network virtualization in the data center; for example, using the popular VXLAN tagging protocol, as well as BGP extensions to carry virtual network information.

- The Open Networking Foundation (ONF) has been developing and enhancing the OpenFlow protocol used for programming switches in an SDN environment, and the associated OF-Config protocol for configuring and initializing switches. The ONF is now also extending OpenFlow to address control of optical networking systems.

- The OpenDaylight collaborative project is an industry initiative to create an SDN controller infrastructure with a standardized northbound API. The controller infrastructure is aimed at multivendor deployment and provides a standardized programming interface to the network.

- The Open Network Operating System (ONOS) is aimed at building a carrier-grade network operating system for SDN. It provides a distributed core of identical controller instances to isolate device and controller instance failures from impacting the rest of the network and a new programming abstraction based on intents, which allows an application to request services without having to know the details of how the service will be performed.

- There is growing industry acceptance of the NETCONF protocol for managing network elements and their configurations and of the related YANG data modeling language for specifying them. Most devices now have an associated YANG model. Also, the OpenDaylight controller uses YANG extensively.

- 400 Gb coherent optics are already being integrated by leading vendors both as router and transport interfaces to create high-scale core networks, with rate tunability. Terabit superchannel systems have been demonstrated in the laboratory as well as in operator field trials.

The fundamental technology shift to cloud computing that happened in the computer industry is now driving networks to transform around the same cloud-computing paradigm of instant adaptability and elasticity.

Summary

The fundamental technology shift to cloud computing that happened in the computer industry is now driving networks to transform around the same cloud-computing paradigm of instant adaptability and elasticity. This transformation requires network fabrics to become tunable to dynamic service needs to support the ability to create the "perfect network" for any application or service on demand. The future tunable network fabric we have described will accomplish this by providing the means to assemble a set of virtual network components to form a custom virtual network with any global-local extent.

These dynamic virtual networks will be controlled by a network OS that provides an abstracted global view of the network to networking apps that use the abstracted view to build customized services. The abstracted view is made possible by adoption of the SDN-based logical centralization of key control plane functions.

New networking apps will drive the development of a bandwidth marketplace for matching consumers and suppliers of on-demand capacity. This marketplace will allow large cloud providers to take advantage of a capacity spot market to drive up average utilizations by absorbing peaks through spot market acquired capacity.

We are truly at the dawn of a new era of IP and optical networking of edge and centralized clouds that will form the basis for transformation of business and consumer lives and a new technological revolution.

References

Bell, C. J., Hogarth, J. M. and Robbins, E., 2009. "U.S. Households' Access to and Use of Electronic Banking, 1989–2007," *Federal Reserve Bulletin* (http://www.federalreserve.gov/pubs/Bulletin/2009/articles/onlinebanking/default.htm#t1).

Bell Labs Consulting, Growth in US Real-Time Bidding Market.

Belobaba, Peter; Hernandez, Kari; Jenkins, Joe; Powell, Robert and Swelbar, William, "Productivity Trends in the US Passenger Airline Industry 1978-2010," Transportation @ MIT Program, MIT Press, MIT, Cambridge, MA.

Bjorklund, M., 2010. "YANG – A Data Modeling Language for the Network Configuration Protocol (NETCONF)," RFC 6020 (https://tools.ietf.org/html/rfc6020).

Broeders, H., and Khanna, S., 2015. "Strategic choices for banks in the digital age," *Insights & Publications*, McKinsey & Company (http://www.mckinsey.com/insights/financial_services/strategic_choices_for_banks_in_the_digital_age).

Brueckner, Jan K. and Spiller, Pablo T., 1994. "Economies of Traffic Density in the Deregulated Airline Industry," *Journal of Law and Economics*.

Chandrasekhar, S.; Liu, X.; Zhu, B. and Peckham, D. W., 2009. "Transmission of a 1.2-Tb/s 24-carrier no-guard-interval coherent OFDM superchannel over 7200 km of ultra-large-area fiber," in *Proceedings of the European Conference on Optical Communication* (ECOC) 2009, Sept., paper PD2.6.

eMarketer 2013, Programmatic Ad Spend Set to Soar: Real-time bidding expected to account for largest chunk of programmatic ad budgets (http://www.emarketer.com/Article/Programmatic-Ad-Spend-Set-Soar/1010343).

Enns, Rob, 2006. "NETCONF Configuration Protocol," RFC 4741 (https://tools.ietf.org/html/rfc4741).

Federal Reserve System 2013. "The 2013 Federal Reserve Payments Study" (https://www.frbservices.org/files/communications/pdf/research/2013_payments_study_summary.pdf).

FierceTelecom 2011. "XO's EoC growth driven by mix of organic growth, partnerships," *FierceTelecom*, May 3 (http://www.fiercetelecom.com/special-reports/competitive-carriers-hone-their-ethernet-over-copper-skills/xos-eoc-growth-driven-mi).

Hao, Fang; Jansen, Arnold; Kodialam, Mualidhara N. and Lakshman, T. V., 2015. "Path computation for IP network optimization," *TechZine*, Alcatel-Lucent, June.

Jain, S. et al, 2013. "B4: Experience with a Globally-Deployed Software Defined WAN," *Proceedings of the ACM SIGCOMM Conference*, 2013.

Kenshoo 2013. New Research Study from Kenshoo Finds Facebook Advertising Boosts Paid Search Return on Ad Spend by 30 Percent, *Press Release*.

Martínez, Germán, 2009. Gemalto, "Mobile Financial Services," slide 2, Banking Channel Trend (http://www.slideshare.net/daniel_b4e/evolucin-de-servicios-financieros-mviles-en-latinoamrica-2487277).

Neely, Michelle Clark, 1997. "What Price Convenience? The ATM Surcharge Debate," *The Regional Economist*, July, Federal Reserve Bank of St. Louis (https://www.stlouisfed.org/publications/regional-economist/july-1997/what-price-convenience-the-atm-surcharge-debate).

ONOS (http://onosproject.org/).

Open Compute Project 2011 (http://www.opencompute.org/).

OpenDaylight (https://www.opendaylight.org/).

Open vSwitch, "Production Quality, Multilayer Open Virtual Switch" (http://www.openvswitch.org/).

Phan, Son, 2006. Cisco Systems Vietnam, "one bank. one architecture" (http://www.slideshare.net/p2045i/one-bank-one-architecture?next_slideshow=1).

Raybon, G.; Randel, S.; Adamiecki, A.; Winzer, P. J.; Salamanca, L.; Urbanke, R.; Chandrasekhar, S.; Konczykowska, A.; Jorge, F.; Dupuy, J.-Y.; Buhl, L. L.; Draving, S.; Grove, M. and Rush, K., 2012. "1-Tb/s dual-carrier 80-Gbaud PDM-16QAM WDM transmission at 5.2 b/s/Hz over 3200 km," in *IEEE Photonics Conf.* (IPC), 2012, paper PD1.2.

Renaudier, J.; Bertran-Pardo, O.; Mardoyan, H.; Tran, P.; Charlet, G.; Bigo, S.(*); Konczykowska, A.; Dupuy, J.-Y.; Jorge, F.; Riet, M. and Godin, J., 2012. "Spectrally Efficient Long-Haul Transmission of 22-Tb/s using 40-Gbaud PDM-16QAM with Coherent Detection," in *Proceedings of the Optical Fiber Communications Conference 2012*, paper OW4C.2.

Skinner, Chris, 2015. "Why the mobile ecosystem just became easy," *Financial Services Club Blog* (http://thefinanser.co.uk/fsclub/2015/03/why-the-mobile-ecosystem-just-became-easy.html).

Tholin, Kathryn, 2006. "Residential Real-Time Pricing: Bringing Home The Potential," *Assessing the Potential for Demand Response Programs,* The Institute for Regulatory Policy Studies, Illinois State University.

Tower Group, McKinsey & Co, Novantas. "Banking Transactions by Channel."

U.S. Bureau of Labor Statistics (http://www.bls.gov/lpc/#tables).

West, M., 2011. "30 Years of Airline Travel," *Contrail Science,* March 9 (http://contrailscience.com/30-years-of-airline-travel/).

Wikipedia 2015, "Metcalfe's Law", (https://en.wikipedia.org/wiki/Metcalfe%27s_law").

Winzer, Peter J., 2014. "Spatial Multiplexing in Fiber Optics: The 10X Scaling of Metro/Core Capacities," in *Bell Labs Technical Journal,* Volume 19, September.

CHAPTER 5

THE FUTURE of the CLOUD

The rise of the new global-local cloud

Mark Clougherty and Volker Hilt

The essential vision

Whether one considers the start of the current cloud era as the 1999 debut of salesforce.com or the introduction of Amazon Web Services (AWS) in 2006, there is little doubt that cloud computing has had a profound impact on our lives over a span of only 10 to 15 years. It is no longer necessary for companies to install complex and expensive equipment with dedicated network connectivity to serve their customers. Cloud computing allows services and content to be offered globally and accessed by anyone with access to the Internet. Furthermore, with the availability of pay-as-you-go cloud infrastructure services like AWS, the barriers to market entry for new services are eliminated. A handful of talented software developers are now able to build service offerings capable of serving millions of users without a huge upfront investment in computing infrastructure. As a result, an entirely new cloud-based economy has emerged.

This first cloud era was built on a regionally centralized cloud, with applications running in a handful of massive warehouse-scale data centers within key geographic regions (North and South America, Europe, Middle East and Africa, and Asia Pacific). While this centralization has enabled economies of scale, it is ill-suited to support services that require latency or bandwidth performance guarantees because distance and the number of network hops generally impairs the ability to offer such control or assurances. Furthermore, geo-political security issues challenge the centralized cloud approach because governments and enterprises require that their data remains within (or avoids) certain political boundaries. As a result, we are on the cusp of a new type of cloud that will address the needs of the "everything digital" era by providing local delivery of services while maintaining the critical aspect of global reach. The emerging cloud will interconnect and federate local resources with the centralized cloud infrastructure to form a new global-local cloud continuum.

This trend is already well underway for content delivery. Over the last decade, the dramatic increase in the delivery of high-quality video via the Internet outpaced the ability of the Internet to meet the associated bandwidth demands economically. As a result, global content delivery networks (CDNs) were created to optimize delivery by allowing content providers to bypass the Internet backbone and deliver content from a network of geographically distributed caches that are closer to users.

As we move into the age of (near) real-time digital automation in which applications and services increasingly require low-latency and/or high-performance interactivity with humans (and devices), functionality must be placed as close to end users or devices as economically possible. As a result, a highly distributed cloud infrastructure will be needed with edge clouds based on new

As we move into the age of (near) real-time digital automation... functionality must be placed as close to end users or devices... with edge clouds based on new cloud technology optimized for high-performance and dynamic operation.

cloud technology optimized for high-performance and dynamic operation. Intelligent algorithms will determine optimal placement of applications within this distributed infrastructure by considering the needs of the services, availability of appropriate resources and location of users and devices.

Applications and services will continue to adapt to changing user needs, shifting usage patterns, including rapid and extreme changes in popularity, the mobility of end users and devices, and changing network conditions. These applications will be updated constantly with new functionalities. This new dynamic in applications, services and demand will require the new global-local cloud to be a learning, automated, and self-optimizing computing and processing digital fabric for the new era.

The past and present

In the fall of 1999, salesforce.com went live and a revolution began. Although the Internet was already entering consumers' daily lives (Netscape was founded in 1994 and Amazon and eBay in 1995), the notion of delivering business services to enterprises via a browser was unknown. With the launch of salesforce.com, businesses no longer needed to purchase and install equipment and software to support enterprise applications; everything could be delivered over the Internet from servers located in third party-owned data centers. Thus, the first era of enterprise cloud services was born and the Internet was transformed from a repository of miscellaneous information to a platform by which high-value services could be created and delivered with global reach and scale. This creation of the underlying cloud infrastructure to support thousands to millions of users distributed across the globe was a non-trivial task requiring significant upfront expenditure and lead time, which hindered the ability to scale new enterprise services.

In 2006, Amazon changed the game with the introduction of AWS, bringing cloud computing infrastructure to the masses. As a result, rather than facing the daunting task of building a massive infrastructure with global reach, entrepreneurs with innovative service ideas could simply deploy their applications on the AWS infrastructure and scale at will. This dramatically reduced the barrier to entry for application providers. AWS provided instant global reach and virtually infinite scale on a pay-as-you-grow basis. Recognizing a strong growth market, many other infrastructure providers then joined the fray, which led to the creation of a global ecosystem from which cloud services could be launched. In 2015, it was estimated that

Amazon operated 1.5 million servers for its AWS, while Microsoft and Google hosted over one million servers each.

While global access to the cloud was critical, cloud operators recognized that a centralized geo-redundant architecture suffers from a number of deficiencies. First, latency is an important factor in quality of experience. Even for applications that are not particularly latency sensitive, delays of many 100s of milliseconds are frustrating to the end user because they are perceptible if any degree of real instantaneous interactivity or response is required. Second, security and trust regulations often require that critical applications and data be served from in country data centers or at least from within countries that are considered trustworthy by the enterprise or regulating government. Finally, from a network cost perspective, it may not be economically viable to serve high capacity applications from a few central sites because the cost of transporting data between the end user and a data center may offset the end user benefits of lower service cost and increased flexibility.

As a result of these deficiencies, major cloud providers have continued to expand their geographical footprint to provide localized resources to meet the needs of their customers efficiently, securely and economically. Today, Amazon operates 11 AWS data centers around the world (Amazon 2015). In addition, Amazon runs edge cloud data centers at 53 locations, 20 of which were added in the past two years. Microsoft claims to operate more than 100 data centers worldwide (Microsoft 2015). Thus, the cloud is beginning to be distributed, although it is not yet truly local.

Latency

Latency is a major issue for global CDN providers. On-demand and time-shifted video has rapidly become the dominant means by which end users consume video, as the video has evolved from short clips of low-quality video streamed to tiny windows on PCs in the early 2000s to the 1080p and even 4k streams of full length professional content delivered in 2015. Netflix movie streaming was introduced in 2008 and by 2011 had 22 million streaming subscribers making up over 32% of peak Internet traffic in the US. This growth continues, with over 62 million subscribers today (20 million outside of the US) and nearly 35% of US peak traffic (Sandvine 2014). As streaming video has become more popular as a primary means of entertainment, consumers have shifted their view to high-resolution devices, which demand high-quality video (1080p or better). The throughput required to deliver these high-quality streams is difficult to achieve and sustain in highly centralized delivery architectures due to the many hops and potential congestion points the streams encounter as they traverse the network. This is a direct result of the well-known transmission control protocol (TCP) bandwidth-delay product that provides a simple rule of thumb for the maximum achievable throughput of a single TCP stream for a given latency and window size:

Maximum throughput (bits/second) = TCP window size (bits)/round-trip latency (seconds)

For a round trip time (RTT) of 100 ms and a typical TCP window size of 64 kbytes, this results in a maximum TCP throughput of just over 5 Mb/s. Given a typical bandwidth of 6 Mb/s for a 1080p stream and over 15 Mb/s for a 4k stream, content providers are challenged to deliver high-quality content without resorting to various TCP optimization techniques. Reducing this latency to 20 ms increases the maximum TCP throughput to over 25 Mb/s. Furthermore, from an economic standpoint, it is much more effective to deliver high-bandwidth unicast streams from nearby caches than to originate millions of streams from a regional data center where the traffic must be carried over the Internet (or a private transport network). For these reasons, global content and CDN providers are continuously expanding their footprint to bring CDNs as close as possible to end users by making arrangements with local Internet service providers (ISPs) to place caches inside of (or directly connected to) public and private networks at localized points of presence (PoPs).

For example, as shown in figure 1, the global CDN provider Akamai operates over 170,000 caches in 1300 public and private networks in 102 countries, allowing them to provide high-quality streaming services to local users. Similarly, Limelight, another CDN provider, operates nearly 20,000 caches globally.

More recently, large content providers have started to establish commercial agreements directly with local and regional network operators to place their own CDNs within local networks to improve performance and reduce delivery cost. In June 2012, Netflix announced the Netflix Open Connect Network through which Netflix provided network operators and service providers with CDN servers called Open Connect Appliances (OCAs), which are placed at locations within the service provider network (Netflix 2012). Netflix reported that a few thousand OCAs were deployed in 2014 (Limer 2014). Google initiated a similar program in 2008 called Google Global Cache, which provides ISPs with Google caches to be placed in their networks at no cost to serve Google content (including YouTube). Neither Google nor their

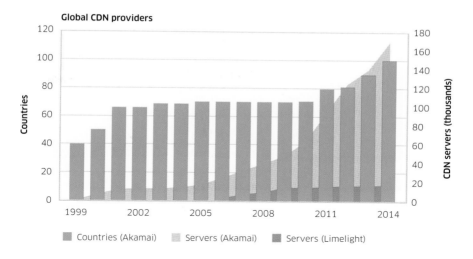

FIGURE 1: Growing localization of global CDN providers (Akamai 1999-2014, Limelight 2008-2014)

partners talk publicly about the extent of the program, but industry experts estimated that deployment rose from a handful of ISPs in 2010 to over 80% of ISPs in North America by July 2013 (MacMillan 2013).

Given these moves, Bell Labs Consulting predicts that an increasing number of global providers will leverage the proximity of local service providers to users to underpin the new global-local cloud paradigm that will emerge in the next five years. As discussed in the next section, this trend will extend beyond just CDN caches and shift the bulk of cloud resources to edge cloud locations closer to end users. As indicated in figure 2, Bell Labs Consulting projects that by 2025, 60% of cloud servers will be located in edge locations as enterprises shift their workloads to the cloud. The distribution of cloud resources between the core and edge tracks the projected traffic growth at each of these locations.

This does not mean that the size and number of centralized clouds will decrease relative to today. Quite the opposite is true. However, Bell Labs Consulting believes that the growth in edge clouds will outpace server growth in all other parts of the network.

Dynamic scalability

Another aspect of cloud services delivery to be considered is dynamic scalability. The continued rise of social media has led to ongoing increases in application popularity at unprecedented rates, which requires operators

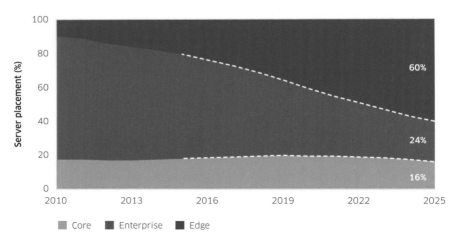

FIGURE 2: The shift to edge clouds. Source: Bell Labs Consulting (Gartner 2011, Gartner 2013, Gartner 2015, IDC 2010)

to scale up if new applications catch on or to phase out rapidly if the anticipated application uptake doesn't materialize or is ephemeral.

As one example of highly dynamic popularity, consider the video creation service Animoto. In 2013, after the Animoto Facebook application was updated to attract more use, the number of users grew from 25,000 to 250,000 within only four days. As a consequence, the Animoto cloud services running on Amazon EC2 had to scale from approximately 50 server instances up to over 3400 server instances – a 68-fold increase in four days.

In a more recent example of this rapid shift in application popularity, the live streaming app Meerkat rose from a download ranking of approximately 1100 to become one of the top 250 downloads in the Apple App Store in six days in March of 2015. It remained there for only 10 days before falling out of the top 500 in just over a day when Twitter released its competing Periscope app, which surpassed the popularity of Meerkat (in terms of Twitter mentions of #meerkat versus #periscope) within three days of its launch (Popper 2015).

Another example of this phenomenon was the Budweiser Puppy Love commercial, which was released four days before the Super Bowl in 2014 and spread rapidly on social media (figure 3a). It received 13 million views on just the first day, racked up more than 35 million views within the first five days and became one of the most shared Super Bowl ads of all time. Similarly, a Samsung commercial for the Galaxy Note 4 had a slow start (12 days!) then gained popularity very rapidly, moving from less than half a million daily

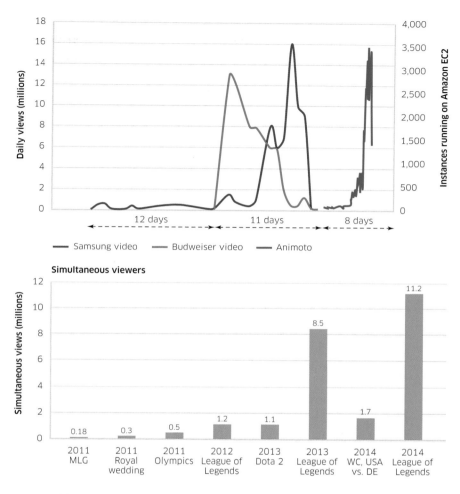

FIGURE 3 (A) AND (B): Illustration of the rapid growth in popularity of online applications, content and services (Rezab 2014, Marshal 2014, Gafford 2014, Redbeard 2014, Panzarino 2014, Makuch 2013, McCormick 2013, Tassi 2011)

views to approximately 16 million daily views within six days. After reaching its peak, the daily views dropped at an even faster pace down to less than 0.1 million daily views within four days.

Along with the extreme popularity cycles of individual items, the overall popularity of online services is growing fast. Figure 3b depicts the viewership of popular online streaming events. In 2011, Major League Gaming events started to reach online audiences of 180,000 simultaneous viewers. Since then, numbers have been constantly increasing for various types of events, with The League of Legends World Championships in 2013 setting another record as online viewership reached 8.5 million unique viewers, only to be exceeded by the 2014 event, which reached a total online viewership of 11.2 million.

Consequently, cloud-based services must be able to handle this highly dynamic traffic profile. This requires advanced monitoring, automation and orchestration capabilities, as discussed below. The network within the data center (DC) must be equally dynamic to connect newly created virtual instances. Thus, software-defined networking (SDN) solutions that automate connectivity with the DC will continue to emerge to enable creation of fully dynamic data centers. Lastly, new cloud technologies (for example, lightweight containers) will appear as the need to support legacy services and applications (for example, based on proprietary OS, customized kernel and specialized hardware) diminishes, allowing improvements in density, modularity and speed.

We are on the cusp of the next era of the cloud. The value of localization is recognized by global cloud providers, though the need has not yet fully materialized. The tools and technologies needed to build this new cloud are beginning to emerge and will be mainstream in the very near future.

The future

As noted in the previous section, some limited localization of the cloud has already begun, but a wholesale shift to the new cloud architecture has not happened. By 2020, the cloud will transform into a true hybrid of global and local resources. It will have the ability to deliver extremely high-performance services to users and adapt these services in real time as demand shifts and users move from place to place. This new cloud will enable users to access services from anywhere without service degradation and, by leveraging the globalization of access discussed in chapter 7, *The future of broadband access*, over any access network provider's infrastructure. The localization revolution will be driven primarily by two factors:

- The continuing explosion of high-quality unicast streaming video
- The emergence of a new class of hyper-local, high-performance, low-latency applications, such as mission-critical enterprise applications, real-time data stream processing applications and security applications.

The low-latency, unicast cloud

As noted previously, consumption of on-demand and time-shifted video has been rapidly increasing in recent years, both in terms of quantity and quality.

Viewers increasingly consume popular content on their own schedule and from sources other than traditional broadcast providers. Again, using Netflix as a proxy for on-demand viewing trends, the average per subscriber Netflix consumption in the US increased to over 1.5 hours per day in 2014, which is a 350% increase from 2011, according to 2014 research from The Diffusion Group (The Diffusion Group 2014). While this increase is impressive, Netflix viewing time remains small compared to traditional broadcast TV viewing, which averages 4 hours 55 minutes per day, per subscriber.

However, as reported by Nielsen, traditional TV viewing is gradually declining. It decreased from 5 hours 11 minutes in 2013 to 4 hours 55 minutes in 2015, or over 4.4% (Nielsen 2015). While this decline can be only partially attributed to the increase in streaming services like Netflix, Amazon, Hulu and others (social networking and other online activities are contributing factors), it does highlight how the network architecture must shift from one optimized for broadcast delivery of live content to one that is better suited for economical on-demand delivery of high-quality content to a highly mobile audience. This drives the need to move content as close as possible to the end user using a global service, local delivery model.

But, there is a new need that is appearing: the local processing of high-bandwidth streams for video analytics, in particular from smart devices as user generated content (UGC) or from IoT devices. Like the streaming video case, real-time video analytics involves sending many unicast streams across the network, but in the opposite direction from streaming video delivery for entertainment, with video from many end sources transported to a single location for processing. Moving intelligence away from the endpoints and aggregating video analytics in the cloud enables low-cost endpoint devices to be used, reduces operational costs and simplifies coordinated analysis across multiple devices (for example, enterprise tracking of a fleet of vehicles or shipping containers). However, carrying many video streams to a single centralized processing function places a heavy traffic load on the transport network. Localization of cloud resources at the edge of the network allows processing to be located near clusters of devices, reducing network traffic while maintaining the benefits of centralized analytics.

This notion of localized processing is also essential to the evolution of the networks themselves. It allows network operators working to virtualize their network functions to achieve the scalable economics required to serve emerging services needs. A primary example of this is the virtualization

The global-local cloud will also be a technology enabler for an entirely new class of applications that are impossible in today's cloud due to performance and latency constraints.

of the mobile network, which will enable operators to respond to changing demands at each location dynamically, with the movement of mobile users and machines, with the time of day and weekday/weekend, and supporting spontaneous clustering due to sporadic events. Separating the baseband unit (BBU) from the base station and instantiating it collocated with other virtual BBUs provides operational cost savings and flexibility (as well as improved spectral efficiency through tight BBU coordination). However, radio-layer performance requirements limit the latency between the radio head and BBU to less than 100 microseconds, and with a propagation delay in optical fiber of 5 microseconds per kilometer of fiber, the BBU can be placed no more than 20 km from the base stations it serves. This architecture is possible only with a highly distributed cloud infrastructure. Even if new split baseband architectures could support timing requirement to ~4 ms, this precludes queuing delay in an intermediate switch or router, which again argues for the need for a highly distributed cloud infrastructure.

The global-local cloud will also be a technology enabler for an entirely new class of applications that are impossible in today's cloud due to performance and latency constraints. With the hyper-local resources of the global-local cloud combined with dramatically reduced access link delay, latency between the application and end user can be reduced to the single digit millisecond (and even sub-millisecond) range. As a result, it is possible to implement applications with very tight feedback control loops in the cloud, including those involving feedback to humans. Figure 4 provides an overview of different applications and their tolerance to latency. As illustrated, application latency is the sum of network latency and the processing time taken by the application. While processing time improves with advances in technology, network latency is driven by the network architecture and physics.

Figure 4 contrasts application requirements: not all applications require the extremely low network latencies possible in the edge cloud. For example, instant messaging is an inherently asynchronous application for which users tolerate many seconds of delay. Similarly, delays of up to approximately one second are tolerable for many web pages to load. These types of asynchronous, non-real-time applications are fairly well supported by today's cloud networks (with localized CDN caches to optimize content delivery and accelerate web page loading). However, there are a set of applications that are not well suited to the 100+ ms latencies of centralized cloud implementations, including those that require human neurological response times to be effective (for example, augmented reality) and some that involve tight control loops to control and/or coordinate machines with very little tolerance for delay (for example, cloud-assisted driving). It is these classes of new applications that will be made practical by the emergence of the global-local cloud.

When considering interaction with humans, a few benchmarks should be considered. The time to blink an eye is approximately 150 ms. A world-class sprinter can react within approximately 120 ms of the starting gun. These numbers define the expectations for applications intended to interact with humans in real time. Clearly, if applications hope to approach this level of responsiveness, network latencies must be much less than 100 ms.

For reference, nerve impulses travel at a maximum speed of approximately 100 m/s in the human body. Therefore, the time required to propagate a signal from the hand to the brain (not counting the time required for the brain to process the signal) is approximately 10 ms. As network latency approaches this same range, it is possible to enable interactions with a distant object with no perceived difference compared to interactions with a local object. Perhaps even more intriguing, is that sufficiently low-latency will permit humans to interact with virtual objects rendered in the global-local cloud as though they were physically co-located.

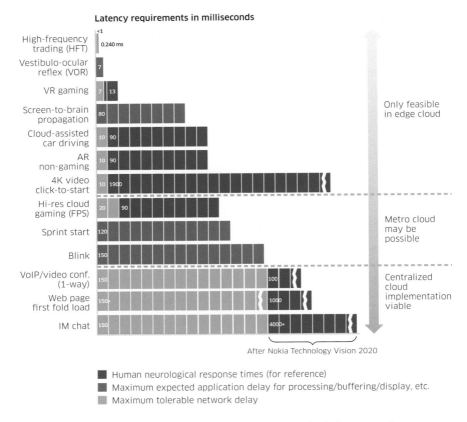

FIGURE 4: Application latency requirements compared to neurological response times (Chioka 2013, Berman 2014, Duffy 2002, Human Benchmark 2014, Microsoft 2015b)

Today, such capabilities are only possible through the use of significant intelligence in the endpoint. For example, while military drones can be controlled from halfway around the world, interaction is possible because they have sophisticated on-board processing and are capable of flying themselves without the need for real-time, low-latency feedback to the pilot thousands of miles away. But, this sophistication comes with a price – the Predator drones flown by the US Air Force cost approximately $4.5 million each (Drew 2009), which excludes this technology from use in more cost-sensitive applications that have the potential to transform human existence and enterprise or vertical industry segments. If the cost were not prohibitive, clusters of low-cost robots could be used to allow remote operators to inspect and repair critical civil infrastructure (for example, bridges, gas and water mains, sewers). This tradeoff suggests that an application in the edge cloud could provide local coordination among them and reduce the requirements for human-machine interaction.

Similarly, localized intelligence could enable sophisticated traffic control applications from which vehicles receive real-time feedback about events close enough to them to require action, but outside of the driver's field of view. Real-time analytics of localized vehicle information with millisecond latency connections to on-board intelligence will revolutionize traffic management without the need for new dedicated infrastructure. These "smart highways" could be operated with higher traffic density and higher speeds, but with fewer accidents.

To understand the delay requirement for transportation, consider an application for coordinating dense clusters of high-speed traffic. As vehicles at the front of the cluster react to events that cause them to reduce speed, the application must process this information, determine which vehicles in the cluster are affected by the event and send instructions to them to allow them to automatically adjust. At 120 km/h, 3 m distance corresponds to 100 ms of delay. With 90% of this budget allocated to the processing required for the application to make the decision and the vehicle to act on the resulting instructions, only 10 ms can be allocated to network latency, with little tolerance for variance and extremely high availability required. This can only be achieved by locating the application in an edge cloud close to the section of highway being managed.

Finally, the low-latency and high-bandwidth service delivery capabilities of the global-local cloud will usher in a new wave of innovative augmented reality (AR) applications. To meet human perception requirements today, augmented reality is dependent on on-board processing and local content. The ability to move this processing and content to the cloud makes AR more portable by allowing appropriate content to be presented to users no matter where they are. However, as the AR content becomes more immersive (for example, a video overlay of reality for instructional purposes), the lag between the user's movements and changes in the AR overlay display become critical in avoiding a poor user experience (including disorientation and nausea). Physiologically, the vestibulo-ocular reflex (VOR) in humans coordinates eye and head movements to stabilize images on the retina. Studies have shown the VOR to be approximately 7 ms. Therefore, to avoid user disorientation, similar latencies must be achieved by AR applications. Localizing resources in the global-local cloud will allow extremely low-latency coordination between the user and AR processing and content, which will in turn enable a nearly imperceptible lag between real and virtual elements. While significant issues exist in this field (not the least of which is a viable unobtrusive display device), enabling AR services to be delivered from the cloud will remove a major obstacle to future innovation.

The secure cloud

As discussed in chapter 3, *The future of security*, security organizations will leverage the localized resources of the global-local cloud to use security appliances and streaming analytics at the edges of the network that are capable of analyzing, storing and replaying all network traffic, and taking action on malicious traffic before it can enter the core. Addressing security at the edges reduces the processing load on individual security functions because traffic at the edge cloud is proportionately lower than that in more centralized clouds. Furthermore, as the cloud becomes more distributed, more traffic will originate and terminate in the edge cloud. As a result, a lower percentage of total traffic will traverse the core, so security functionality will have to be implemented at the edge.

The high-scale cloud

With massive distribution of resources in the global-local cloud, new technologies will be required for service monitoring, automation and optimization. Typically, services consist of many components interacting to support the overall service. For example, a CDN can consist of content delivery nodes, request routing/control functions, origin nodes and management functions. Each of these components is composed of multiple service functions that implement the required functionality. The functions of a service often have different performance requirements and can be placed at multiple different data centers. For example, content delivery nodes and request routing/control functions of a CDN can leverage the high-bandwidth and ultra-short latencies provided by edge data centers, while origin nodes and management functions can leverage the scale of centralized data centers.

The section below outlines critical cloud technologies that will be required for a high-scale cloud.

1. To virtualize or to containerize?

Virtualization of computing hardware was first introduced in the 1960s to provide mainframe resources to different applications running simultaneously on the same machine. Today, the virtualization of server hardware is a key enabler of cloud computing systems. Virtual machines (VMs) provide an emulation of a full computing system that executes software in a similar way as a physical system. A VM runs a complete operating system (OS) − the guest OS − and the desired applications and services. Multiple VMs are able to run on the same server and each VM is able to run a different OS. Even though VMs share the resources of the server (for example, network interface, disk, CPU) software applications running inside a VM are not aware of this sharing. VM images contain the operating system, application files, and executables needed by the VM and can be booted on any server that has the required resources to create an instance of the VM.

Containers provide a lightweight method for the virtualization of services and have emerged recently as an alternative to VMs. Like VMs, containers enable the execution of multiple isolated services on a single host. Unlike VMs, all containers on a host run on the same operating system and do not contain an OS of their own. Namespaces are used to provide the illusion of isolation to services in a container. Different namespaces provide containers with isolated "slices" of different aspects of the host environment, including processes, networking, users and file systems. The executable code of the application, the required functional libraries and configuration files are located in the host file system where they can be shared among containers using common files and libraries. Figure 5 illustrates the key elements of VMs and containers.

Both containers and VMs provide virtualization and control over access to resources of the underlying hardware. However, containers typically boot in a few seconds, compared with 10s of seconds for VMs. Also, containers create only minimal CPU and memory overhead, both during startup and runtime (Boden 2014). These features enable containers to scale instances up and down substantially faster than VMs and provide a significantly better total system performance than VMs. Containers can also run services at near-hardware speed, with high predictability and with improvements of between 10% for typical workloads and a factor of 10 for certain types of disk I/O sensitive workloads (Felter 2014 and internal Bell Labs investigations). Finally, because software images for containers do not contain an OS and can use shared libraries and files on the host, they are much smaller in size than comparable VM images.

From the perspective of interference, performance and security, VMs provide much stronger isolation between the VMs running on the same host than the isolation offered by containers. This is particularly important when services from different developers and users are executed on the same host.

Also, all containers running on a host must be compatible with the same OS kernel. VMs carry their own OS and do not suffer this restriction, so preparing legacy software for execution in a VM is, typically, much simpler than porting legacy software into a containerized environment.

Classical virtualization

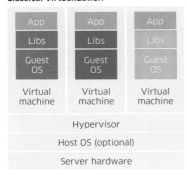

Lightweight containers

FIGURE 5: Comparison of virtual machine and container architectures

Criteria	Virtual machines	Containers
Relative image size	>3x	1x
Boot time	10s of seconds	Few seconds
Performance overhead (typical workload)	>10%	<5%
Performance overhead (disk I/O)	>50%	Negligible
Isolation of applications	Good	Fair
Security concerns	Low-medium	Medium-high
OS flexibility	Excellent	Poor
Management ecosystem	Mature	Evolving
Impact on legacy application architectures	Low-medium	High

TABLE 1: Comparison of the salient attributes of VMs and containers

As shown in Table 1, the attributes of containers make them an attractive choice for environments with many small workloads or micro-service architectures (Fowler 2014), for performance-sensitive workloads, and for environments requiring fast boot-up and flexible deployment strategies. VMs are well suited for environments in which services from different vendors or operators must share the same infrastructure (for example, in a public cloud), increased security requirements must be met, or legacy software must be virtualized.

VMs have been in use for many years and are relatively mature. Container technology is still new, but it is making quick progress because of the attention it is receiving from the open source and research community, with recent projects like Docker highlighting the potential. VMs and containers address different virtualization needs and will both continue to evolve and play an important role in local edge clouds and centralized data centers.

2. Dynamic distributed service management

Managing all the elements of a complex service across multiple data centers is non-trivial and requires new algorithms to drive placement and scale-out of VMs or containers. The algorithms must consider multiple optimization goals, such as current demand, resource availability, maximum latency to customers, redundancy and cost to compute the optimal scale and placement of service elements (Hao 2014).

New orchestration techniques are being developed from the understanding of models that capture dependencies between elements of a service and scaling functions that relate elements to each other and to application-level performance metrics. These models enable an orchestrator to manage a service as a whole instead of treating each service element independently. Application-level traffic predictions can be translated into specific virtualized service configurations so that services can meet even stringent service level agreements (SLAs) under dynamic conditions.

In the future, configuration parameters and models for services will be inferred automatically by new analytics algorithms that can detect the structure of a service and identify dependencies between service elements using cross-correlation analysis of behavioral similarities. Additional techniques, such as transfer entropy identification, can augment the results of the correlation analysis by measuring the amount of information transferred among different elements over time and enable an orchestrator to generate highly efficient new service configurations (Arjona Aroca 2015).

Traffic prediction algorithms can anticipate future demands for a service as a function of time-of-day, events that are happening (or are scheduled) and even the weather. These algorithms anticipate demand changes before they manifest themselves in the system and adjust services automatically so they are ready to handle the coming traffic.

Traffic predictions will also leverage past traffic patterns by analyzing traffic traces, identifying recurring characteristics and clustering them into regular patterns, such as weekday, weekend or holiday traffic. Based on these patterns, a capacity profile can be calculated that describes the resources required by a service to handle the anticipated load and can directly generate the required growth and de-growth decisions, as shown in figure 6. In figure 6 (a), the red line represents the capacity profile calculated to address a weekday traffic pattern with sufficient margin specified by the operator. Advanced anomaly detection techniques are used

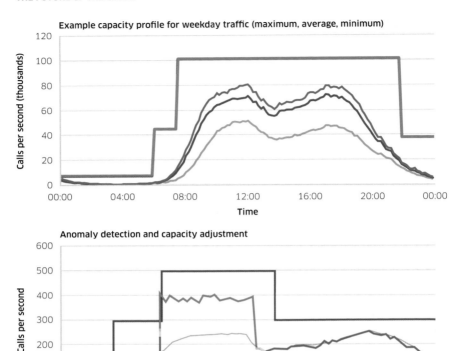

FIGURE 6 (a) AND (b): Illustration of capacity prediction and anomaly detection for a communications application

to identify unforeseen events that could not be predicted and enable a service to quickly respond to them. For example, a severe accident or natural disaster will lead to a very different communication pattern than that expected for a regular day. This is shown in figure 6 (b) where actual demand increases suddenly (green/yellow lines) and deviates from the expected traffic pattern (red line). Once an anomaly has been detected, prediction algorithms can calculate the traffic expected during the anomaly based on past experience and linear regression to extrapolate to future demand and create a new capacity profile (blue line) to handle the demand. These techniques enable a system to detect anomalies early, predict the capacity needed to handle them and quickly scale a service to this capacity, thereby substantially reducing the time it takes for the system to respond to these events.

Traffic prediction mechanisms will also take external data sources into consideration, which are outside of the service, but highly relevant for service usage. For example, severe weather alerts can be leveraged to anticipate an increase in communications traffic. Mobile network information about a sudden clustering of users can be used to trigger service capacity scaling-up around a gathering crowd and Twitter feeds and other social media can be leveraged to anticipate video streaming of instantaneously popular content.

These techniques enable a cloud orchestrator to learn about events that impact a service, its environment and its users. The challenge will be to understand the real meaning of the data observed and the corresponding life events, and to translate this understanding into specific orchestration actions to provide optimal service to customers.

Data centers, and edge clouds in particular, will not only support general purpose computing resources, but will also provide an array of heterogeneous hardware resources with different capabilities to address different performance needs of various application classes. For example, a data center may offer servers with network I/O acceleration, other servers with high-performance CPUs and hardware blades with field-programmable gate arrays (FPGAs). Advanced orchestration systems will consider the specific hardware capabilities in the management process and place service instances on resources that provide optimal support for the functions implemented, which will increase service performance substantially.

Given the number of decision dimensions outlined in this section, intelligent automation will clearly be a key factor in the orchestration and optimization of services. Human oversight will be needed at the beginning to prevent and manage unexpected consequential interactions. But it will become increasingly rare and limited to exceptional events, such as the creation of new data centers (or growth of existing ones) and recovery from major disasters.

3. Networking the distributed cloud

The highly dynamic nature of applications in the global-local cloud will necessitate the use of SDN control of the network. The network must automatically adapt connectivity between the components of cloud applications, as well as between cooperating applications, to avoid service interruption as applications scale and adjust themselves to shifting demand and network conditions. SDN provides this automation within and between the distributed data centers of the global-local cloud. Figure 7 illustrates the networking of two components of an application within a single data center.

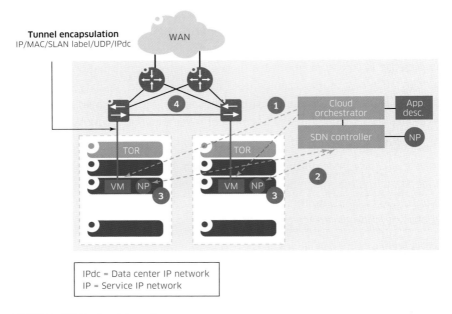

Tunnel encapsulation
IP/MAC/SLAN label/UDP/IPdc

IPdc = Data center IP network
IP = Service IP network

FIGURE 7: SDN in the data center

In step 1, the cloud orchestrator instructs the infrastructure manager to instantiate the VM components of the application, as specified by the application descriptor file, which describes how the application is constructed, the number and type of VMs required, the software images to be run on the VMs and the connectivity among the components. At the same time, the cloud platform constructs the network policies (NP in the figure) that describe the required networking and passes them to the SDN controller. As each VM is instantiated, a policy query is issued to the SDN controller to obtain networking policies for the VM (step 2). The SDN controller supplies the policies (step 3) and the virtual switch creates virtual tunnels between the VM components as indicated by the network policy (step 4). The figure also shows the use of virtual extensible local area network (VXLAN) tunneling (RFC7348 2014) within the data center, though other encapsulations are possible. The VXLAN is an evolution of virtual LAN (VLAN) encapsulation that provides isolation of traffic on a common network and allows VMs residing in two different IP networks to appear as though they are on the same private LAN.

Typically, a network service is comprised of multiple functions (for example, load balancers, web servers, security appliances), known as a service chain. In the global-local cloud, functions in a service chain may be distributed

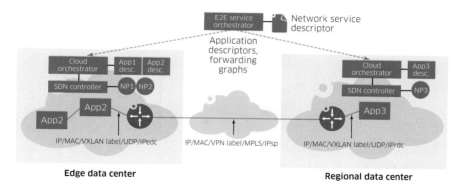

FIGURE 8: Constructing distributed services

across the cloud, with the performance critical functions instantiated in an edge cloud close to end users and less critical applications located in the regional cloud, as shown in figure 8.

In figure 8, a network service orchestrator (NSO) function is invoked to configure the service chains based on a network service descriptor. The service descriptor includes the set of functions that make up the chain as well as a description of how these functions are to be interconnected. The service orchestrator establishes the service chain by passing application descriptor files, which describe each function in the chain, and forwarding graphs, which describe the interconnections between the functions, to the cloud orchestrators of the impacted cloud domains (edge and regional in this example). The functions are instantiated in the respective clouds and interconnected using network policies derived from the forwarding graph that describes the service. Edge routers are used to provide interconnectivity among data centers across the wide area network (WAN), with a mapping between the VXLAN in the data center and WAN virtual private network (VPN). Mapping of the VXLAN to the VPN extends the LAN across the regional and edge data centers, but requires a separate VPN for every VXLAN carried between the edge and the regional data center.

The new cloud service business model

The new global-local cloud will provide local delivery of high-bandwidth, high-performance, low-latency services with global accessibility, enabling

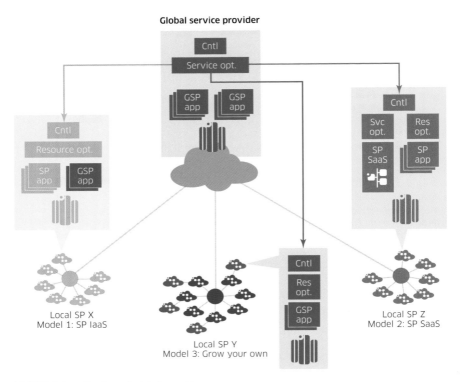

FIGURE 9: Localization of global providers

the same high-quality services to be delivered to users regardless of their location. From an industry evolution perspective, this will likely play out in one of two ways.

In the first scenario, global service providers (GSPs) will enhance their local presence through local service providers (LSPs) or via acquisition of local infrastructure as shown in figure 9.

In the second scenario, traditionally local or regional service providers will choose to expand to a global presence through mergers, acquisitions, and partnerships as shown in figure 10.

The relationship between the local and global service providers may follow one of four basic models, which will be chosen by considering the business goals of the providers involved, as outlined below.

1. *Local service provider provides infrastructure for global cloud services* – In this model, the LSP will operate as an Infrastructure-as-a-Service (IaaS)

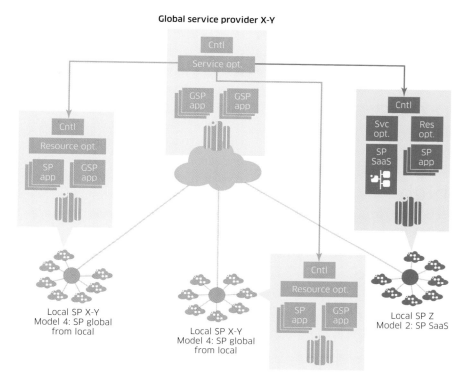

FIGURE 10: Globalization of local providers

provider, offering multi-tenant infrastructure services to GSPs. The GSP provides the application software instances and manages the service whether the components are instantiated in the LSP's edge cloud, or in the GSP's regional/centralized cloud. Service level optimization, including placement and scaling of application instances, is the responsibility of the GSP, with the LSP providing only localized resource optimization (for example, the ability to match applications with appropriate resources based on local cloud and network conditions).

Using a global video delivery service as an example, the GSP will instantiate its own CDN software instances in LSP edge clouds and determine placement and scaling of these instances. This model enables the LSP to monetize the ability to provide resources in close proximity to end users, but limits the LSP's ability to participate in the full service value chain because it is not otherwise involved in service delivery to end users. Likewise, the GSP is limited in its ability to adapt the service to local network conditions and resource status, relying entirely on the information that the LSP chooses to expose.

The new global-local cloud will provide local delivery of high-bandwidth, high-performance, low-latency services with global accessibility, enabling the same high-quality services to be delivered to users regardless of their location.

2. *Local service provider provides hosted "as-a-service" functions for global cloud services* – In this model, the LSP will operate a set of services in a Platform-as-a-Service (PaaS) or Software-as-a-Service (SaaS) model, offering generic functions or complete service components to GSPs. Here, a GSP will manage the end-to-end service, but will rely on the LSP to provide functions and services that are fully optimized and automatically scaled for the dynamic local demand, resource status and network conditions.

Again using the global video delivery service example, the LSP will offer fully locally optimized CDN software instances that adapt to changing conditions and demand dynamically within the local network with full knowledge of activity at the service, network and resource levels. The global video delivery operator will use these CDNs to deliver video to end users served by the LSP. This model will enable the LSP to participate fully in the service value chain because it will serve as an optimized local delivery service for the GSP. In effect, the LSP is able to leverage its detailed knowledge of resources and network conditions to more efficiently optimize services than would be possible in the IaaS model above. However, some GSPs will be reluctant to rely on a service they do not fully control or to expose details of their end-to-end service operations to LSPs, who may be offering a competing service of their own, albeit on a more local or regional scale.

3. *Global operators provide the local infrastructure* – In this model, GSPs will address the local delivery issue by growing their own local infrastructure from which to deliver their services by extending their own infrastructure into the local network. This may be accomplished by constructing infrastructure in leased or owned sites near existing LSP edge cloud sites, in sites acquired from LSPs (old central offices, for example), or in space actually collocated with LSP edge cloud sites (for example, leased space within LSP edge cloud sites). This model will enable GSPs to offer a fully optimized global-local solution without relying on resources from the LSP. Of course, such a model will probably prove to be prohibitively expensive for all but the very largest of GSPs and will probably first materialize for very high-demand local areas (for example, very densely populated metro areas). For other areas, this approach will need to leverage models one and/or two described above to create a truly global-local cloud.

4. *The local operator provides global services* – In this model (shown in figure 10), LSPs will create a global-local cloud through mergers, acquisitions, and partnerships (perhaps including one or more GSPs). As a result, the edge clouds of LSPs will be combined for increased local reach and to add regional/centralized resources. This model is the inverse of model three, in that it is driven by the desire of LSPs to become more global, rather than the converse.

Models three and four result in the creation of a similar global-local cloud infrastructure, but both will probably be augmented with models one and two to achieve full local coverage. However, model four relies on the ability of the LSP federated consortium to create or attract a viable global service to host on the global-local cloud and compete effectively with services provided or supported by GSPs, such as today's web scale providers.

It is likely that by 2020, we will see a mix of all of these scenarios, depending on regional specifics and regulatory environments, as well as the ambitions of the different players involved. However, due to the significant economic outlay required to build global infrastructure, it is also likely that there will be a limited number (for example, 10) of GSPs who will leverage a large set of affiliated LSPs in a relationship akin to that of the airlines today.

The future begins now

Today, there is increasing consensus among service providers in North America, Europe and Asia that the cloud must expand to the edge to support the flexibility and requisite performance requirements for the "everything digital" era. Consequently, equipment vendors are engaging in many proofs

Recognizing the benefits of increased localization, global cloud providers are steadily increasing the distribution of their cloud infrastructures to address the bandwidth and performance needs of their customers.

of concept and releasing first generation products in support of highly dynamic distributed clouds (Technavio 2014, Wagner 2015a), including:

- Cloud orchestration solutions capable of managing a widely distributed cloud infrastructure with automated life cycle management of applications
- SDN control of intra- and inter-data center networking (BCN 2015, Wagner 2015b)

Open source projects are already beginning to dominate the IaaS space (for example, the OpenStack, OpenDaylight, KVM, XEN initiatives). OpenStack is quickly emerging as the "cloud operating system" that provides management functions for compute, storage, and networking resources. The OpenStack community is also starting to address the monitoring and orchestration needs of advanced deployments in the OpenStack Ceilometer, Monasca and Heat projects. The ETSI Industry Specification Group for Network Functions Virtualization (ETSI ISG NFV) is defining an architecture with one of its key objectives being the creation of an infrastructure that "allows software to be located at the most appropriate places", explicitly including central offices (ETSI 2014). Recent projects like Open NFV are working to adapt these open source systems to the specific needs of network service providers as defined by ETSI NFV. Equipment vendors are also leveraging these solutions and adding additional capabilities that further improve distributed system performance, and service delivery and assurance (Verge 2015, Alcatel-Lucent 2014, Redapt 2014).

In anticipation of the need for highly modular and flexible micro-services architectures, equipment vendors, LSPs, and GSPs are exploring the use of the emerging containers ecosystem (for example, Docker, Marathon, Mesos) for the realization of efficient cloud-based software designs. Research organizations are also working to create new algorithms and technologies that support highly dynamic, high-performance cloud applications, automate the life cycle management of services, enable efficient distributed cloud designs and create innovative new applications that are enabled by the global-local cloud (like augmented reality and assisted remote control).

Recognizing the benefits of increased localization, global cloud providers are steadily increasing the distribution of their cloud infrastructures to address the bandwidth and performance needs of their customers, as evidenced by Amazon's discussion of their AWS global infrastructure:

> "Amazon Web Services serves over a million active customers in more than 190 countries. We are steadily expanding global infrastructure to help our customers achieve lower latency and higher throughput, and to ensure that their data resides only in the region they specify. As our customers grow their businesses, AWS will continue to provide infrastructure that meets their global requirements." (Amazon 2015).

Also, as noted earlier in this chapter, in recognition of the need for more localized delivery of content, major CDN and content providers are placing caches inside local networks and providing direct peering connections to ISPs (Netflix 2102, MacMillan 2013). This trend will continue to blur the line between global cloud providers and LSPs. and further support the notion of the new global-local cloud paradigm.

In the local/regional network service provider space, we are beginning to see evidence of operators exploring the creation of highly distributed clouds in support of their own network functions. For example, a major North American operator announced plans to convert many of its central offices into data centers that will host its own virtualized network functions. The same operator is participating in the Open Networking Lab's central office re-architected as data center (CORD) proof of concept. Both moves are clear evidence that the operator is moving in this direction (Le Maistre 2015). For the time being, this effort is focused on the operator's own network functions. But, ultimately, this will provide the basis for the global-local cloud infrastructure described in this chapter. Similar initiatives by European and Asia-Pacific operators are underway.

Summary

In the first decade of the 21st century, the emergence of the global cloud revolutionized how services and content were delivered to users and enterprises. Seemingly overnight, it was possible to access a huge array of innovative services with nothing more than Internet connectivity and a browser. These new services offered virtually unlimited capacity to serve millions of users, regardless of their location, by removing the barrier of building infrastructure for a potentially large customer base. As a result, the global cloud unlocked the potential of innovators who lacked sufficient scale in funding to realize their visions. It became possible for entrepreneurs to deploy a service with very little upfront investment and scale quickly when demand materialized and, if demand didn't materialize, the service could be quietly retired with very little cost. This removal of risk made the kinds of service innovations we enjoy today possible. Many of these services would never have seen the light of day if investors had to risk large sums to get them off the ground.

Although highly centralized cloud architectures lent themselves very nicely to global accessibility and economies of scale, it has become apparent that such architectures suffer from performance predictability, and latency issues due to the presence of significant numbers of network hops between the application servers and end users. Furthermore, limited choices of sites sometimes ran counter to national regulations for sensitive data storage and concerns over the security of stored data. To address these concerns, the centralized cloud has begun to regionalize, expanding its footprint to include sites in different geo-political regions. In addition, to address performance, latency and transport economics issues, CDN networks have provided even more distribution to bring content ever closer to end users, up to the point of deploying global provider CDN appliances within local networks.

We are now at the cusp of a second cloud revolution that will result in further distribution of the cloud toward the network edge. As user consumption of content finally makes the transition to being dominated by on-demand streaming, there will be a need to move very high-quality content very close to end users for performance and delivery cost optimization. At the same time, new applications are emerging that provide high-performance real-time services. Virtualized mobile networks, vehicular control and augmented reality services require single digit millisecond latency and cannot be realized by applications running in centralized or regional data centers. These factors are driving the need for a cloud that provides global access to services, but supports the delivery of these services from data centers placed at the edge of the network: the global-local cloud. To enable this new reality, new technologies are being developed that support the delivery and scaling of unpredictable and highly dynamic applications with highly mobile users via localized, optimized resource pools.

As we move toward 2020 and beyond, the majority of services must globalize their reach, but localize delivery or contextualization. Services and content available only within a specific localized network will not be acceptable to highly mobile consumers who will expect their services to adapt to their lives, not vice versa. At the same time, for high-performance mission-critical services, local delivery will be essential to provide users with a high-quality of experience and, in some cases, to make the services possible at all. This requires a combination of global and local capabilities not possessed by any single provider today. Therefore, a new provider landscape will emerge comprised of mergers, partnerships and acquisitions that will enable the infrastructure and control systems necessary to meet these needs to be built. The endpoint of this revolution will likely be the formation of an ecosystem that will consist of a handful of major global-local service providers who will leverage the resources and services of 100s of local/regional service providers and enable a new human and business era.

References

Akamai 1999-2014. "Annual Reports," *Akamai web site*, (http://www4.akamai. com/html/investor/financial_reports.html).

Alcatel-Lucent and RedHat, 2014. "Cloudband With Openstack As NFV Platform," *TMCnet*, (http://www.tmcnet.com/tmc/whitepapers/documents/ whitepapers/2014/10694-cloudband-with-openstack-as-nfv-platform.pdf).

Amazon Web Services 2015. "Global Infrastructure," *Amazon Web Services web site*, (http://aws.amazon.com/about-aws/global-infrastructure/).

Arjona Aroca, J., Bauer, M., Braun, S., Hilt, V., Lugones, D., Omana, J., Riemer, J., Sienel, J., Voith, T., 2015. "Insight-Driven NFV Orchestration," in preparation.

Berman, B., 2014. "Our Bodies' Velocities, By the Numbers," Discover, June 04 (http://discovermagazine.com/2014/julyaug/18-body-of-work).

Boden, R. 2014. "KVM and Docker LXC Benchmarking with OpenStack," *Software Defined Boden blog*, May 1 (http://bodenr.blogspot.be/2014/05/ kvm-and-docker-lxc-benchmarking-with.html).

Business Cloud News 2015. "China Telecom cloud subsidiary taps Nuage Networks for SDN deployment," *Business Cloud News*, February 5 (http:// www.businesscloudnews.com/2015/02/05/china-telecom-cloud-subsidiary- taps-nuage-networks-for-sdn-deployment/).

Chioka, 2013. "What is Motion-To-Photon Latency?," *Garbled Notes web site*, November 3 (http://www.chioka.in/what-is-motion-to-photon-latency/).

Drew, C.2009. "Drones Are Weapons of Choice in Fighting Qaeda," *The New York Times*, March 16 (http://www.nytimes.com/2009/03/17/business/17uav. html?pagewanted=all&_r=0).

Duffy, K., 2002. "Reaction Times and Sprint False Starts," *Kevin Duffy's web site*, September 21 (http://condellpark.com/kd/reactiontime.htm).

ETSI 2014. "Network Functions Virtualisation (NFV); Architectural Framework ETSI GS NFV 002 V1.2.1," *ETSI Group Specification*, December (http://www.etsi. org/deliver/etsi_gs/nfv/001_099/002/01.02.01_60/gs_nfv002v010201p.pdf).

Felter, W., Ferreira, A., Rajamony, R., and Rubio, J., 2014. "An Updated Performance Comparison of Virtual Machines and Linux Containers," *IBM Research Report*, RC25482 (AUS1407-001) July 21 (http://domino.research.ibm.com/library/cyberdig.nsf/papers/0929052195DD819C85257D2300681E7B/$File/rc25482.pdf).

Fowler, M. and Lewis, J. 2014. "Microservices," *Martin Fowler web site*, March 25 (http://martinfowler.com/articles/microservices.html).

Gafford, T. 2014. "League of Legends 2014 World Championship Viewer Numbers (Infograph)," *onGamers News*, December 1 (http://www.ongamers.com/articles/league-of-legends-2014-world-championship-viewer-n/1100-2365/).

Gartner 2011. "Forecast: Servers, All Countries, 2008-2015, 4Q11 Update," *Gartner*, December 14 (https://www.gartner.com/doc/1875715/forecast-servers-countries--q).

Gartner 2013. "Forecast: Servers, All Countries, 2010-2017, 4Q13 Update," *Gartner*, December 20 (https://www.gartner.com/doc/2640537/forecast-servers-countries--q).

Gartner 2015. "Forecast: Servers, All Countries, 2012-2019, 2Q15 Update," *Gartner*, June 23 (https://www.gartner.com/doc/3080717/forecast-servers-countries--q).

Hao, F., Kodialam, M., Lakshman, T.V., Mukherjee, S., 2014. "Online allocation of virtual machines in a distributed cloud," IEEE Infocom.

Human Benchmark 2014. "Reaction Time Test," *Human Benchmark web site*, August 18 (http://www.humanbenchmark.com/tests/reactiontime).

IDC 2010. "Server Refresh: Meeting the Changing Needs of Enterprise IT with Hardware/Software Optimization," *Oracle web site*, July (http://www.oracle.com/us/corporate/analystreports/corporate/idc-server-refresh-359223.pdf).

Le Maistre, R., 2015. "AT&T to Show Off Next-Gen Central Office," *LightReading*, June 15 (http://www.lightreading.com/gigabit/fttx/atandt-to-show-off-next-gen-central-office/d/d-id/716307).

Limelight Networks 2008-2014. "Form 10-K Annual Reports," *United States Securities And Exchange Commission filings*, (http://www.wikinvest.com/stock/Limelight_Networks_(LLNW)/Filing/10-K/).

Limer, E., 2014. "This Box Can Hold an Entire Netflix," *Gizmodo*, July 23 (http://gizmodo.com/this-box-can-hold-an-entire-netflix-1592590450).

Makuch, E., 2013. "The International hits 1 million concurrent viewers," *GameSpot*, August 13 (http://www.gamespot.com/articles/the-international-hits-1-million-concurrent-viewers/1100-6412911/).

Marshal, C., 2014. "Top Ten Video Ads on YouTube, Apparently: September 2014," *ReelSEO web site*, October 8 (http://www.reelseo.com/top-video-ads-youtube-sep-2014/).

McCormick, R., 2013. "'League of Legends' eSports finals watched by 32 million people," *The Verge*, November 19 (http://www.theverge.com/2013/11/19/5123724/league-of-legends-world-championship-32-million-viewers).

McMillan, R. 2013. "Google Serves 25 Percent of North American Internet Traffic," *Wired*, July 22 (http://www.wired.com/2013/07/google-internet-traffic/).

Microsoft 2015. "We Power the Microsoft Cloud," *Microsoft Datacenter web site*, (http://www.microsoft.com/en-us/server-cloud/cloud-os/global-datacenters.aspx).

Microsoft 2015b. "Plan network requirements for Skype for Business," *Microsoft Technet web site*, June 18 (https://technet.microsoft.com/en-us/library/gg425841.aspx).

Netflix 2012. "Announcing the Netflix Open Connect Network ," *Netflix US and Canada Blog*, June 4 (http://blog.netflix.com/2012/06/announcing-netflix-open-connect-network.html).

Nielsen 2015. "The Total Audience Report Q1 2015," *Nielsen web site*, Q1 (http://www.nielsen.com/content/dam/corporate/us/en/reports-downloads/2015-reports/total-audience-report-q1-2015.pdf).

Nokia 2014. "Technology Vision 2020 Reducing network latency to milliseconds," *Nokia White paper*, July 2014 (http://networks.nokia.com/sites/default/files/document/technology_vision_2020_reduce_latency_white_paper_1.pdf).

Panzarino, M., 2014. "WatchESPN's Record 1.7M Concurrent World Cup Viewers Beats The Super Bowl," *TechCrunch*, June 26 (http://techcrunch. com/2014/06/26/watchespns-record-1-7m-concurrent-world-cup-viewers-beats-the-super-bowl/).

Popper, B., 2015. "Here's why Meerkat's popularity is suddenly plunging," *The Verge*, April 1 (http://www.theverge.com/2015/4/1/8319043/why-is-meerkats-popularity-suddenly-plunging).

Redapt, 2014. "Integrated OpenStack™ Cloud Solution with Service Assurance," *Intel web site*, (http://www.intel.com/content/dam/www/ public/us/en/documents/white-papers/intel-service-assurance-redapt-white-paper.pdf).

Redbeard 2014. "One World Championship, 32 million viewers," *League of Legends web site* (http://na.leagueoflegends.com/en/news/esports/esports-editorial/one-world-championship-32-million-viewers).

Rezab, J. 2015. "Welcome to the era of social video," *Presentation at Social Media Week NYC*, February (http://www.slideshare.net/socialbakers/social-media-week-the-era-of-social-video).

RFC7348 2014. "Virtual eXtensible Local Area Network (VXLAN): A Framework for Overlaying Virtualized Layer 2 Networks over Layer 3 Networks," *IEEE RFC7348*, August (https://tools.ietf.org/html/rfc7348).

Sandvine 2014. "Global Internet Phenomena Report: 2H 2014," *Sandvine web site*, November 20 (https://www.sandvine.com/downloads/general/global-internet-phenomena/2014/2h-2014-global-internet-phenomena-report.pdf).

Tassi, P., 2011. "Major League Gaming Breaks Online Viewer Record, Approaches TV Numbers," *Forbes*, October 20 (http://www.forbes.com/sites/insertcoin/2011/10/20/major-league-gaming-breaks-online-viewer-record-approaches-tv-numbers/).

Technavio 2014. "Top 26 Companies in the Global NFV Market," *Technavio web site*, April 10 (http://www.technavio.com/blog/top-26-companies-in-the-global-nfv-market).

The Diffusion Group 2014. "TDG: Netflix Streaming Volume Up 350% in 10 Quarters," *TDG web site*, September 25 (http://tdgresearch.com/tdg-netflix-streaming-volume-up-350-in-10-quarters/).

Verge, J., 2015. "Vendors Push Telco Cloud NFVs Atop OpenStack," *Data Center Knowledge*, March 27 (http://www.datacenterknowledge.com/archives/2015/03/27/vendors-push-telco-cloud-nfvs-atop-openstack/).

Wagner, M., 2015a. "PoCs Pave the Way for NFV," *LightReading*, April 14 (http://www.lightreading.com/nfv/nfv-tests-and-trials/pocs-pave-the-way-for-nfv/d/d-id/715061).

Wagner, M., 2015b. "Leading Lights Finalists: Most Innovative SDN Deployment (Network/Data Center Operator)," *LightReading*, May 29 (http://www.lightreading.com/carrier-sdn/sdn-architectures/leading-lights-finalists-most-innovative-sdn-deployment-(network-data-center-operator)/d/d-id/716009).

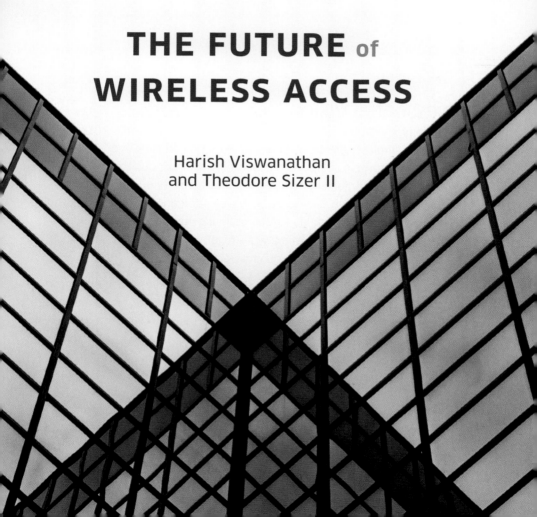

CHAPTER 6

THE FUTURE of
WIRELESS ACCESS

Harish Viswanathan
and Theodore Sizer II

The essential vision

Wireless communications began when Marconi invented the radio in 1894. Innovation in wireless communications has been breathtaking ever since, starting with microwave transmission over long distances and then moving into the mobile access domain. The two major systems for wireless network access, namely mobile cellular and Wi-Fi, have undergone many generations of technology evolution, each marked by dramatic improvements in network efficiency and user experience, and have – together with advances in mobile computing – arguably transformed human existence.

Today the expectation is that we should be wirelessly connected at all times and in all places in order to increase our social interactions, as well as to continuously access information and services. But the next decade will witness an even more dramatic change in wireless access as the technology transforms, yet again, to enable new modes of human communication and content consumption. The wireless connectivity of ever-more powerful devices, with digital interfaces to physical objects, goods and infrastructure in many industries, will enable the digitization and connection of everything and everyone, everywhere.

We believe the essential ingredients of the future wireless access network can be described as follows:

1. *The delivery of seemingly infinite capacity* – The future wireless system will seek to provide seemingly infinite capacity everywhere through the use of contextual information. By understanding the specific user-service needs, it will provide the optimal capacity for each personalized service couplet. It will require a transformation of the network, with massive network densification and integration of multiple access technologies. The system will need to use frequencies above 6 GHz and new and re-farmed licensed, unlicensed and shared spectrum below 6 GHz, as well as large-scale, multiple antennas. This seemingly infinite capacity will be delivered both in the data plane (user or application data transfer), as well as in the control plane (signaling required to establish sessions for user data transfer), facilitating what can be considered ultra-broadband and ultra-narrowband applications simultaneously.

2. *The realization of continuous performance optimization* – The future wireless system will be optimized for performance across multiple dimensions, in addition to capacity. The system will be able to achieve ultra-low-latency, high reliability, and massive user and service scalability. It will become more distributed, virtualized and "cell-less," with connectivity to spectrum provided by multiple radios and air interfaces simultaneously. This will support the emergence of new use cases in enterprise and industry segments, driven by the emergence of new digitized devices and interfaces to physical objects and infrastructure.

3. *The creation of an extremely energy efficient network* – The future wireless system will substantially reduce energy consumption in mobile devices and the network. The network will be energy optimized for human communications services of all types, by utilizing smaller cells and sophisticated beamforming techniques, as well as local power and energy harvesting. The network design will also ensure that short-burst communications, which are typical of Internet of Things (IoT) devices, consume the minimal amount of energy, thus ensuring battery life in excess of 10 years for devices with sporadic communication.

In the rest of this chapter, we will outline these major trends and the key emerging technologies in more detail, as well as where we stand today.

The past and the present

Capacity

The evolution of mobile cellular networks has been characterized by the introduction of a new technology generation every decade or so, with a step change improvement in performance compared to the previous generation. Similarly, the short range unlicensed band Wi-Fi technology (IEEE 802.11) has improved successively with the introduction every few years of new specifications. The technology advance has been driven by a persistent increase in user demand for mobile access, which in turn has been growing because of advances in device capabilities and declining communication costs (due to Moore's Law type gains). In general, increase in demand comes about because of increasing numbers of devices on the network, as well as user activity. In many countries, device penetration has substantially exceeded 100%, with many users having more than one device per person. User behaviors are also changing with ever-more reliance on data communications for multiple forms of communication (personal, media and control, as described in chapter 9, *The future of communications*).

To illustrate this ever-increasing dependence on mobile interactivity, figure 1 shows the growth in the number of instant messages per minute worldwide over time, and in particular the large variation in the peak to average ratio over the course of the year, based on data for the popular "WhatsApp" messaging application.

We forecast a rise in mobile messaging of 14x between 2012 and 2020 (left side of figure 1). In addition to the challenge this represents in terms of short-burst wireless capacity, the large peak to average ratio reflects the need for dynamic scalability or capacity on demand.

We expect this trend in mobile data capacity demand to continue into the future, and to be accentuated by a dramatic increase in control plane capacity of the network, driven by a majority of the new devices on the network being IoT devices, which generate a disproportionately large amount of control traffic relative to data traffic. This is in contrast to consumer broadband devices, such as smartphones, which typically generate more data plane than control plane (signaling) load. That said, existing consumer mobile services will also likely evolve to stream 4K video in the near term, delivering ever-higher concurrent capacities in dense geographical areas. This will continuously drive the need for significant additional data capacity that future wireless access systems will have to support.

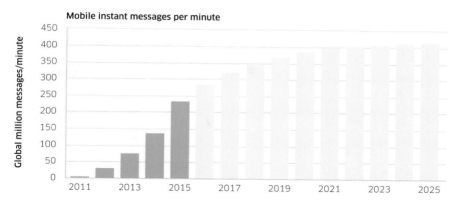

Mobile instant messages per minute

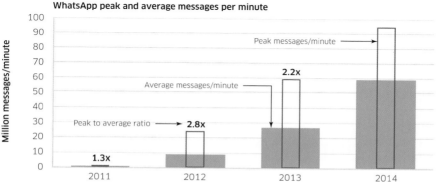

WhatsApp peak and average messages per minute

Peak messages as on New Year's Eve (31st December) each year

FIGURE 1: Growth in number of all mobile messages per minute worldwide and, specifically, peak and average of WhatsApp messages[1]

Figures 2 and 3 show current control and data plane demand with our projections for the future evolution of the two. The control plane load is estimated using the global, daily number of connection requests, each of which corresponds to device transitioning from idle state to active state. In LTE technology, each such state transition involves more than 10 control plane signaling messages between the device, the base station, and the core network control element.

We predict that there will be between a 31x and 127x increase in control plane traffic in 2025, relative to 2015 (figure 2), and a 61x to 115x increase in bearer traffic over the same time period (figure 3). Consequently, the

---

1 - These graphs were based on analysis done by Bell Labs Consulting, drawing from multiple data sources, including Ovum 2015, Pyramid 2015, Statista 2014, 2015a, 2015b, and Whitfield 2014.

The future wireless access system must be capable of providing seemingly infinite capacity... with market economics that are not significantly different than in 2015.

future wireless access system must be capable of providing seemingly infinite capacity for both ultra-broadband and ultra-narrowband traffic with market economics that are not significantly different than in 2015.

It is already clear that future mobile cellular access will evolve from fourth generation (4G) long term evolution (LTE) to fifth generation (5G) cellular technology around 2020, while on the unlicensed wireless access front, IEEE will introduce a new generation of more powerful technology beyond today's 802.11ac at 5 GHz and 802.11ad at 60 GHz (Wi-Gig) in the same timeframe. We expect future wireless access to integrate both licensed and unlicensed access into a single seamless wireless access network, which will enable the massive growth in capacity outlined above, with the optimum economics. It is likely, therefore, that the era when users had to decide when to connect to licensed and unlicensed technologies based on a finite amount of monthly data capacity will be over. All radio resources will be managed as part of a seamless whole, with the optimum air interface selected based on service needs and minimum cost.

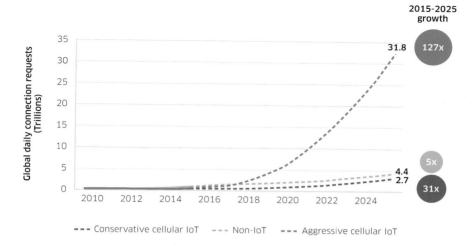

FIGURE 2: Growth in control plane demand (Bell Labs, 2015)

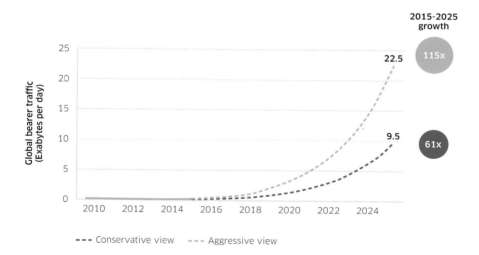

FIGURE 3: Growth in data plane demand (Bell Labs, 2015)[2]

2 - Figures 2 and 3 graphs were taken from an analysis done by Bell Labs Consulting based on LTE traffic models and drawing from multiple data sources, including Alcatel-Lucent field data, Beecham 2008, GSMA 2015, Machina 2011, Pyramid 2011, 2014.

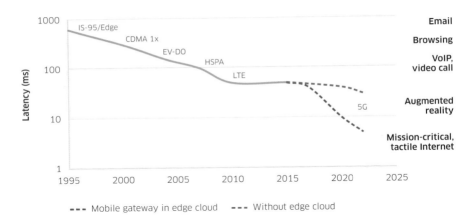

FIGURE 4: End-to-end latency performance of cellular networks, based on 4G 2012 and Bell Labs projections

Performance

In addition to system capacity, wireless systems have seen a steady improvement in performance across multiple dimensions over the years, resulting in enhancements in voice quality and broadband data experience. Higher throughputs, lower call drop or data timeout rates, and lower transactional latencies have enabled an enormous number of Internet-connected mobile applications, video streaming and electronic commerce to and from mobile devices. Figure 4 shows the average end-to-end latency of mobile networks over successive technology generations since the dawn of the mobile data era in second-generation (2G) cellular systems, together with new applications that are enabled as latency decreases. Latency has reduced over one order of magnitude and is further expected to reduce by another order of magnitude in future wireless access.

This latency reduction has been achieved by improvements in signaling and new data capabilities, including smaller frame durations, increases in communications bandwidth and improvements in device processing power, leading to reduced packet roundtrip times over the air. As we move to 5G, the end-to-end latency will also critically depend on whether the mobile edge and the application reside in an edge cloud close to the user, or not. While the air-interface latency can be minimized independently of the service latency, having the application or service functions close to the user is also necessary to reduce the end-to-end round trip time for the service as a whole, which again argues for the placement of these functions in a common edge cloud.

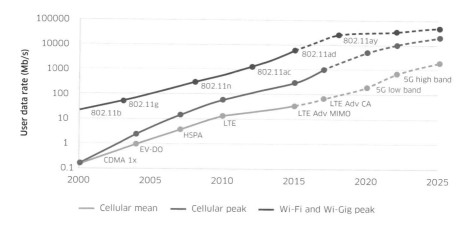

FIGURE 5: User data rates have increased over several orders of magnitude for Wi-Fi, Wi-Gig and cellular access technologies. Times shown roughly correspond to when the technology became broadly available in the market (from IEEE and 3GPP standards documentation)

Figure 5 shows the increase in throughput over the past 15 years for both cellular and Wi-Fi and Wi-Gig technologies. Peak refers to the theoretical peak data rate that is typically achieved only when the user is very close to the cell site and is the only one being served by the base station or access point. Mean refers to the average of the rates achieved as the user moves across different locations within the cell. In both cases, the throughput relates to a single active user in the cell sector. It is clear that throughputs have increased by several orders of magnitude over the decades, enabling the growth in applications, such as mobile web services and video streaming. We expect the trend to continue in the future, with greater emphasis placed on improving the mean user-experienced data rate, rather than the theoretical peak data rate.

Also shown in figure 5, the 802.11 family of technologies (Wi-Fi, Wi-Gig) have exhibited higher peak rates than cellular systems, since 802.11 technologies are designed for much smaller cell sizes compared to cellular and can operate with wider bandwidths, even with less transmit power. However, the performance degrades significantly with increases in the number of users, due to the use of a collision sense multiple access (CSMA) protocol for gaining access to the wireless channel. As cellular evolves to include higher carrier frequencies with larger bandwidth swathes (for example, in the 20 to 100 GHz range), we expect theoretical peak rates to become comparable to those for unlicensed technologies, such as 802.11, as shown by the extrapolation to 2025 in figure 5.

User experience in mobile systems is also governed by other performance criteria, such as coverage, call drop rate, handoff failure rate and other radio link failure events. Wireless systems are designed and optimized to maintain target values for such performance indicators. For example, the call drop rate typically is maintained between 0.5% and 1% by tuning various parameters, such as transmit power, handoff thresholds and cell site antenna orientation, as well as the addition of cell sites. Improvements in technology that impact these key performance indicators (KPIs) are not immediately reflected in the KPIs themselves, but more in the ease with which the targets are achieved.

In the early days of cellular systems, performance optimization involved drive tests in which a vehicle would drive around each cell location to make measurements in the field. The data helped determine the best values for various system parameters to be used by the network planners. With the introduction of self-optimizing network (SON) technology in fourth generation (4G) cellular systems, performance optimization became dynamic and automated. In future wireless networks, devices will have multiple connections to the network through various technologies, air interfaces and sites. There is an inherent redundancy in this multi-connectivity that will result in significantly reduced call and session drops for devices, even when individual technology links may be dropped. We can therefore expect a significant reduction in the effort expended to meet target KPIs as the network becomes inherently more robust.

Energy efficiency

With the clear threat of uncontrolled global warming due to excess fossil fuel usage, the pressure to reduce energy consumption and carbon footprint is universal across all industries. The entire information and communications technology (ICT) industry causes only 2.3% of the world's CO_2 emissions (GeSI, 2015, p. 8), which on the surface would seem to suggest that the energy impact of telecommunication networks is relatively minor. But telecommunication providers have a social responsibility, as well as an added incentive to significantly reduce their operating expenses. Typically, network energy consumption represents 75% of the energy cost; and energy costs comprise 8% of the overall operational expenses in developed economies, and 30% in developing economies. There is a clear economic, as well as social, imperative to lower network energy consumption.

This need for energy efficiency has driven major improvements in network energy over the last few years. Figure 6(a) shows energy efficiency of various LTE cellular technologies relative to the 3G baseline technology

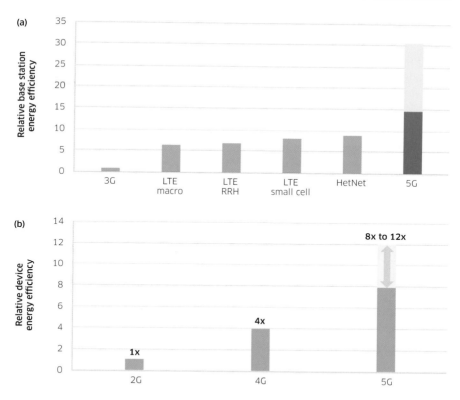

FIGURE 6: Energy efficiency of cellular access (a) base station (b) IoT device for short burst access, based on Bell Labs estimates from the Gwatt tool (www.gwatt.net)

obtained from the Bell Labs GWATT[3] tool. *LTE macro cell* refers to the traditional macro cell where the base band processing unit and the radio are located in the cabinet below the mast, and only the antenna is on the tower. *LTE RRH* refers to the remote radio head architecture, where the radio head is located on the tower, thereby eliminating loss of power in the cable and coupling. *LTE small cell* refers to a low power, single sector base station with a footprint much smaller than a typical macro cell. Finally, *HetNet* refers to a network architecture consisting of a mix of these macro cells and small cells in a given region. As can be seen, wireless technology evolution has generally resulted in increasing energy efficiency. We expect that 5G will more dramatically improve energy efficiency by using a larger number of antenna elements, minimizing transmission of broadcast signals unrelated to traffic and allowing some cells to be powered down when not in use.

3 - Available at www.gwatt.net.

It is important to note that as we enter the era of IoT, *device* energy consumption will also become a significant factor in the design of wireless access networks. Figure 6(b) shows the improvement in energy efficiency for transmission of small bursts, which LTE has achieved compared to 2G (currently the two leading technologies for cellular IoT devices), as well as the expected improvement for 5G. For massive deployment of IoT devices to occur, a device battery life is needed in excess of 10 years based on 10s of transactions per day. The industry is responding to the challenge with improvements introduced in Release 13 of LTE, specifically to extend battery life. However, we expect future wireless access to help dramatically improve device battery life through introduction of new air-interface technologies, as described in the following section.

The future

In the preceding section, we alluded to the future of wireless access and the introduction of new technologies and architectural elements to dramatically improve network performance and cost, as well as device battery life. They will enable the network to support a wide variety of applications, including ultra-narrowband IoT applications (driven by the instrumentation and automation of industries and enterprises) and ultra-broadband, video-intensive applications. In this section, we describe the essential technologies that we believe will become an integral part of the wireless network of 2020 and beyond.

Ultra-narrowband: New air-interface for below 6 GHz

LTE is based on an orthogonal frequency division multiplexing (OFDM) waveform that maintains orthogonality between multiple-device-originating signals simultaneously transmitted to the base station, as long as the signals are synchronized, and all signals are based on the same basic symbol period. Tight synchronization is achieved through a mechanism called *closed loop timing advance*, in which the base station signals transmit timing information to the devices prior to transmission of a data burst.

This process has the following implications:
- The signaling overhead incurred in this procedure is acceptable for large volume flows, such as video or large data files, but it has a significant impact for devices that send only small bursts of data, such as instant messages, application updates or "keep alives" for TCP connections. This type of short burst communication will be the norm for future IoT devices.

- This synchronization protocol has an impact on the device energy consumption that will significantly reduce the battery life of IoT devices that only sporadically send a small amount of data.

- The minimum latency that can be achieved is limited, since the procedure introduces an extra round trip for the timing advance signaling before the actual transmission of information. This will limit the ability to support low-latency in real-time-responsive applications.

What prevents optimization of today's systems for the different latencies required is that all signals need to have the same symbol period for orthogonality. Future wireless systems will need to allow both synchronous and asynchronous transmission modes, as well as symbol periods that are dynamically tuned to the different types of applications. This will be enabled through a modification of the basic OFDM waveform to include filtering, which will eliminate the inter-user interference. Figure 7(a) shows the interference suppression benefits, visible through the sharper roll off of the power spectral density for one instance of a filtered OFDM signal, known as the universal filtered OFDM (UF-OFDM) waveform. This results in less signal power leaking into adjacent frequency resources used by other signals. While the ultimate choice of the waveform will depend on what is adopted in the 3rd Generation Partnership Project (3GPP) standards discussions, we expect some form of such filtering to enable asynchronous communication and co-existence of multiple symbol periods.

With regard to the need to support both synchronous and asynchronous modes, figure 7(b) shows a new frame structure with smaller symbol periods for low-latency traffic, normal periods for video streaming, and larger periods for more delay-tolerant broadband traffic. Multiple symbol periods will also increase the efficiency of the system.

Finally, with the future wireless system expected to support a large number of IoT devices that only transmit short bursts, the future air interface will be further optimized for such traffic by introduction of a so-called connectionless service. This air interface service option will allow the devices to wake up and send data efficiently, without closed-loop synchronization overhead and without significant signaling overheads, thereby maximizing battery life. As a result, we can expect a significant improvement in spectral efficiency of transmitting bursts that are smaller than 300 bytes.

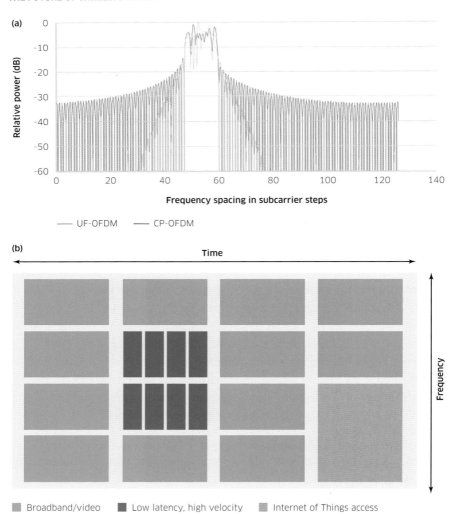

FIGURE 7: (a) The interference suppression capability of UF-OFDM and OFDM waveforms and (b) application of this waveform to flexible air-interface design with different symbol periods

Ultra-broadband: Spatial expansion and the high-band air interface

As discussed in chapter 1, *The future X network.* we are approaching the limit of the low-band spectrum and spectral efficiency dimensions. Consequently, we need to turn to the spatial dimension and increase the number of cells deployed per unit area using the same spectrum, thus reducing the size of each cell. Such cell division has been used for increasing the wireless network capacity since the beginning of cellular networks, but rather than dividing by

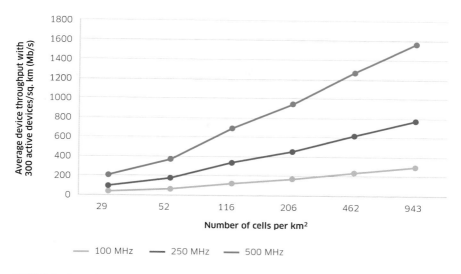

FIGURE 8: Throughput increases linearly with increasing small cell density and signal bandwidth (Lopez 2015)

a factor of two, the cell division must be by a factor of 10 or more. Figure 8 shows that throughput per device increases linearly as the density of cells is increased, with simulation results (Lopez, 2015) shown for three different bandwidths in an interference limited scenario (for example, the signal to interference and noise ratio is limited by out of cell interference rather than noise power). In this regime, increase in bandwidth also results in a linear increase in throughput, making the high band (>6 GHz) attractive, as there is a significant amount of additional spectrum available (up to 10 GHz) that could be allocated for future cellular communications.

Frequencies above 10 GHz, such as the millimeter wave bands (mmwave) have traditionally been used for fixed wireless communications in applications such as backhaul and fixed access. But due to the significant attenuation of these bands by atmospheric elements, such as air and rain, they require line-of-sight propagation and beamforming to be employed, making them unsuitable for conventional macro-cell based wireless architectures. However, with the impending saturation of licensed spectrum availability in low bands discussed, there is significant momentum in the industry to utilize the high-frequency bands for mobile networks. In small cell configurations they will provide very high user throughputs and system capacity in very densely populated areas. We thus expect network densification in future wireless access systems to include such high-band

Future wireless systems will need to allow both synchronous and asynchronous transmission modes, as well as symbol periods that are dynamically tuned to the different types of applications.

small cells. But there are a number of significant challenges that must be addressed in order to make high-band small cells a reality:

- *Robust control signaling* – Because of the harsh propagation conditions in high frequencies stemming from increased blocking effects of the human body and other obstructions, and reduced diffraction, connectivity will likely be intermittent. Therefore, to maintain service continuity, a robust control channel and fallback data channels will be required. The use of a low-band link as an anchor link for the control channel offers a promising solution; future high-band access will therefore most likely be tightly coupled with a low-band channel.

- *Beam acquisition and tracking* – The use of beamforming is critical for achieving a reasonable range (for example, 100 m) in high bands because of significantly greater path loss at higher frequencies (path loss increases as the square of the frequency). Furthermore, as the user location, orientation or the environment changes, it is necessary to make adjustments to beam directions during the data transmission. This will require the invention and implementation of reduced sampling beam search algorithms and signaling procedures, relying on the low-band control channel to reduce beam search time.

- *Radio component power and cost* – The use of more integrated radio frequency chips will allow a larger number of radio paths and antennas with both power and cost efficiency. They will be required to form the narrow beams needed to achieve cell sizes comparable to that of small cells in lower frequencies.

- *Backhaul* – In a dense network of high-band small cells, access to backhaul connections could potentially be an issue. Therefore, in-band backhaul, involving the use of the same system for both access and backhaul, could be used to reduce the number of nodes connected to wired backhaul. Massive MIMO could also be used to direct beams from macro cells to each of the small cells in the macro umbrella. In addition, high-band, line-of-sight, microwave solutions will continue to be used for backhaul of low-band solutions.

Ultra-broadband: Massive MIMO for massive throughput

Twenty years ago, Jerry Foschini, working at Bell Labs, first described the benefits of the use of multiple antennas at both the base station and at the mobile device. Now this multiple antenna approach is used in every major wireless access standard and handset available today. LTE systems currently have solutions that use up to 16 antennas at the b ase station. Theoretically, increasing the number of antennas shows great benefits by expanding capacity from a single site through greater use of beamforming, as well as decreasing energy usage. Simulation studies have shown capacity increases of nearly an order of magnitude, when one considers numbers of antennas of a hundred or more (Marzetta 2015 and Rusek et al 2013), as shown in figure 9.

While the use of antenna arrays with traditional beamforming is required for high-band to achieve longer range, as discussed, at low-band frequencies Massive MIMO can be used in a heavy, scatter-propagation environment and nonline-of-sight configuration, to achieve much higher throughput at a given range. However, the antenna arrays are correspondingly larger, since they scale with the wavelength, which is inversely proportional to the frequency used.

Architecture: Scalability and flexibility with a virtualized radio access network (vRAN)

Network functions virtualization (NFV) also has a role in macro cell architecture. Some of the baseband processing functions can be moved onto general purpose processors (GPP) located in a data center, rather than performing all processing at the cell site itself. The new architecture, called vRAN for virtualized RAN, is illustrated in figure 10. In conventional distributed and integrated radio base stations, Layer 2 (L2) – medium access control and radio link control functions – and Layer 3 (L3) – call control, packed data convergence protocol, encryption and header

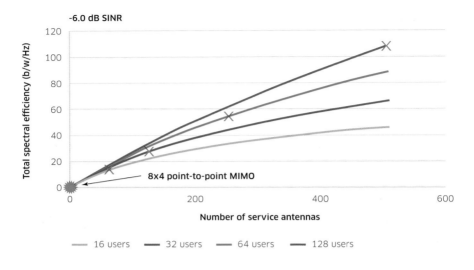

FIGURE 9: Massive MIMO theoretical spectral efficiency. Crosses indicate points where the number of antennas is four times that of the number of users served (Marzetta 2015).

compression – are all performed at the cell site. In a vRAN, they are performed in the data center.

In either case, specialized hardware is used to perform Layer 1 (L1) functions, such as intensive digital signal processing. It is only the location that differs, whether located at the base station site or in the data center. The capacity of the link between the cell site and the base station depends on whether the L1 processing is performed at the cell site or in the data center. The former approach is referred to as *Dual-site vRAN* and the link between cell site and data center is called a *mid-haul link;* whereas the latter approach is called *Centralized vRAN* and the link is called a *front-haul link.* A conventional IP/ Ethernet network can be used for mid-haul links, whereas for a front-haul link, the transmit digital (so-called, CPRI) samples of the waveform require a multi-gigabit-per-second link, typically, wavelength division multiplexing (WDM) optical transport.

In either case low-latency (~6 ms for mid-haul and ~300 ms for front-haul) is required in order that the tight timing requirements of the radio layer are met. This essentially mandates the use of an edge-cloud data center, as outlined in chapter 5, *The future of the cloud.*

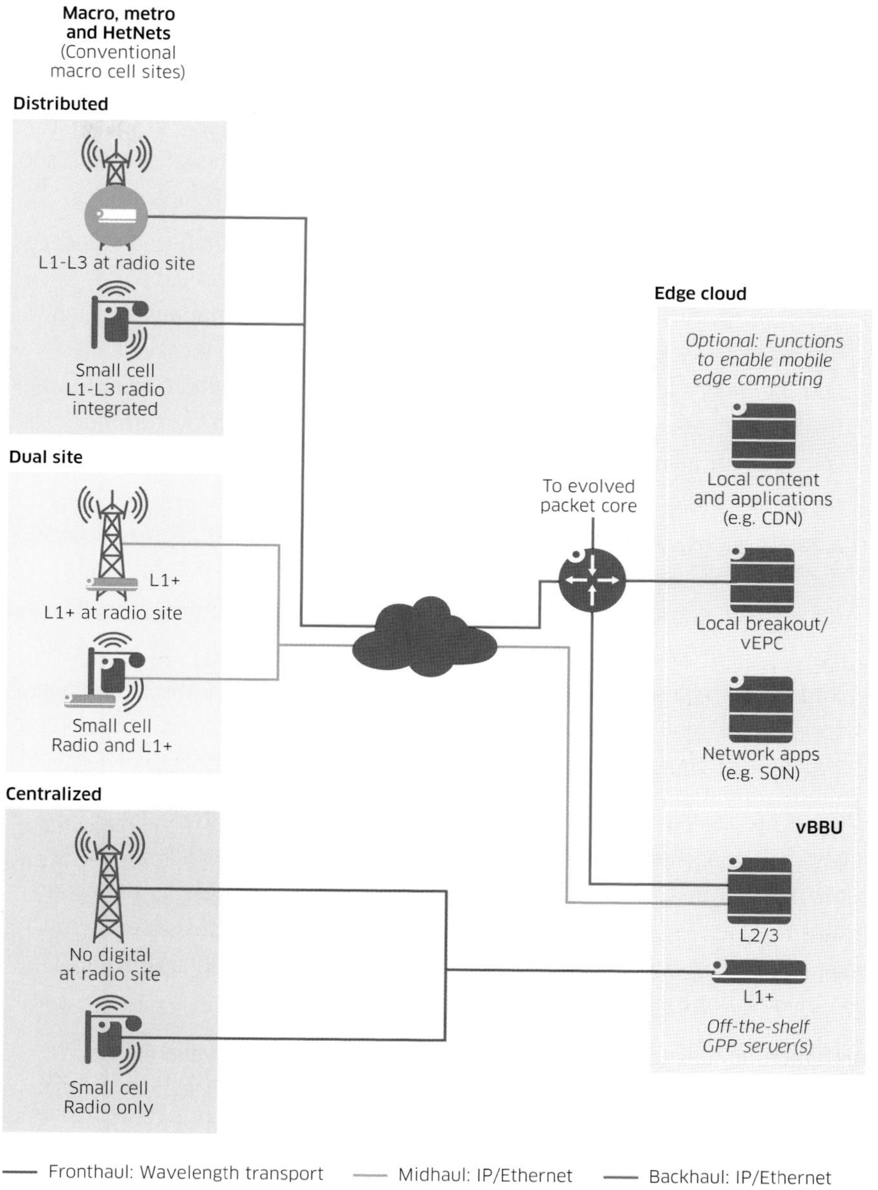

**Macro, metro
and HetNets**
(Conventional
macro cell sites)

Distributed

L1-L3 at radio site

Small cell
L1-L3 radio
integrated

Dual site

L1+

L1+ at radio site

Small cell
Radio and L1+

Centralized

No digital
at radio site

Small cell
Radio only

To evolved
packet core

Edge cloud

*Optional: Functions
to enable mobile
edge computing*

Local content
and applications
(e.g. CDN)

Local breakout/
vEPC

Network apps
(e.g. SON)

vBBU

L2/3

L1+

*Off-the-shelf
GPP server(s)*

—— Fronthaul: Wavelength transport   —— Midhaul: IP/Ethernet   —— Backhaul: IP/Ethernet

FIGURE 10: Illustration of dual-site and centralized vRAN architectures.

Some of the key benefits of the vRAN architectural approach are:

- *Flexibility for future enhancements* – The vRAN architecture allows for easy enhancements of L2/3 algorithms for radio resource management (RRM) and distributed self-optimizing networks (SON). It is also easy to scale the number of simultaneously active users (SAUs), since additional compute and storage can be added to the processing through virtual machine scaling.

- *Cross-site carrier aggregation / dual connectivity* – LTE-Advanced standards already support aggregation of carriers to serve a user device from different bands and different sites. In the case of cross-site aggregation, the lower layer processing is generally performed at the local cell site, and the higher layer, at a central cell site. The vRAN architecture naturally facilitates cross-site aggregation, since L2/3 processing is performed centrally for all cell sites.

- *Pooling* – During periods of a very high ratio of peak-to-average demand across regions, the vRAN architecture will be better able to satisfy the peak demand using pooled baseband computing resources. Furthermore, the deployment economics are improved by re-using the pooled resources for non-real time applications during long periods of inactivity (for example, at night).

- *Radio performance (uplink)* – Simulations show ~60% uplink throughput gain for coherent multipoint processing (CoMP) of signals, by allowing subtraction of interfering signals from each other – either on a single site or between sites.

With vRAN implemented in distributed edge cloud data centers, there is also the opportunity to deploy local core network functions from the same data centers, such as, serving-gateway and packet-gateway nodes, or future 5G-gateway nodes. This approach, which could be considered mobile edge computing, could also include applications and content caches for continuously optimized wireless service delivery.

Architecture: "Cell-less" architecture for high reliability and availability

The preceding vRAN architecture also facilitates a user-centric, "cell-less" architecture in which devices are simultaneously connected to multiple radio access technologies – such as LTE, 5G low-band, 5G high-band and Wi-Fi – in heterogeneous macro- and small-cell network layers. At any given time, the device is served by multiple radio heads, each carrying application traffic with particular characteristics for which it is best suited, as shown in figure 11. As illustrated, mobile devices are no longer connected to a single cell, but rather different sets of radio end points are connected to different devices. The upper layer (L2/3) processing for different devices is at different

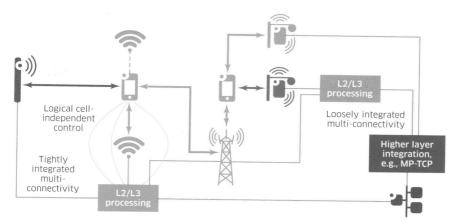

FIGURE 11: Illustration of cell-less, multi-connectivity architecture

locations, even though they share some common links. In other words, the mobile service is provided at any given time through a set of links with processing disassociated from any particular cell maintaining the radio links, thus creating the "cell-less" architecture. Furthermore, a persistent logical control channel can be maintained independent of the specific cell site, to facilitate adding and dropping links.

The benefits of a multi-connectivity approach include:

- Higher instantaneous throughput by aggregating resources across multiple radio links

- Improved application performance by routing flows via the most appropriate radio links, taking into account the target latency, throughput, and burst requirements of each flow

- Reduced call drop rates because of handoffs, since the device is connected through other links when migrating across cells

Architecture: Unlicensed and licensed spectrum

Traditional cellular systems have been based only on licensed spectrum. Unlicensed spectrum, such as the ISM band spectrum, has transmit power limits that prevents its use for wide area macro cells. Thus, it has been utilized primarily by Wi-Fi systems that operate indoors, or over shorter distances. However, with the advent of mobile network small cells, which are well below the transmit power limits, unlicensed spectrum becomes attractive for use by cellular technologies. Aggregation of somewhat unreliable, unlicensed spectrum, together with licensed spectrum, can increase throughput substantially, as long as the unlicensed spectrum is not overloaded. For example, throughput is doubled when 20 MHz unlicensed spectrum is combined with 20 MHz of LTE licensed spectrum. Small cells that combine licensed and unlicensed spectrum will be part of future wireless access, with unlicensed spectrum used either as a supplementary downlink-only channel, or for both uplink and downlink using time division duplex. Standardization for use of unlicensed spectrum for LTE is ongoing in 3GPP Release 13, as license-assisted access (LAA), and similar concepts will be part of the future 5G specifications.

Architecture: A new core for optimum adaptability

NFV is already transforming core network elements that were previously on dedicated hardware, promising OPEX and CAPEX savings for operators. Similarly, software defined networking (SDN) is enabling dynamic control of network services in the data center and wide area networks between data centers. 5G will offer the first opportunity to design a mobile core network that embraces SDN and NFV as a fundamental aspect of the architecture, instead of merely an alternative implementation approach for a mobile core. The new core architecture will allow the benefits of SDN and NFV to be fully realized in ways not possible with today's mobile network standards.

We expect the new mobile core to incorporate a simpler, logically centralized, network controller that will interface to SDN-enabled routers, serving as mobile gateways. The architecture will substantially increase the flexibility to assign mobile network edge nodes on the data path, facilitate local breakout for enterprise or local CDN nodes, unify fixed and mobile access/edge data plane nodes, as well as reduce the complexity of the control plane. We expect the core network to become access-type independent, allowing the device to connect to the core through any wireless access technology, using independent signaling methods. Figure 12 shows the new core network architecture we envision.

Other features of the new core network would include:

- A unified, policy-based service management framework where control applications establish per-user and per-flow policies for charging, mobility, security, QoS and optimization using the network controller

- Additional mobility modes in active and idle states, which would result in fewer signaling messages for low-end IoT or machine-to-machine (M2M) devices, thereby increasing their battery life

- Network slicing, where each virtual network slice has a different set of requirements, for instance, enabling multiple enterprise services to be carried independently or serve a specific set of devices

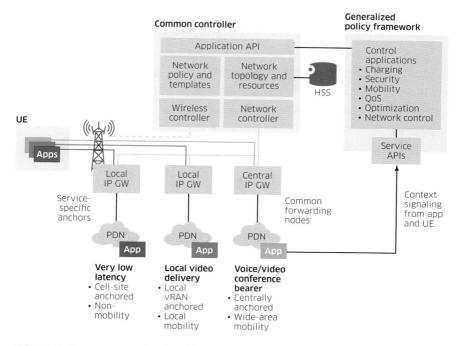

FIGURE 12: The new core network architecture. Each device or user equipment (UE) has multiple IP addresses corresponding to different anchors and packet data networks (PDN) for different services

Operational optimization: Minimizing energy consumption

Integration of different types of access points and radio access technologies allows for greater coordination and the many benefits described for cell-less access solutions. However, this control can also be used to maximize network energy efficiency by dynamically balancing the powering of access points to match the geographic distribution of users and their data needs. Diurnal variations in wireless usage are well known, as are geographical variations. The ability to power down access points when a user can be equivalently served by another node (for example, a macro-cellular coverage layer instead of a small cell).

In addition, by using Massive MIMO to direct energy only to the individual users for which data is served, the overall energy efficiency of base stations is enhanced. And because energy is not sent toward other users in a cell, interference is reduced, for further energy savings.

Putting it all together: End-to-end architecture of future wireless accesss

The end-to-end architecture of the future wireless access system that we envision is summarized in figure 12. It incorporates multiple radio access technologies, multiple spectrum bands, potentially a new core network separating control and data, a vRAN distributed edge cloud and low-latency services. The radio access network is a combination of macro cells based on vRAN and fully integrated small cells, with each handling a subset of access technologies: 4G LTE, 5G low-band, 5G high-band, Wi-Fi and Wi-Gig, in multiple spectrum bands. Devices will use multiple technologies to connect through multiple sites, with low-band used for small data transfers and the control signals for both low and high bands.

The future begins now

Although the future wireless network is a massive advance in network architecture and technology, the path to this network begins with the deployment of LTE-advanced and upcoming releases of 3GPP specifications. For example, LTE-advanced already incorporates features, such as carrier aggregation and dual connectivity, which will be essential technologies of 5G. Release 13 of the 3GPP specifications will include features that allow tight integration of LTE and Wi-Fi, called LTE and Wi-Fi Link Aggregation (LWA), as well as the use of LTE in unlicensed spectrum, or LAA – as discussed above. Equipment providers and operators are in the process of implementing proofs of concept and trial systems of these features. Similarly, the migration to vRAN architecture and virtual packet core are beginning with early deployments and trials. Finally, a number of industry forums on 5G have facilitated discussion and demonstration of 5G technologies.

Here are a few examples of activities in 2015:

- In February 2015, a lab-based proof of concept for an LTE unlicensed system was demonstrated that delivered a 450 Mb/s downlink (TeleGeography 2015). This system combined 20 MHz of licensed spectrum with 40 MHz of unlicensed spectrum in the 5 GHz band (as a supplementary downlink) to achieve this rate.

- In March 2015, there was an announcement of a plan to deploy LTE in unlicensed spectrum in a North American operator's mobile network (Qualcomm 2015).

- In May 2015, a proof of concept demonstrated – with extensive lab and campus-wide test results – that with minor modifications to include channel sensing, LTE can coexist with Wi-Fi in the same spectrum, without any degradation to Wi-Fi performance. An LTE-U open industry forum has been created for the ecosystem to meet and determine operational and compliance specifications (see www.lteforum.org).

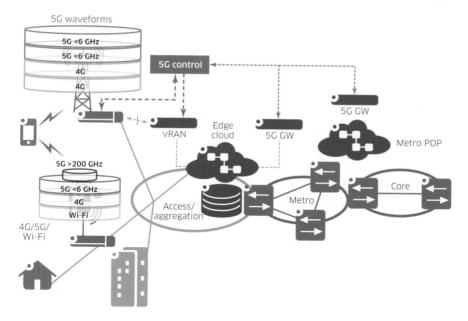

FIGURE 13: End-to-end architecture of future wireless access.

- At Mobile World Congress in 2014, there were demonstrations of proof-of-concept vRAN solutions with voice-over-LTE (VoLTE) calls made through the vRAN and vEPC demonstrator (Alcatel-Lucent 2014).

- Technology advances for 5G have been ongoing in industry and university research labs since 2013. In addition to academic conferences, there are a number of 5G industry forums that serve as meeting grounds where many of these new technologies are discussed and demonstrated. All major vendors and operators are participating in these forums to drive consensus on key 5G technologies through these pre-standards discussions. The Next Generation Mobile Network Forum of leading mobile operators is also shaping the future of wireless access with release of a 5G white paper and requirements (NGMN 2015).

- Demonstrations from different vendors have included at least the following 5G technologies: UF-OFDM waveform, multi-RAT networking, mmWave access and vRAN.

Summary

Over the last decade, wireless communication has moved from being a luxury to a necessity with users now expecting to be always connected and always untethered. In the next decade, we will see these expectations expand as all communication[4] goes wireless, whether in the home, the office or anywhere else on the planet, as part of the new digital reality. This digitization and connection of everything and everyone will lead to expectations of infinite capacity and immediate response, as well as of high reliability and robustness. The notion of the Internet will shift from being something to which we connect from our smartphones, tablets and PCs, to something that envelopes us with myriad devices, wirelessly connected together, monitoring, supporting and automating our lives.

Enabling this vision requires a network that is capable of delivering seemingly infinite ultra-narrowband and ultra-broadband capacity, when and where it is required, optimized for throughput, latency, reliability, scalability and energy efficiency. The key technologies and architectural elements for the wireless network of 2020 and beyond are already in the process of being invented and implemented, but much remains to be done to realize this wireless future and the fully digital human condition it will make possible.

4 - We are using communication in the broadest sense, see Chapter on the Future of Communications.

References

4G Americas 2012. "Mobile Broadband Explosion: The 3GPP Wireless Evolution."

Alcatel-Lucent 2014. "China Mobile shows VoLTE via virtualized network at Mobile World Congress using Alcatel-Lucent's virtualized proof of concept LTE RAN BBU and virtualized evolved packet core solutions," Press Release, Feb 19 (https://www.alcatel-lucent.com/press/2014/china-mobile-and-alcatel-lucent-demonstrate-progress-mobile-nfv-development-voice-and#sthash.N8ba2iwN.dpuf).

Beecham 2008. "Internet of Things: worldwide M2M market forecast."

Bell Labs Consulting 2015. "Mobile Demand Consumption Model – NAR forecast 2015-2025," March 27, internal tool and presentation.

GeSI 2015. Global e-Sustainability Initiative, *#Smarter2030 ICT Solutions for 21st Century challenges* (http://smarter2030.gesi.org/downloads/Full_report2.pdf).

GSMA Intelligence 2015. Current year-end data except interpolated subscribers and connections (https://gsmaintelligence.com).

Lopez-Perez, D., Ming Ding, Claussen, H. and Jafari, A. H., 2015. "Toward 1 Gb/s/UE in Cellular Systems: Understanding Ultra-Dense Small Cell Deployments," *IEEE Communications Surveys & Tutorials*.

Machina 2011, "M2M Communications in Automotive 2010-2020," July, "M2M Communications in Healthcare 2010-2020," May, "M2M Communications in Consumer Electronics 2010-2020," July, "M2M Communications in Utilities 2010-2020," July, " M2M Global Forecast and Analysis 2010-20," October.

Marzetta, T. 2015. "Massive MIMO: An Introduction," *Bell Labs Technical Journal*, March.

NGMN 2015. "5G White Paper," NGMN Alliance, Feb 17 (https://www.ngmn.org/uploads/media/ NGMN_5G_White_Paper_V1_0.pdf).

Ovum and Bell Labs Consulting 2015. "Peak Messages as on New Year's Eve (31st Dec) Each Year." OTT Communications Tracker: 4Q14, March.

Pyramid Research 2011. "Mobile data total region for NA, EMEA, CALA and APAC," December.

Pyramid Research 2014. "Mobile data total global," December.

Pyramid Research 2015, "Mobile data," June (http://pctechmag.com/wp-content/uploads/2014/01/ WhatsApp-1.jpg).

Qualcomm 2015. "Qualcomm and Alcatel-Lucent Join Forces with T-Mobile to Accelerate Licensed Assisted Access Adoption," Press Release, March 15 (https://www.qualcomm.com/news/releases/2015/03/01/ qualcomm-and-alcatel-lucent-join-forces-t-mobile-accelerate-licensed).

Rusek, F., Persson, D., Lau, B. K., Larsson, E. G., Marzetta, T. L., Edfors, O., and Tufvesson, F., 2013. "Scaling Up MIMO: Opportunities and Challenges with Very Large Arrays", IEEE Signal Processing Magazine, January.

TeleGeography 2015., "Ericsson, Qualcomm achieve 450 Mb/s downlink in LTE-U trial," 11 Feb (https://www.telegeography.com/products/commsupdate/articles/2015/02/11/ericsson-qualcomm-achieve-450Mb/s-downlink-in-lte-u-trial).

Statista 2014, Number of LINE App's registered users from December 2011 to October 2014 (in millions), October.

Statista 2015a, Number of monthly active WhatsApp users worldwide from April 2013 to April 2015 (in millions), April.

Statista 2015b, Number of monthly active WeChat users from 2nd quarter 2010 to 1st quarter 2015 (in millions), May.

Whitfield, K. 2014. "The 1.5 Trillion $ dollar story of SMS," Portio Research, December (http://www.portioresearch.com/en/blog/2013/the-15-trillion-$-dollar-story-of-sms.aspx).

CHAPTER 7

THE FUTURE of BROADBAND ACCESS

Peter Vetter

The essential vision

For the past two decades, broadband access has been provided by traditional telecommunication and cable TV operators. These operators have leveraged their existing twisted pair and hybrid fiber-coaxial (HFC) wireline networks to offer triple play voice, data and video services. Typically, they have operated with a national market focus while serving the needs of their original domestic markets. With the increasing importance of mobile communications and networking in the last decade, many of these operators purchased mobile networks in other markets. Then, with the move toward quadruple-play services, including wireless services with triple play offers, a convergence of fixed and mobile operators in each domestic market began to occur. However, given the hyper-competitive nature of the marketplace, this convergence often came at the expense of further global expansion. As a result, a relatively localized set of converged operators emerged.

In the coming years the future of broadband will be affected by three dramatic changes in business models and enabling technologies:

1. *A new global digital democracy* – In the future, users will have access to broadband services anywhere in the world. Service will be offered by global providers independent of region or local access networks. These

global service providers will offer the same service package and level of service to any user and enable access from any location – at home, work, or on the move – by leveraging any affiliated local access provider.

2. *A new global-local service provider paradigm* – In the next access era, global service providers will offer multiple tiers of services by partnering with regional access providers to create a new global-local service provider paradigm. Specific service packages will be built depending on the technological capabilities of the local access provider and related commercial agreements with the global service provider. Local access providers in the same region will compete to offer global service providers with the most compelling hosting and digital delivery infrastructure, which will be comprised of a seamless mix of wireline and wireless connectivity.

3. *A new universal access experience* – As the new paradigm unfolds, users will become access agnostic. They will expect to be able to move seamlessly between fiber, copper and wireless connections. This will require operators to shift resources dynamically between different access technologies. Therefore, the virtualization of network functions in the edge cloud will be an important enabler of this flexible, converged infrastructure. Additional flexibility will come from the installation of universal remote nodes that can be reconfigured to support any combination of access technologies at any time.

This chapter outlines the evolution of broadband access and highlights key indicators for the paradigm shift that will happen. It elaborates our vision of a future in which the technology, architecture and business models that enable broadband access will differ significantly from what they are today.

The past and present

The beginning of the digital democracy

The first digital democracy in which affordable broadband access was accessible in nearly all markets emerged over the past two decades. It was driven by aggressive commercial competition among network service providers and a remarkable series of innovations in technology that reduced the cost per bit dramatically.

In the early 1990s, the deregulation of the telecommunication industry in most markets and the advent of the Internet marked the start of broadband access services. Then, the World Wide Web provided a platform for sharing

The first digital democracy in which affordable broadband access was accessible in nearly all markets emerged over the past two decades. It was driven by aggressive commercial competition among network service providers and a remarkable series of innovations in technology that reduced the cost per bit dramatically.

content at scale efficiently. Before that time, service providers offered a limited portfolio of content and services, built on proprietary platforms and limited to the walled garden of their network realm. Fiber to the home (FTTH) was seen as the ultimate solution for broadband access and PayTV video services were considered to be the killer application that would fund the cost of deploying a new optical access infrastructure.

However, this utopian ideal was hampered by the enormous cost associated with deploying new wired infrastructure to every home, the amount of time required to build out the new networks (decades) and the realization that video services offered limited additional revenue potential. Combined, these factors made the return on investment (ROI) period for FTTH deployments a decade or more. This was deemed unacceptable to investors and shareholders of publicly traded access providers. Consequently, access network providers started looking for alternative technologies that would reuse their existing infrastructure to enable deployment of broadband services faster and with an acceptable ROI.

In 1997, incumbent telecom operators in North America started using new digital subscriber line (DSL) technology over their twisted pair copper wires. This technology used 1 MHz of bandwidth above the voice channel (4 kHz) to modulate a digital data signal to homes where it was demodulated by a DSL modem. At the same time, the community antenna television (CATV) providers became multi-service operators

(MSOs) by using cable modem technology over their coaxial cable plant, using so-called hybrid fiber-coaxial (HFC) technology that used a related digital modulation and demodulation scheme to transmit data to homes using a small amount of spectrum above the video spectrum (above 550 MHz). These technologies were relatively economical to deploy and offered speeds up to 10 Mb/s (peak rate).

As a result, FTTH was restricted to greenfield deployments for which new infrastructure had to be deployed anyway and the relative economics were comparable to those of copper-based technologies. The most notable exceptions to this evolution were Japan and Korea where fiber deployments in metropolitan areas and buildings was considered a long-term economic imperative and the relative density of apartment buildings made the economics more attractive. Chinese operators were also deploying FTTH, due to the lack of adequate twisted pair infrastructure and a desire to create a future-proof solution. On the east coast of North America, the low-cost associated with the aerial installation of fiber, combined with higher demand for video services, created a positive business case for FTTH. To minimize the investment risk, Google Fiber changed the business model by signing subscribers before deploying FTTH in select North American cities.

Historically, because they started as local telephone or video service operators, telecom service providers and MSOs were national businesses. Telecom networks were interconnected through exchanges that created a global service, but geographic restrictions on video content rights limited connections. This changed in the early 2000s. Increasing competition from web-scale service providers who offered global services on top of any access network infrastructure, combined with the associated value shift from connectivity services to content and contextual services discussed elsewhere in this book, led network service providers to develop different strategies to become more global players. Many looked at successful global-local expansion strategies in other industries, which showed that it was possible to build very profitable businesses by expanding beyond a national base.

As shown in figure 1, global companies in different industry segments have leveraged their brand and the assets of local partners to be successful worldwide. These companies derive more than 50% of their revenues from foreign markets. For example, Coca-Cola pursues a global marketing and product strategy, but relies on local companies for the bottling and distribution of its drinks. Similarly, global airlines like Lufthansa have evolved to be global carriers with worldwide coverage, but achieve this by

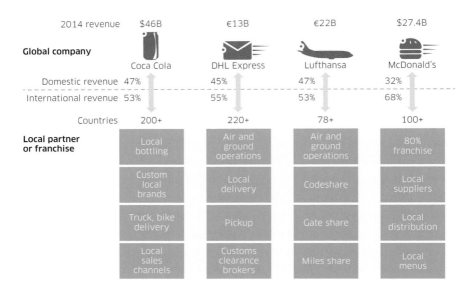

2014 revenue	$46B	€13B	€22B	$27.4B
Global company				
	Coca Cola	DHL Express	Lufthansa	McDonald's
Domestic revenue	47%	45%	47%	32%
International revenue	53%	55%	53%	68%
Countries	200+	220+	78+	100+
Local partner or franchise	Local bottling	Air and ground operations	Air and ground operations	80% franchise
	Custom local brands	Local delivery	Codeshare	Local suppliers
	Truck, bike delivery	Pickup	Gate share	Local distribution
	Local sales channels	Customs clearance brokers	Miles share	Local menus

FIGURE 1: Examples of global companies in different industries that collaborate successfully with local partners (Bell Labs Consulting based on references from annual report data)[1]

sharing codes and facilities with regional airlines and relying on partners for air control and ground operations in local airports.

Taking a cue from these and other successful global-local companies, some service providers made initial forays into international markets where they formed joint ventures. For example, BT and AT&T joined forces to create Concert Communications Services and Vodafone and Verizon created Verizon Wireless. Others, such as Orange and Vodafone, acquired local operators. However, these ventures had limited success because they could not achieve economies of scale by combining geographically dispersed and architecturally dissimilar networks (Manta 2015 and Wikipedia).

The era of local access providers
The current market dominance enjoyed by national and regional broadband access providers is the result of historical incumbency based on owned physical infrastructure. But with the emergence of virtualized network functions (NFV) and software-defined networking (SDN), providers have an opportunity to extend their reach beyond historical geo-physical boundaries and achieve greater scale and a larger market share.

1 - Analysis and modeling by Bell Labs Consulting based on data from various sources (Coca-Cola 2014, Deutsche Post 2014, Lufthansa 2014, Passenger 2014, McDonald's 2014).

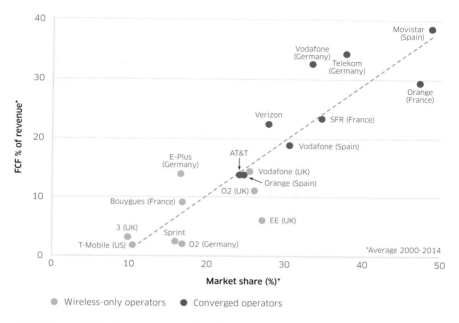

FIGURE 2: Correlation between market share (*average for 2000 to 2014) and free cash flow (FCF) for different operators[2]

Figure 2 shows how a larger market share creates a better free cash flow (FCF) to revenue ratio for converged wireline and wireless service providers in their domestic markets. It also emphasizes that providers can extend their reach beyond national borders and leverage global branding and marketing to attract more subscribers than would be possible with a local brand and its attendant smaller marketing budget. These operators have the advantage of more financial clout in the procurement of content, equipment from system vendors, connectivity with transport network providers and, as we will see below, connectivity to local access providers.

The advantages of going global have continued to drive mergers and acquisitions in the service provider space between wireline and wireless providers (table 1). In an era where end users expect seamless access to broadband services, regardless of where they are and whether they are connected via a fixed or mobile network, converged global service providers have the advantage in terms of service delivery and the superior economics associated with sharing network costs across wireline and wireless networks. However, the ability to achieve the scale and profitability advantages seen

2 - Bell Labs Consulting based on data from GSMA 2015.

by truly global brands has been very limited. But, Bell Labs believes the industry is on the verge of a new era of global-local expansion, as described later in this chapter.

ACQUIRER	TARGET
VF (Germany) – mobile	KDG – fixed
VF (Spain) – mobile	Ono – fixed
Orange (Spain) – converged	Jazztel – fixed
Numericable (France) – fixed	SFR – converged

TABLE 1: Wireless and wireline access provider acquisitions in 2014 to 2015

The evolution of broadband access speed

As shown in figure 3, competition and user demand have driven access providers to deliver higher-speed broadband services to users at affordable prices and, with it, they have enabled the evolution of the digital democracy.

The first broadband access service in the 1990s delivered rates ranging from a few hundred kb/s to a couple of Mb/s, which was a huge improvement compared to the tens of kb/s possible with a dial-up modem. Access capacity for DSL services delivered over twisted pair grew exponentially over the years, as shown in figure 3 (Maes 2015). Every new technological step enabled providers to deliver new services: asymmetric digital subscriber line (ADSL) was well suited for early web browsing on the Internet, while very-high-bit-rate DSL (VDSL) was ideal for the delivery of video. While many believed that this would be the endpoint for DSL, the introduction of vectoring (Oksman, Schenck et al 2010) to cancel crosstalk between twisted pairs in the same bundle, which was inspired by interference cancellation techniques used in wireless communication, resulted in a significant improvement of the signal quality over VDSL. This gave providers the ability to support up to 100 Mb/s and multiple simultaneous streams of high-definition content per household. Where multiple pairs to the home were available, it was also possible to combine their capacity through bonding across pairs, which combined the capacity of the pairs into a single, logical, high-capacity transmission link. And this remarkable technological feat – delivering near-fiber-like speeds over copper loops that were designed for 56 kb/s and, in some cases, were approaching 100 years of age – continues today, thereby turning copper into commercial gold.

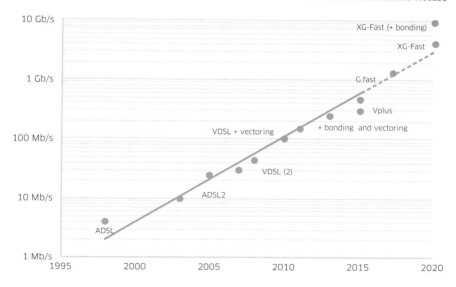

FIGURE 3: Evolution of the speed achievable over twisted pair copper using DSL technology (Maes 2015)

The latest DSL standards, Vplus and G.fast, are about to be deployed. Vplus achieves up to 300 Mb/s over a single twisted pair and features a vectoring scheme that is compatible with existing VDSL2. G.fast further boosts ultra-broadband transmission speeds up to one Gb/s (figure 3). As we will see below, evolution to even higher rates can be expected and has already been demonstrated in research (Coomans 2014).

Likewise, the demand for higher bandwidth and higher video resolution has driven innovation in cable networks. Figure 4 shows the evolution of the downstream capacity on a cable network for different generations of the data over cable service interface specification (DOCSIS) standard. Unlike the point-to-point rate per subscriber shown for DSL in figure 3, DOCSIS capacity is shared among hundreds of homes attached to the same cable tree (figure 7). Therefore, the capacity shown is the instantaneous peak rate available to subscribers. What this means is that one subscriber has access to all the data bandwidth, but the achievable *sustained* rate per subscriber is significantly lower (for example, only about 10% of this rate when 10 subscribers are simultaneously active).

In 1997, DOCSIS 1.0 specified the first, non-proprietary, high-speed data service infrastructure capable of providing Internet web browsing services. The ability to differentiate traffic flows for the improvement of service

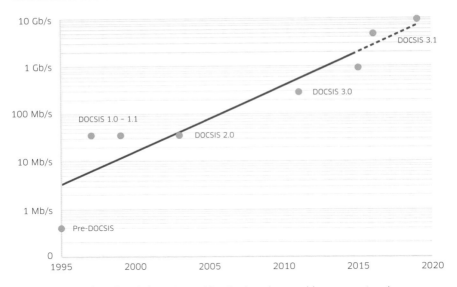

FIGURE 4: Evolution of total downstream bit rate shared on a cable access network
(Bell Labs Consulting 2015)[2]

quality appeared with DOCSIS 1.1. Enhancement of the upstream bandwidth was introduced with DOCSIS 2.0, so that new services like IP telephony could be supported. DOCSIS 3.0, which is used worldwide today, significantly increased capacity by bonding channels (for example, combining 24 or 32 channels for downstream data and 4 or 8 for upstream data), effectively using a wider spectrum for interactive services. With a USA spectral plan, each additional channel provides 6MHz bandwidth. The latest DOCSIS 3.1 standard enables up to 10 Gb/s downstream and 1 Gb/s upstream thanks to the use of a wider spectrum for downstream (1 GHz or higher) and upstream (up to 200 MHz), as well as the introduction of orthogonal frequency division multiplexing (OFDM), a signal format with improved spectral efficiency that is already widely used in DSL and wireless.

There has been a similar evolution in the data rates supported by optical access (figure 5). A passive optical network (PON) has proven to be the most economical choice because it allows multiple subscribers (typically 32) to share the expensive downstream laser and it supports the use of a signal distribution fiber, which is then passively split to each home with individual drop fibers (FTTH in figure 6). The bandwidth is shared through time division multiplexing (TDM) on one wavelength channel for downstream

2 - Based on Bell Labs modeling and the following standards: "DOCSIS 1 - ITU-T Recommendation J.112 Annex B (1998)"; "DOCSIS 1.1 - ITU-T Recommendation J.112 Annex B (2001)"; "DOCSIS 2 - ITU-T Recommendation J.122, December 2007"; "DOCSIS 3 - ITU-T Recommendation J.222, December 2007"; "DOCSIS 3.1 Physical Layer Specification (CableLabs), CM-SP-PHYv3.1," June.

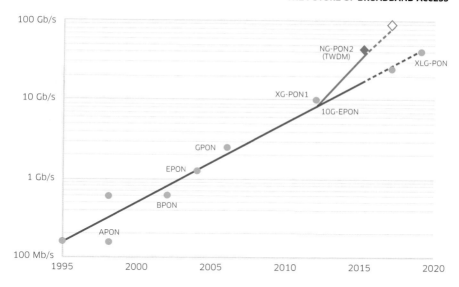

FIGURE 5: Evolution of downstream capacity on a passive optical network (Vetter 2012)

and another wavelength channel for upstream. The early, pre-standard PON systems based on asynchronous transfer mode (ATM) – APONs– supported symmetrical rates of 155/155 Mb/s and asymmetrical rates of 622/155 Mb/s (downstream/upstream). They were the baseline for a broadband PON (BPON) that was standardized in 1998 and which included an additional radio frequency (RF) overlay wavelength for the distribution of video channels. The subsequent standard Ethernet PON (EPON) and gigabit PON (GPON) provide a shared bandwidth of 1 Gb/s and 2.5 Gb/s respectively and are both still widely deployed today.

A 10 Gb/s capable PON and 40 Gb/s PON have recently been defined in standards. They will be deployed over the next five years to allow multiplexing of more subscribers over a PON (for example, 128) or allow simultaneous support for backhaul and enterprise applications, without compromising the bandwidth available for residential services. The latest next-generation PON (NG-PON2) standard achieves 40 Gb/s by stacking four wavelength pairs of 10 Gb/s in a time and wavelength division multiplexing (TWDM) scheme (Poehlmann, Deppisch et al 2013). A 40 Gb/s-capable PON (XLG-PON) with pure TDM serial bit rates can be expected, if cost-effective approaches can be found (van Veen, Houtsma et al 2015).

The evolution of the broadband access architecture
The continuous increase in data rates has had a significant impact on the

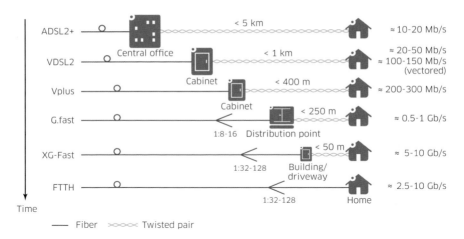

FIGURE 6: Telecom access architecture evolution to deep fiber solutions and remote nodes that reuse existing copper (with the exception of the FTTH case)

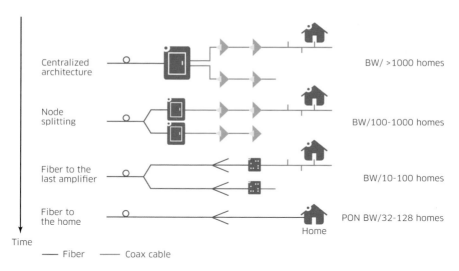

FIGURE 7: Cable access architecture evolution with increasing share of bandwidth (BW) per home

access network architecture. The higher DSL speeds require transmission over a wider frequency band (for example, 1 MHz for ADSL, 8 or 17 MHz for VDSL, 34 MHz for Vplus, and 106 MHz for G.fast), which due to propagation physics, is only possible over shorter distances of twisted pair copper (figure 6). Therefore ADSL can run from a central office over a distance of 5 km because of the low frequency band used, but because it uses higher frequencies, VDSL can only be delivered from a street cabinet close to the end user (typically 1 km or less). Vplus is a recent addition to the DSL family that offers higher rates for homes closer to the cabinet by doubling the frequency band. The even higher frequency bands used by G.fast necessitate deployment from a distribution point at the street curb or basement of a building. As a result, fiber is increasingly being installed to G.fast remote nodes that are moving ever closer to the end users.

Likewise, the bandwidth per subscriber in a cable network can be improved by reducing the node size and using deeper fiber penetration, ultimately bypassing the last amplifier, as shown in figure 7.

As discussed in the chapter on *The future of wireless*, the same evolution is occurring in wireless networks with the move from macro cells (~1 km cell radius) to small cells (<500 m cell radius). Consequently, ultra-broadband copper (whether DSL or cable-based) access and ultra-broadband wireless access are evolving to very similar architectures and distances from subscribers, a trend which is pointing the way to a new converged future, as described in the next section.

The future

Changing the game in broadband access

As outlined earlier, we believe that a new global-local paradigm will emerge in which global service providers (GSPs) will seek partnerships with regional or local service providers (LSPs) to become tenants of the access infrastructure and provide local delivery of digital access to end users. Through these partnerships, GSPs will expect to have the flexibility and ability

...to deliver the desired bandwidth in the future, high-speed wireless and wireline technologies will be located at distances of 100 m or less to the end users, pointing the way to a new converged access network.

to instantiate and configure required capabilities in the access infrastructure as needed. This includes everything from basic subscriber management and configuration of bandwidth and quality of service (QoS), to more advanced functions like mobility, security, storage of data, or even virtualization of a user's customer premises equipment (CPE). To the greatest extent possible, GSPs will want to offer the same set of service packages anywhere in the world, regardless of the location or the type of access network to which the subscriber is connected at any given time. Of course, this will depend on the technological capabilities of the local access infrastructure. But one of the goals of the future access network will be to create the illusion of seemingly infinite capacity using the optimum set of access technologies in each location, with seamless interworking between the technologies.

To support these requirements, LSPs will transform their infrastructure to host different GSPs, as well as their own domestic subscribers, with a simple interface that provides an abstraction of underlying network capabilities. The local access infrastructure will support multi-tenancy to enable sharing of the access network by multiple GSPs, on demand and with the required profitability to sustain access network growth. Given that competition between LSPs in the same region will continue to exist, each will try to attract GSPs by offering the most compelling hosting infrastructure. To meet these requirements, today's dedicated fixed and mobile access architecture will evolve to a new, fully converged, cloud-integrated broadband access network, as shown in figure 8 (a) and (b).

The new architecture will contain three key enabling technology building blocks:

- Small, localized data centers serving as an edge cloud infrastructure, which will host virtualized access network functions and support flexible programming of services by GSP tenants

- Universal remote access nodes, which will extend access programmability to the physical layer, so that operators can configure any combination of wireless and wireline access technology, as needed, based on demand patterns

- A network operating system (Network OS), which will control the creation, modification and removal of functions, and schedule the usage of available resources across different global service agreements

In addition to the introduction of these key technologies, the evolution to higher capacity in the access infrastructure will continue to fiber-like speeds,

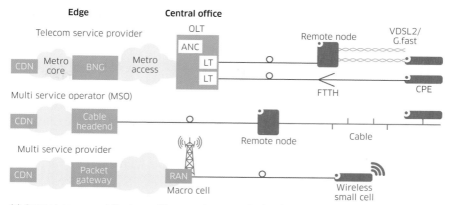

(a) Current access architecture with separate networks for different service providers

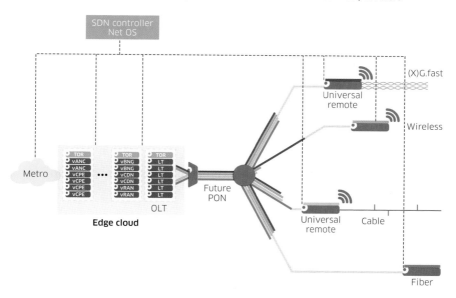

(b) Future converged access architecture with functions consolidated in a unified edge cloud

FIGURE 8 (a) AND (b): Current and future architectures for the broadband access network

but not just over fiber. Beyond G.fast, 10 Gb/s capable DSL (called XG-Fast) over tens of meters of twisted pair has already been demonstrated in research labs, with products likely to emerge by 2020 (figure 3) (Coomans 2014). At the same time, cable will move beyond DOCSIS 3.1 with the expected synergistic use of technology similar to XG-Fast that will enable XG-Cable solutions capable of symmetrical 10 Gb/s rates. In addition, wireless data rates will

Collocating
wireless
and wireline
physical
layers in the
same remote
unit offers
significant
operational
and capital
cost benefits
because of the
shared site
acquisition,
fiber backhaul,
power
supply and
maintenance
opportunities.

evolve to multiple Gb/s with the evolution to 5G, as described in chapter 6, *The future of wireless access*. In each case, these ultra-high access rates will be possible by reducing distances between the access nodes/cells and end users to less than a hundred meters.

This vision of the future raises the obvious challenge of backhauling traffic to and from remote nodes. To support the high capacity of these nodes (up to 10s Gb/s), high-speed optical access technologies will be required in the feeder network (Pfeiffer 2015). The feeder network will be based on next-generation passive optical networks (NG-PON2) or, possibly, XLG-PON. And pure wavelength-division multiplexing (WDM) could be used if very high bit rates (>10 Gb/s sustained per remote node) and minimal latency are required.

The central office in the edge cloud

The global-local cloud introduced in chapter 5, *The future of the cloud*, which will have compute and storage resources distributed to the edge of the network, is well suited to host key enabling functionality for the future access network. In essence, the edge cloud will move the data center to the central office (CO). For access network operators, the edge cloud will be created by consolidating multiple current COs into a more central location at the metro edge, as shown in figure 8, which will create an access network with an extended reach of approximately 20 to 40 km. The maximum distance between the edge cloud and end users will be dictated by latency requirements, if latency-sensitive physical layer processes are implemented in the edge cloud, and by the preference for a fully passive optical feeder network without locally powered repeaters or amplifiers.

The edge cloud will consist of commodity server blades with general purpose compute and storage resources connected via top-of-the-rack (TOR) switches. The optical line terminal (OLT) function that drives the PON and has, until now, been implemented in dedicated hardware platforms (figure 8 (a)), will be virtualized and decomposed into line termination (LT) server blades. These blades will contain the physical layer and medium access control (MAC) layer functions of the PON and the control plane of the OLT will be implemented as a virtual access network controller (vANC) (figure 8(b)). This will enable separate scaling of the functions that depend more on the (fixed) physical interface (the Layer 1 and basic MAC functions) from the functions that depend on the number of subscribers and flows (the Layer 2 (Ethernet) and Layer 3 (IP) forwarding functions).

This new architecture will integrate access network resources in a data center with an SDN controller to enable rapid configuration of new broadband services as service chains between constituent functions and with lower operational costs compared to a conventional access network (Kozicki, Oberle et al 2014).

Using this new architecture, LSPs will create separate vANC instances for each hosted GSP and the domestic service, while SDN will be used to connect these instances to the underlying (common) physical layer functions. This can be achieved in any edge cloud of any LSP, depending on needs and subscriber location. In addition, GSPs will also be able to instantiate virtual CPE (vCPE) and virtual broadband network gateway (vBNG) subscriber management functions to provide nomadic subscribers with the same set of capabilities they have at home. In fact, the virtual BNG or CPE functions will be further decomposed into their constituent functions, which can be service-chained together, as needed, to meet the service requirements for each subscriber and with the optimum economics. The edge cloud will also host the virtualized radio access network (vRAN) function described in chapter 6, *The future of wireless access*. Finally, it will be possible to leverage available local storage resources for virtualized content distribution network (vCDN) functions to enable providers to optimize video content delivery for each local subscriber.

Toward the universal remote

As noted earlier, to deliver the desired bandwidth in the future, high-speed wireless and wireline technologies will be located at distances of 100 m or less to end users. Furthermore, they will both be backhauled by PON-based solutions, wherever possible. Therefore, it will be beneficial to collocate these technologies at the same site, so they can share power and backhaul connectivity. But, the Bell Labs vision for broadband access goes one step further: there will be convergence at the component level as well.

The analysis in figure 9 shows that the total capacity per wireless node has grown faster than that of wireline nodes. This has been driven by the desire of users to have everything available on mobile devices that is available over wireline networks. And, this coincidence becomes convergence if we recognize that wireline and wireless access are now based on the same digital signal modulation format (orthogonal frequency-division multiplexing (OFDM)), similar signal processing technologies and comparable spectral bandwidth (approximately 100 MHz to 1 GHz). Combined with the same deployment location (distance), this convergence suggests that a common digital signal processing platform could be used for all access technologies in future. It creates a clear opportunity for a truly converged universal remote that can be configured for any wireless or wireline access network deployment scenario and adapted to dynamic changes in demand on each interface.

To illustrate this potential, compare the OFDM signal flow for G.fast and LTE in figure 10. It is clear that there is substantial functional similarity between the two at the digital signal processing layer (Layer 1 or L1). In general, the requirements of this layer (forward error correction (FEC), processing for cross-talk cancellation, and Fast Fourier Transform (FFT)) are not performed optimally on a general purpose processor, even with the most advanced multi-core processors. In addition, such an approach would consume too much power to be acceptable for a remote node deployment. Therefore, the best approach is to make use of specialized hardware for the digital signal processing and to virtualize the L2 and L3 functions (vANC functions) to run in the edge data center on general purpose servers.

We believe that the future ultra-broadband access network will leverage this split architecture (Doetsch et al. 2013). It will have vANC functionality running in the edge cloud and connected to a universal remote node that will contain programmable, reconfigurable hardware accelerators that have the ability to continuously adapt to changing technologies and user demands over time. Once again, as with the PON example, vANC instances will be created for each GSP. They will be connected to the universal remote, which will be programmed according to the requirements of the set of hosted GSPs

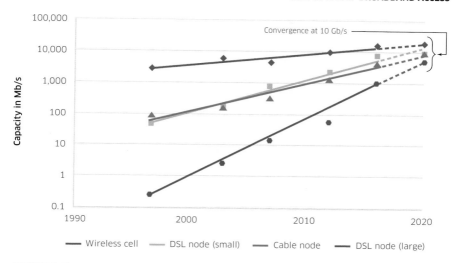

FIGURE 9: The total capacity of access nodes for different wireline and wireless access technologies is converging (Bell Labs Consulting)

and the LSPs' own local subscribers. And they will use SDN and a Network OS to efficiently and dynamically share physical layer resources across multiple physical access technologies, as described in the following section.

Collocating wireless and wireline physical layer functions in the same remote unit offers significant operational and capital cost benefits because of the shared site acquisition, fiber backhaul, power supply, and maintenance opportunities. In addition, the programmability of the new universal remote offers much greater flexibility for the deployment of a converged access network that can adapt to:

- *Geographical variations:* Because the universal remotes are placed so close to end users, the specific needs for wireless/wireline capacity, the number, type and reach of copper connections required or the use of line-of-sight versus nonline-of-sight transmission for wireless will vary at every location. A programmable remote can be configured to accommodate each specific location.

- *Long timescale variations:* As mobile traffic grows more rapidly than fixed, or subscribers churn between different access technologies, the LSP can reprogram node resources to better reflect these changes and achieve higher throughput (by more extensive signal processing) on the desired interfaces.

- *Short timescale variations:* Traffic patterns vary over short timescales. For example, mobile usage typically increases during the day when users

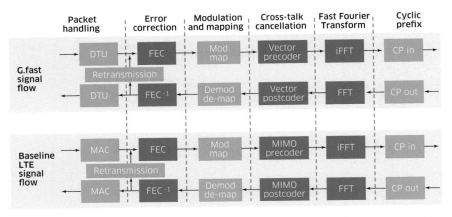

■ Function executed on general purpose processor   ■ Function benefiting from HW acceleration

FIGURE 10: Comparison of digital signal processing flow of G.fast DSL and LTE wireless with indication of common blocks (data transfer unit (DTU), medium access control (MAC), forward error correction (FEC), multiple input multiple output (MIMO), FFT, inverse FFT (iFFT) and cyclic prefix (CP))

are moving outside the home. But, the reverse is true in the evening. The universal remote can adapt to these diurnal patterns to provide optimal throughput at all times.

- *Rapid introduction of new capabilities:* A programmable remote node can adapt to support new technology standards or features that can differentiate an LSP from competitors by enabling the LSP to offer a superior level of service delivery.

In summary, a universal remote extends the programmability of the future broadband access infrastructure to the physical layer and will enable a truly end-to-end SDN to be created. The high capacity and flexibility of the new architecture will enable deployment of multiple types of services that required separate networks in the past (residential or business, fixed or mobile) on a single flexible infrastructure, which will support multiple GSPs, as well as the LSP's local subscriber base.

The role of a Network OS

As discussed in chapter 4, *The future of wide area networks*, a Network OS will also be needed to enable dynamic scheduling of network and processing resources among different functions and services in the broadband access infrastructure (figure 8). Similar to the role of an OS in a computing or mobile device, a Network OS will allow for rapid installation or removal of functions without interrupting ongoing services. The multi-tenancy requirement for multiple GSPs will be supported by the creation of network slices by the Network OS. These slices are similar to the virtual machines on computing hardware. But, in addition to slicing computing resources, they partition programmable physical resources, such as capacity and wavelengths on the fiber network or the digital signal processing resources in the universal remote, to instantiate the desired wireline or wireless access capacity.

In the future ultra-broadband access network, the Network OS will leverage SDN control functions to:

- Provide GSPs with an end-to-end service control by hiding the details of the access network through technology-agnostic abstractions and integrating network control with other network domains (for example, the metro aggregation network or the macro wireless network)

- Act as a network hypervisor that allows independent GSPs to program network functions dynamically within their network slice, without modifying resources allocated to other tenants

- Resolve the contention between multiple network services sharing the same infrastructure, while satisfying different service level agreements (SLAs)

- Optimize the use of resources automatically based on an up-to-date view of network conditions and demand from each tenant network based on sophisticated monitoring and analytics that allow network slices to be modified before any degradation of quality of service occurs

A new universal remote extends the programmability of the future broadband access infrastructure to the physical layer and will enable a truly end-to-end SDN to be created.

From a technology perspective, SDN and NFV are being increasingly embraced by the broadband access community as enabling technologies for multi-tenancy and rapid configuration of new services.

In addition to the functions it will enable for GSPs, the Network OS will provide the host LSP with a unified mechanism for optimizing its access infrastructure cost, revenue, performance, and reliability. The network abstraction provided by SDN will also enable the LSP to install new capabilities rapidly, without affecting existing user services offered by the tenant GSPs.

The future begins now

The broadband access industry is already moving toward a new era of global convergence. Service providers are merging to become fully converged operators that can increase their market share (figure 2 and table 1). Also, web-scale players are offering broadband access services by leveraging Wi-Fi and FTTH deployments in some cities. In particular, in addition to the gigabit services offered by Google Fiber in selected cities, Google has started large-scale trials of a free Wi-Fi access service (supported by ads) using 10,000 Wi-Fi pylons in New York City (Payton 2015). And Google's Project Fi, announced in March of 2015, is perhaps the clearest evidence to date of the upcoming global-local revolution, with subscribers obtaining wireless service from Google, based on local access provided through an ecosystem of local access providers (Google 2015).

From a technology perspective, SDN and NFV are being increasingly embraced by the broadband access community as enabling technologies for multi-tenancy and rapid configuration of new services. For example, in North America, an operator is promoting a new architecture and a prototype with central office functions in an edge cloud (Le Maistre 2015), using the open network operating system (ONOS) (ONOS 2015). The architecture instantiates and controls virtualized subscriber management functions for the provisioning of access services via G.fast. Virtual CPE services are also being pioneered and trialed on a large scale with some early products already announced (Morris and LightReading 2015).

Meanwhile, the European Telecommunications Standards Institute (ETSI) industry specification group (ISG) on NFV has created dedicated technical committees for broadband wireless and broadband cable access (ETSI 2015). The latter is in collaboration with CableLabs, which also runs an open networking project that is exploring how to virtualize cable access headend systems (Donley and CableLabs 2015). And the Broadband Forum (BBF) is expected to release technical reports on fixed access network sharing (FANS), architectures for access network virtualization that are consistent with ETSI NFV by the end of 2016, and access programmability through SDN by the end of 2017 (Broadband Forum 2015).

Clearly, we are also at the beginning of Gb/s services for residential access and the requisite accelerated roll-out of fiber deeper in the network, either to the home or, in many cases, to remote nodes closer to the home. Driven by governments in Asia (for example, South Korea and Japan), or by competition by Google Fiber in the US, gigabit to the home initiatives are important catalysts for this evolution.

Looking at the evolution toward remote nodes, there are also clear early moves in the market:

- Several equipment vendors have announced their first G.fast products that feature near-gigabit speeds via twisted pair copper from a distribution point unit (DPU) at a 100 m from the home (Alcatel-Lucent 2015).

- Researchers have recently demonstrated a prototype of 10 Gb/s over DSL over tens of meters of twisted pair copper using XG-Fast (Coomans 2014).

- Cable system vendors have announced the availability of products compatible with DOCSIS 3.1, which supports downstream bit rates up to 10 Gb/s (CableLabs and Arris 2014).

- Some operators have recognized the opportunity of collocating wireless access points with DSL access nodes (Iannone 2013), which is an important precursor to a fully converged universal access node. System and component vendors have also announced agreements to jointly develop a system on a chip for a programmable universal remote node (Joosting 2015).

Summary

In 2020 and beyond, the future broadband access network will see a shift from local and regional service providers to a more GSP model. To achieve global coverage and optimize local services delivery, GSPs will form partnerships with LSPs to offer ubiquitous access and multiple tiers of services to end users. This model has been successful in other industries, driven by the branding and scale advantages of global companies. We predict that this model will also be applied to the global telecommunications market. Early signs of this trend are apparent with large service providers merging or acquiring other players and achieving superior financial performance as a result. Furthermore, service providers with a converged fixed and mobile portfolio show stronger business performance than providers that only focus on either fixed or mobile.

Consequently, the role of the LSP is changing and this change will have a profound impact on the technologies deployed in the future broadband access network. Local access providers will compete for business partnerships with GSPs, with a more limited focus on their own local service offers.

Due to the virtualization of central office functions in the edge cloud, multiple tenant GSPs will be able to obtain a slice of the LSP-owned access network. This will allow them to configure multiple tiers of service packages for their subscribers. In this market, LSPs with a combination of wireline and wireless access capabilities will have a clear advantage in terms of the range of service offers they can provide relative to those limited to either wireless or fixed access networks. Driven by user demand and competition, local operators will continue to increase broadband access speeds to gigabit rates over FTTH in an increasing number of areas. In a significant portion of the network where FTTH deployments are expected to remain economically unattractive, they will offer fiber-like services over existing copper or high-speed air interfaces.

Since the ultra-high-speed copper and wireless networks will require short transmission distances of less than 100 m, they will become collocated to benefit from the economics of shared optical feeder network, site acquisition and installation, power supply and maintenance opportunities. Furthermore, Bell Labs believes that there will be convergence at the chipset level as well, where there will be an opportunity to make a universal remote node that is fully programmable for any combination of wireless and wireline technologies to enhance the deployment flexibility of a truly converged access network.

What will emerge from the above changes is a new digital democracy in which anyone will have access to broadband services at fiber speeds and competitive prices from anywhere in the world by leveraging the optimum local access provider. Users will be able to seamlessly shift between fixed and mobile networks and between wireline and wireless technology, experiencing a fully converged broadband access service that delivers an optimal digital experience anywhere, anytime.

References

Alcatel-Lucent 2015. "G.fast: Clear a New Path to Ultra-Broadband," *Alcatel-Lucent web site*, (https://www.alcatel-lucent.com/solutions/G.fast).

Arris 2014. "SCTE 2014 CCAP interoperability demo highlights DOCSIS 3.1 leadership," *Arris web site*, September (http://www.arriseverywhere. com/?s=Docsis+3.1).

Broadband Forum 2015. "Technical Working Groups Mission Statements," *Broadband Forum web site* (https://www.broadband-forum.org/technical/ technicalworkinggroups.php).

CableLabs 2014. "New Generation of DOCSIS Technology," *CableLabs press release* (http://www.cablelabs.com/news/new-generation-of-docsis-technology).

CableLabs 2015. "SDN Architecture for Cable Access Network Technical Report," *CableLabs web site*, June 25 (http://www.cablelabs.com/specification/ sdn-architecture-technical-report).

Coca-Cola Company 2014. "Annual Report 2014 Form 10-K 2014; Net Operating Revenues by Operating Segment; North America vs. Total," *Coca-Cola web site* (http://assets.coca-colacompany.com/c0/ba/0bdf18014074a1ec a3cae0a8ea39/2014-annual-report-on-form-10-k.pdf)

Coomans W., Moraes, R. B., Hooghe, K., Duque, A., Galaro, J., Timmers, M., van Wijngaarden, A. J., Guenach, M., and Maes, J., 2014. "XG-FAST: Toward 10 Gb/s Copper Access," proceedings of IEEE Globecom, December.

Deutsche Post AG 2014. "Annual Report:, "DHL Express Revenues: Europe vs. Total," (from key figures by operating division table) *Deutsche Post Annual Report web site* (http://annualreport2014.dpdhl.com).

Doetsch, U., et al. 2013. "Quantitative Analysis of Split Base Station Processing and Determination of Advantageous Architectures for LTE," *Bell Labs Technical Journal*, June.

Donley, C., 2015. "SDN and NFV: Moving the Network into the Cloud," *CableLabs web site*, (http://www.cablelabs.com/sdn-nfv).

ETSI 2015. "Network Functions Virtualization," *ETSI web site* (http://www.etsi. org/technologies-clusters/technologies/nfv).

Google 2015. "Project Fi uses new technology to give you fast speed in more places and better connections to Wi-Fi," *Project Fi web site* (https://fi.google.com/about/network).

GSMA 2015. "GSMA Intelligence: Wireless Market Data by Country," *GSMA Intelligence web site*, July. (https://gsmaintelligence.com).

Kozicki B., Oberle K., et al 2014 "Software-Defined Networks and Network Functions Virtualization in Wireline Access Networks," From Research to Standards Workshop, *Globecom* 2014, pp. 680–685, December.

Iannone, P. et al, 2013. "A Small Cell Augmentation to a Wireless Network Leveraging Fiber-to-the-Node Access Infrastructure for Backhaul and Power," *Optical Fiber Communication Conference and Exposition and the National Fiber Optic Engineers Conference* (OFC/NFOEC), March.

Joosting, J-P., 2015. "Freescale and Bell Labs explore wireline and wireless virtualization on the road to 5G," *Microwave Engineering Europe*, February 19 (http://microwave-eetimes.com/en/freescale-and-bell-labs-explore-wireline-and-wireless-virtualization-on-the-road-to-5g.html?cmp_id=7&news_id=222905842).

Le Maistre R. 2015. "AT&T to Show Off Next-Gen Central Office," *Lightreading*, June 15 (http://www.lightreading.com/gigabit/fttx/atandt-to-show-off-next-gen-central-office/d/d-id/716307).

Lightreading 2015. "Telefónica, NEC Prep Virtual CPE Trial," *Lightreading* Oct 10 (http://www.lightreading.com/carrier-sdn/nfv-%28network-functions-virtualization%29/telefonica-nec-prep-virtual-cpe-trial/d/d-id/706032).

Lufthansa 2014. "First Choice Annual Report 2014: Passenger Airline Group 2014, "Net traffic revenue in €m external revenue; Europe vs. Total," *Lufthansa web site* (http://investor-relations.lufthansagroup.com/fileadmin/downloads/en/financial-reports/annual-reports/LH-AR-2014-e.pdf).

Maes, J., and Nuzman, C., 2015. "The Past, Present and Future of Copper Access," *Bell Labs Technical Journal*, vol. 20, pp. 1-10.

Manta 2015. "Verizon Communications Inc (Verizon)." April (http://www.manta.com/c/mmlg0rt/verizon-wireless).

McDonald's 2014. "Annual Report 2014 Revenue by Segment and Geographical Information; US vs. Total," *McDonald's web site* (http://www. aboutmcdonalds.com/content/dam/AboutMcDonalds/Investors/McDonald's 2014 Annual Report.PDF).

Morris, I., 2015. "Orange Unveils NFV-based Offering for SMBs," *Lightreading*, March 18 (http://www.lightreading.com/nfv/orange-unveils-nfv-based-offering-for-smbs/d/d-id/714503).

Oksman, V., Schenk, H., Clausen, A., Cioffi, J. M., Mohseni, M., Ginis, G., Nuzman, C., Maes, J., Peeters, M., Fisher, K., and Eriksson, P.-E., 2010. "The ITU-T's New G.Vector Standard Proliferates 100 Mb/s DSL," *IEEE Communications Magazine*, vol. 48, no. 10, pp. 140–148, October.

ONOS 2015. *ONOS web site* (http://onosproject.org)

Payton M. 2015. "Google wants to bring free Wi-Fi" to the world…. and it's starting NOW," *Metro UK*. http://metro.co.uk/2015/06/25/google-wants-to-bring-free-wifi-to-the-world-and-its-starting-now-5265352

Pfeiffer T. 2015. "Next Generation Mobile Fronthaul Architectures," *Invited paper*, M2J.7, OFC.

Poehlmann W., Deppisch B., Pfeiffer T., Ferrari C., Earnshaw M., Duque A., Farah B., Galaro J., Kotch J., van Veen D., Vetter P. 2013. "Low-cost TWDM by Wavelength-Set Division Multiplexing," *Bell Labs Technical Journal*, Vol. 18, No. 3, pp. 173-193, November.

van Veen, D., Houtsma, V., Gnauck, A., and Iannone, P. 2015. "Demonstration of 40-Gb/s TDM-PON over 42-km with 31 dB Optical Power Budget using an APD-based Receiver", *Journal of Lightwave Technology*," vol. PP, Issue: 99, DOI: 10.1109/JLT.2015.2399271.

Vetter, P. 2012. "Next Generation Optical Access Technologies," *Tutorial at ECOC*, Amsterdam, p. Tu.3.G.1, September.

Wikipedia 2015. "Concert Communications Services," (https://en.wikipedia.org/wiki/Concert_Communications_Services).

Wikipedia 2015. "Orange SA," (https://en.wikipedia.org/wiki/Orange_S.A.)

Wikpedia 2015. "Vodafone," (https://en.wikipedia.org/wiki/Vodafone).

CHAPTER 8

THE FUTURE of the ENTERPRISE LAN

The invisible LAN

Colin Kahn

The essential vision

Over the last decade, wireless technology has become the dominant means for consumer device connectivity. The days of wired Ethernet are effectively over for consumers, however, wireless has yet to transform enterprise device connectivity to the same extent. As a consequence, the level of innovation that has characterized the consumer mobile device ecosystem has been largely absent in the enterprise space. The result is an environment that remains rooted in the original *wired* telephony paradigm of a desktop phone and workstation, with network connectivity provided using Ethernet over twisted pair. The rise of *bring your own device* (BYOD) – a consumer-employee-driven phenomenon – has started a revolution in enterprise networking that will only be further fueled by technological advances in consumer wireless telecommunications, ultimately bringing about the demise of wired networking in the enterprise.

The future enterprise LAN will be an invisible all-wireless network, transparently bringing secure, wireless connectivity to enterprise users and visitors to the space. This new network will remove the tether between the employee and the physical space (office, meeting room, building or campus), and simultaneously enable the connection of digital "things" that manage and monitor the enterprise infrastructure and processes, automatically tailoring services to the needs of each application and device.

There are three key elements of this vision:

1. *Virtualized converged radios beat single wires* – The fixed LAN will be replaced by a confederation of wireless access radio technologies, both licensed and unlicensed – and everything in between. These radio resources will be dynamically organized and re-organized into network slices that are abstracted and invisible to the user, with devices multiply or singly connected at times according to the instantaneous needs for throughput, latency and reliability. The resources will be marshaled and directed to each campus or building location, with spatial resolution on the order of individual people.

2. *Zero-touch secure services* – Access to enterprise applications and services will be available both inside and outside enterprise premises, with zero-touch simplicity that requires no user interaction. Setup of virtual private network (VPN) connections, entry of secure-codes, and selection of access connectivity will be automated. Enterprise network controllers will create multi-tenant, virtual workgroups, with secure virtual workspaces and associated services, including the requisite permissions.

3. *The emergence of the in-building operator* – Leveraging new spectrum ownership regimes, independent providers will service the enterprise as wireless LAN providers. They will partner with wide-area operators for shared use of spectral assets to provide seamless connectivity, independent of location within, or outside, the enterprise location. Mobile access will effectively become part of an integrated local and wide-area information networking service.

The past and present

It is important to consider the reasons for this delayed adoption in order to understand what must be overcome to complete the transition to the all-wireless enterprise. As shown in figure 1, this can be attributed broadly to the perceived shortcomings of current unlicensed spectrum wireless solutions in an enterprise context:

- *Inadequate and inconvenient security compared to wired connectivity* – Only in the last few years has Wi-Fi network security been sufficiently strong (128 bit-AES with secure key exchange) to prevent rogue or malicious intrusion. However, in many cases, the process of authenticating or authorizing users remains much more onerous than on wired or cellular networks.

Wi-Fi security adaptation

YEAR	PROTOCOL	SECURITY
Pre-1997	OSA	None
1997	WEP	Insecure protocol - deprecated
2003	WPA	Interim fix for WEP – temporal key integrity protocol not secure
2004	WPA2/802.11i	Current protocol with AES-based encryption
2007	WPS	WPS PIN vulnerability

Wired LAN speeds: ~80 x Wi-Fi

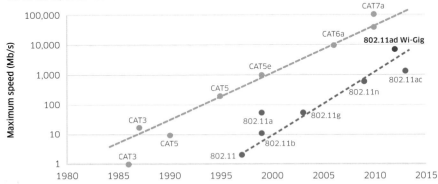

Wired LANs upgrades lag smart device explosion

LAN	Cable refresh	15 years
	Equipment upgrade	5 years
Wireless devices	Refresh rate	2 years
	5-year smartphone sales	3.7 Billion

FIGURE 1: Shortcomings of current unlicensed spectrum wireless solutions in an enterprise (Canalys 2008; Custom Cable 2011; Entner 2011, 2013, 2015; Hoelzel 2013; Kauffman 2013; Pyramid 2012, 2013, 2015b; Security Uncorked 2008; Wikipedia 2015a–c)

- *Insufficient capacity compared to wired LANs* – Category 7 cable-based LANs can support 1 Gb/s continuous throughput, compared to the average throughput – well below 100 Mb/s – of most Wi-Fi access points, due to interference, power and protocol limitations.

- *Solutions that pre-date the explosion of wireless-enabled devices* – Most current LAN infrastructure was upgraded over the past decade when the dominant devices were still desktop phones (TDM and then IP-based) and desktop workstations with Ethernet LAN ports. Wi-Fi was an optional add-on, not designed as a primary method for obtaining connectivity for the applications and devices in use today.

For the above reasons, the adoption of wireless technologies in the enterprise has been relatively slow compared to consumer and public hotspot adoption, as shown in figure 2. In the last few years – roughly starting in 2010 – we are seeing a shift, driven by the BYOD phenomenon and:

- The evolution in wireless LAN technology to 802.11n, which has both 2.4 and 5 GHz air interfaces for a manifold increase in capacity

- The emerging deployment of enterprise small cells for improved coverage (cellular technology)

- Automated, secure authentication of users, using cellular, SIM-based approaches and identity management systems

This has begun a transformation in enterprise networking, from wireless as a simple but unreliable overlay provided for convenience, to a primary mode of connection and an essential enabler of productivity (figure 2). To continue and complete this transformation, further evolution in wireless LANs is required to support the throughput, latency and high availability required for mission-critical, enterprise services.

The present and impending shift

In today's enterprise, Wi-Fi and cellular technologies are both being deployed (figure 2), but often for different applications, as service-specific wireless overlay networks. Today, Wi-Fi can be configured to provide continuous service as nomadic users move about a campus or around a building, and it works well for

Most current LAN infrastructure was upgraded over the past decade when the dominant devices were still desktop phones and workstations... Wi-Fi was an optional add-on.

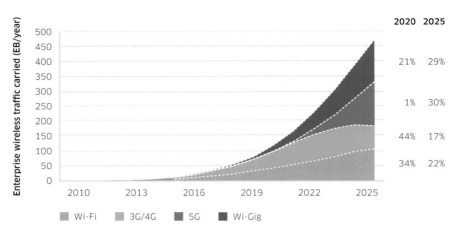

FIGURE 2: Enterprise wireless network traffic in EB/yr (Bell Labs 2015a–b; Korotky 2013; Dell'Oro 2015; Cisco 2008–2013)

most delay-tolerant, high volume data services, such as web services and file transfers. However, though IEEE 802.11e-based quality of service (QoS) techniques have been implemented in access points, the end-to-end quality in actual deployments is generally not sufficient for reliable, real-time voice and video communications services. There are three primary reasons for this:

1. Smartphone Wi-Fi receiver sensitivity is significantly worse than that of laptops (10 to 15 dB), which results in poor Wi-Fi coverage for smartphones in networks designed for laptops. This is problematic for communications applications that require session continuity, which in turn requires complete and continuous in-building coverage.

2. Since Wi-Fi utilizes unlicensed spectrum bands, external interference to the Wi-Fi network is uncontrolled and unpredictable, making radio frequency planning difficult and uncertain, even with continuous monitoring to identify, localize and mitigate interference sources.

3. Lack of LAN-WAN continuity. When the user leaves campus Wi-Fi coverage, a new IP network is encountered, resulting in a new IP address being assigned – as the session is dropped on the wireless LAN and reestablished on the macro cellular network – disrupting the current service activity.

Therefore, for applications that require a consistent quality of experience or session continuity while mobile, a cellular service is used. The macro cellular network or a distributed antenna system (DAS) – acting as a local repeater for an externally originated macro cellular signal – most often provides enterprise cellular coverage today. However, to meet the capacity and coverage requirements of the entire set of enterprise applications – as well as rising device density – what is needed are enterprise small cells, using licensed cellular spectrum. These small cells are typically deployed by mobile network operators and play a hybrid private/public access role in the enterprise, allowing private access to enterprise users within the enterprise location, but seamless hand-off to the public macro-cellular network, outside the enterprise. Initially, starting around 2010, small cells were deployed to provide indoor coverage for mobile communications due to the inferior penetration of the macro radio signal inside buildings. But with the increasing prevalence of mobile broadband data services, the data capacity they provide has become an important complement.

Recognizing the increasing importance of both licensed and unlicensed spectrum – and their need to work together – there are a number of enhancements underway in the areas of unlicensed, cellular and shared spectrum access.

Unlicensed spectrum

In the area of unlicensed access, IEEE 802.11ad (Wi-Gig) has been standardized to provide peak data rates of 7 Gb/s using spectrum in the 60 GHz band. Propagation at 60 GHz is primarily restricted to line-of-sight communications, since reflections from surfaces degrade signal strength, and penetration through walls, floors or the human body is poor. In the enterprise, Wi-Gig will be used over short distances to provide network connectivity from nearby Wi-Gig access points, and for docking – connecting PCs, tablets, and mobile devices to displays and other devices.

Also in the area of unlicensed access, the most recent Wi-Fi specification, 802.11ac, is now widely available in commercial devices, providing device peak rates up to 1.3 Gb/s. IEEE standardization work has also begun on its successor, 802.11ax, which will increase average user throughput by at least a factor of 4 in dense deployment scenarios. Other pending amendments to 802.11 include,

- 802.11ah will provide extended range and longer battery life by supporting operation at frequencies below 1 GHz, necessary for machine-to-machine communications
- 802.11aq will provide a means to advertise services available through the access prior to association
- 802.11ai will shorten the time needed to establish initial connectivity

Cellular spectrum

In the area of cellular access, 4G/LTE licensed-assisted access (LAA) will allow aggregation of licensed LTE carriers with LTE unlicensed carriers (LTE-U), particularly in the 5 GHz band. It is being defined by 3rd Generation Partnership Project (3GPP) for Release 13, expected sometime in 2016. LTE-U/LAA will supplant Wi-Fi to some extent, as it is two to three times as efficient, but both will proliferate to meet the demands of devices with varying radio capabilities.

In the area of 5G cellular access,[1] a consensus is building within the industry for both mm-wave (20-100 GHz) and low-band (sub-6 GHz) carriers. Compared with 4G, low-band 5G carriers are expected to provide better optimization for diverse traffic profiles, supporting connectionless access for short data bursts, and higher throughputs and lower latency if required by applications. Starting around 2020, 5G mm-wave or high-band carriers will be able to provide massive capacity due to the large spectrum bandwidth available at high frequencies, enhancing both device throughput and network capacity. Using smart-beam steering and an ability to track a user through direct and reflected paths, the 5G high-band carrier will be able to work in concert with a low-band carrier to provide service to fixed and mobile devices.

Shared spectrum

In the area of shared spectrum, there are significant blocks of spectrum around the world in the sub-6 GHz frequency bands that are ideally suited for mobile communications. However, because they were allocated to commercial or government use prior to the explosion in mobile connectivity, they remain under-utilized. Attempts to relocate incumbent users of this inefficiently used spectrum have proven expensive and time-consuming.

A more viable approach is to share spectrum in these bands by establishing multiple usage tiers. Incumbent users are considered primary and are given unimpeded access, whereas secondary or tertiary users only have access at places and times when a higher tier user is not active. They must immediately cease usage when a higher tier user pre-empts them. This is orchestrated in real-time by a spectrum access system that uses a database and dynamic software controls to ensure interference is avoided.

The US Federal Communications Commission (FCC) voted in April 2015 to adopt three-tier sharing in the 3.5 GHz band, which contains 150 MHz

---

1 - When we refer to 5G, we mean IMT-2020.

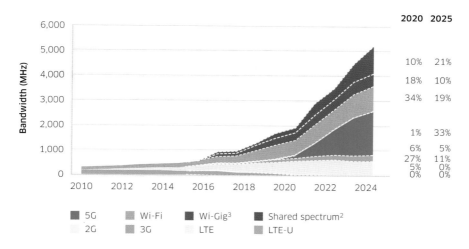

FIGURE 3: US potential spectrum for enterprise traffic in MHz (Bell Labs Consulting)[4]

of spectrum currently used for military and satellite operations. Shared spectrum in the 3.5 GHz band is particularly well suited for indoor small cells. The propagation characteristics are good for enterprise-scale coverage, and the likelihood of interference with the outdoor primary user is low. In addition, the FCC and the National Telecommunications & Information Administration (NTIA) is considering sharing spectrum in the 1750–1850 MHz range (NTIA 2015a-b). In late 2014, the FCC auctioned AWS-3, which contains the 1755-1780 MHz paired band, with mandated provisions for sharing with federal systems for an indefinite period (FCC 2014b). In the 2020–2025 timeframe, the 2700–2900 MHz and 4200–4400 MHz bands, currently used by the Federal Aviation Administration, have also been identified as candidates for spectrum sharing.

As a result of these evolutions, we predict that the spectrum available for enterprise usage will massively increase and be spread across these technologies, as shown in figure 3.

As mentioned, today enterprise wireless connectivity is often dependent on users successfully executing a sequence of one or more steps to select an access technology, choose an access point, enter security credentials and set up a VPN. Then, once connected, action is again required by the user

2 - Used for network connectivity
3 - Multiple technologies
4 - Analysis of data by Bell Labs Consulting with reference to various sources (Cisco 2008; FCC 2012, 2014; Goldstein 2011, 2015; Neville 2014a-b; Rayment 2014; Sprint 2012; T-Mobile 2014; US Cellular 2012; Verizon 2010) and specifically for shared spectrum (FCC 2014b, 2015; NTIA 2015a-b; PCAST 2012).

when coverage drops, or there is insufficient support for QoS, or the limited support for mobility on Wi-Fi affects service. For users, the value of wireless connectivity is based on convenience and transparent access: it saves time and boosts productivity. So when manual intervention is required – to set up connectivity or reconnect after failures – this value is diminished and the wired LAN becomes more attractive.

Fortunately, the advent of a new type of VPN connectivity, based on software-defined networking (SDN) principles, is making progress toward reducing complexity, and the configuration and reconfiguration required. Hardware-based VPNs have existed for several decades to support secure, multi-tenant access to the same IP network infrastructure, allowing the various locations of an enterprise to be connected together to form their own (virtual) private network. However, due to the relative cost and complexity of deploying and managing the required VPN router functionality in each enterprise location, this service has been restricted to only the largest sites of the largest enterprises. The new, SDN-enabled approach creates a software instance of the VPN for each location, leveraging virtualization technologies and cloud infrastructure. A software-defined VPN (SD-VPN) can, in this way, be created for any and every enterprise location, radically simplifying secure connectivity management.

In a fully implemented SD-VPN, a mobile device at an enterprise location can connect over either Wi-Fi or the legacy wired LAN to enterprise applications deployed in centralized data centers. The application flows are routed to a virtual router in the enterprise data cloud and then forwarded to the application server, with no configuration required on the part of the end user, or installation of a security client (for example, an IPSec client).

Unfortunately, when the device leaves the enterprise and attaches to the LTE network, or an external Wi-Fi network, this SD-VPN is not accessible. A dedicated VPN client has to establish a secure connection between the device and an enterprise VPN gateway that resides in an enterprise site, or in a separate data center. IPSec is most often used for these VPNs, as shown in figure 4.

There are two issues with this approach:

1. The VPN client must be installed and maintained on the device, requiring user interaction and entering of credentials each time the device connects.

2. Traffic is not directly routed to the enterprise applications from the external network. Instead traffic has to first be sent through an IPSec

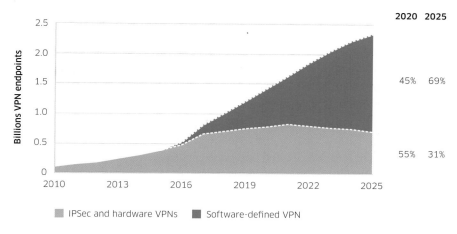

FIGURE 4: VPNs to secure enterprise traffic: IoT and user devices (Bell Labs Consulting)[5]

tunnel across the public Internet to an IPSec tunnel-termination router at another enterprise site. It is then sent to the application data center through another VPN tunnel. This results in non-optimal routing, the inability to support QoS due to the traversal of the public Internet, and the frequent scaling limitations characteristic of IPSec, termination routers.

As indicated in figure 4, we believe that in the next decade, the vast majority of growth in secure connections will be in SD-VPNs in order to solve this problem, as discussed in the next section.

Today, enterprise connectivity is bifurcated between two service providers:

1. A cellular service provider for reliable, seamless access for mobile devices

2. An IT provider of Wi-Fi and wired access connectivity for laptops, copiers, IP phones and other user devices

This split is not driven by the enterprise business needs, but by the independent evolution of information networks into the wireless domain (adding Wi-Fi to fixed LAN access) and service provider networks (adding smart devices to the original wired voice service).

The problem goes even deeper than this. Virtually every mobile network operator would like to provide enterprise services to enterprises, but the (locked) devices provided by many operators prevent usage on other operators' networks, or there is spectrum incompatibility with different bands deployed by different operators.

5 - Analysis of data by Bell Labs Consulting with reference to various sources (Bell Labs 2015c; Frost 2013; Metro 2011; Ovum 2012, 2014; Pyramid 2015a; World Bank 2015a-b).

Although multi-operator networking is clearly essential to support employee-owned (BYOD) devices, as well as the multitude of IoT devices that we expect to be deployed over the coming decade, multi-operator networking can only be achieved today by installing multiple sets of operator-specific small cells, which multiplies equipment, installation and maintenance costs, and adds to the inconvenience of coordinating multiple deployments and contracts. The problem is addressed currently by using an enterprise controlled host/operator-neutral distributed antenna system (DAS), that supports all frequency bands, but with limited baseband capacity (throughput), as it is shared over all the floors and cells in the building.

The future

The "hard" division between networks, described in the previous section, cannot continue to exist, given the expected rise in the numbers and importance of mobile user devices, as well as the spread of IoT for a wide range of business applications (see chapter 11, *The future of the Internet of Things*). In the future, new business models will emerge that naturally support coordination among multiple cellular operators and IT providers. These new models will simplify in-enterprise deployment with converged multi-operator, multi-technology cells, and manage allocation of connectivity resources to users across all available licensed and unlicensed spectrum. In this section, we will lay out our vision of the future wireless enterprise network and how it will accomplish these objectives.

Virtualized, converged radios beat single wires

In 2015, virtualization technologies and SDN are only beginning to transform the networking industry. Network components have historically been implemented on dedicated hardware platforms that were optimized for a single function or set of functions. They are already beginning to be virtualized and instantiated on general-purpose processors in data centers. This approach will yield significant advantages in performance, economics and scalability, as well as adaptability to changing demands and needs. These virtual network instances will be dynamically connected together to form a service chain using SDN.

This industry-wide evolution to SDN and network functions virtualization (NFV) will make it possible to create enterprise- and application-specific virtual private wireless networks (VPWNs) consisting of radio access networks (RANs) and associated mobile core networks. SDN/NFV will automate the allocation of in-building network resources using management and control functions that reside in an edge cloud and can be instantiated and scaled, based on demand. This new, SDN/NFV-enabled VPWN will allow enterprises:

- Optimal connectivity, spanning all access technologies and frequency bands
- A unified multi-technology enterprise LAN dedicated to their needs
- Complete control over policy and security

The architecture that provides this VPWN is shown in figure 5. The essential components are:

1. *Converged enterprise cells* (CEC) deployed by the in-building operator to provide service to enterprise subscribers and transparently provide access to visitors from outdoor mobile networks. These multi-technology, multi-band, multi-tenant-capable nodes will greatly simplify enterprise wireless LAN deployment by removing the need for many different technology-specific access points and the associated cost and complexity of deploying and managing them.

2. *A virtual control plane* containing packet core control functions and an SDN controller. The latter creates and connects the required SD-VPNs between sites, and manages connectivity within the enterprise network. In the future, as the user connects to different enterprise applications and services, the SDN controller will dynamically update paths between the different wireless cells serving the user's device, and the gateways in the edge cloud that connect to applications and services.

3. *A virtual user plane*, which will include the radio access network (RAN) and the core network gateway functions that control the paths to each application and service for each user. The gateways will be able to be created locally in the enterprise location for low-latency and efficient local content access, or in a local data center, either owned by the enterprise or by their in-building VPWN operator.

4. *Virtual access* for any employee or visitor, with continuous connectivity from the in-building wireless access network to the wide-area wireless network. This will be ensured by using the optimum set of radio network resources and path(s) to the appropriate core network functions.

CECs are one of the essential technologies that will drive the migration to an all-wireless enterprise LAN. They have the capacity advantages of small cells and the host-neutral advantages of DAS. These new types of wireless cell will support:

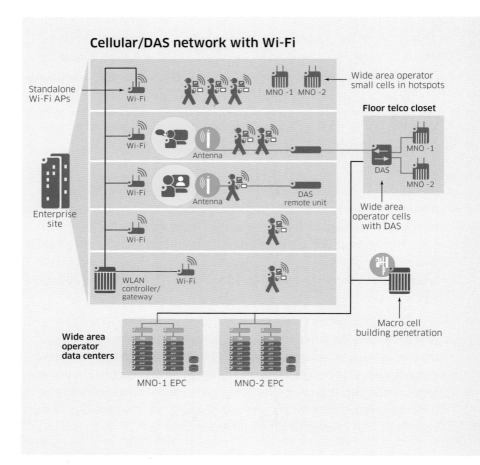

FIGURE 5: Wireless enterprise LAN architecture

- *Multiple bands, multiple technologies, multiple operators* – A diverse array of frequency bands will be supported (for example, < 2 GHz, 2–6 GHz, and 20–80 GHz), using spectrum from multiple, wide-area operators. Also supported will be unlicensed spectrum in the current Wi-Fi bands (2.4 and 5 GHz) and Wi-Gig (60 GHz) bands, as well as shared spectrum bands. All bands will utilize multiple cellular technologies, starting with LTE and the evolutions of LTE (LTE-A, LTE-U, LTE-M), as well as 5G – from 2020 and beyond.

- *Interfaces to multiple core networks for "operator neutrality"* – To any mobile operator subscriber who visits the enterprise location, the CECs will appear to belong to their mobile network operator, and service will be provided transparently through that operator's core network, using the compatible set of radio resources. To enterprise users, the CECs will appear

Virtualized private wireless network

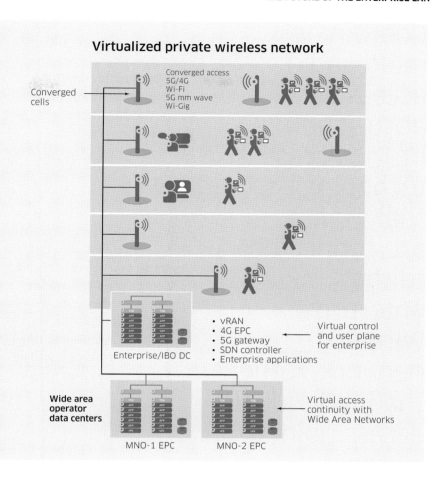

to belong to the in-building operator and the network service will be provided using the fully aggregated set of spectrum resources.

This future is beginning to be realized today. Coordinated multi-radio access allows abstraction of the access technology so that devices are connected to the enterprise wireless network, rather than individual access points or cells. This is already supported, in part, as 3GPP dual connectivity allows a 4G device to be simultaneously connected to more than one geographically distinct base station. In Release 13 of the 3GPP standards, LTE-WLAN radio level integration will allow for LTE and Wi-Fi downlink capacity aggregation for both co-located and non-co-located cases. In addition, today the 3GPP multi-operator core networks (MOCN) feature allows sharing of

Coordinated multi-radio access allows abstraction of the access technology so that devices are connected to the enterprise wireless network, rather than individual access points or cells.

base-station resources among multiple operator packet cores, with subscriber signaling and user-plane traffic directed to the applicable operator core. Dedicated core networks (DÉCOR) extends this in 3GPP Release 13 to allow sharing of a RAN among multiple core networks of a single mobile operator.

This trend toward multi-radio access integration will continue in the future, with routing directed by an enterprise core network controller that will have the perspective to coordinate among the access options. This will provide the flexibility to allow devices to be served based on the technologies they support and the needs of the applications in use, while balancing resources and ensuring fairness among subscribers. At any instant, compatible devices may be multiply connected through different access technologies on distributed or co-located access nodes. Connected devices may have flows routed along multiple parallel paths, and uplink and downlink data may be routed over different accesses. The controller will ensure mobility by adding and dropping connections as they become available, and updating routing instructions. Thus, connectivity will be completely abstracted from a specific access technology, and QoS characterized by aggregated link capabilities, rather than individual access technology characteristics. This will create a "cell-less" and "radio-technology-agnostic," VPWN.

It is also important to note that the VPWN core architecture is also flexible and scalable. Small enterprises will only have the CEC deployed locally and connected via an SD-VPN to off-site (edge cloud) data center resources, which can be shared with other small enterprises. For large enterprises, the mobile core, local gateways (GWs) and applications may be hosted using cloud resources in the enterprise-owned data center, or in an operator-hosted (edge cloud) data center.

Zero-touch secure services

With the advent of virtualized packet cores, mobile network operators are starting to offer private network services over LTE as an alternative to traditional VPN clients. To deliver a secure connection to a specific enterprise, a dedicated packet core gateway and an access point name (APN) are configured in the LTE network. Traffic is routed to this dedicated core gateway, and cellular air-interface encryption is used to secure the enterprise's traffic. For BYOD users, however, the operator must configure specific entries in the home subscriber system (HSS) to authorize access to the enterprise APN, a resource-heavy process that is undesirable. In addition, a VPN is still needed to secure the connection between the APN and the enterprise's own private network, as this connection may well be routed over a public IP network (for example, the Internet). End-to-end security requires installation and management of IPSec clients in each mobile device. While cumbersome for BYOD smartphones, this is all but impossible for many IoT devices, which typically do not support such capabilities.

SDN will enable a better solution by extending the SD-VPN connectivity, which is beginning to be used for enterprise site connectivity, all the way to the mobile device, as illustrated in figure 6. This requires no enterprise user configuration by the operator, and does not require installation and configuration of an IPSec client in each mobile device.

Here the mobile device has a simple virtual switch element (vPE-F) that is controlled by an SDN controller. The SDN controller configures the flow table, establishing a tunnel to securely route IP packets directly to the appropriate enterprise data center, where the target enterprise application server is running. A separate tunnel is not required for each application, as the same

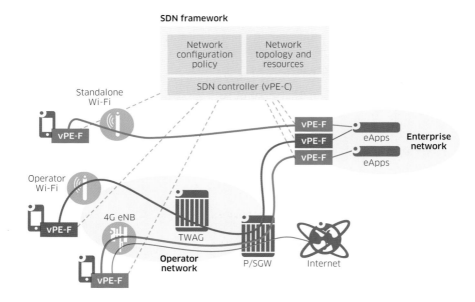

FIGURE 6: SD-VPN service

policy can be applied to all application flows with the same destination data center. Flows that are not enterprise flows, which are common for BYOD devices, are associated with a different flow table entry to directly route to the Internet. This new approach provides zero-touch secure services for any device that can support a simple switch table entry.

This zero-touch approach will also support per-service, QoS characteristics –such as latency and bandwidth – or system characteristics – such as reliability and resiliency. This is achieved by associating different devices with different enterprise gateway functions. To take the example of a typical manufacturing enterprise, IoT sensors on a factory floor and employee smart devices may be served by different instantiations of the packet core network functions, with different network connectivity guarantees (capacity vs. latency).

The emergence of the in-building operator

As highlighted earlier, a new business model structure is called for that eliminates today's divisions among wide area mobile operators and providers of enterprise LAN connectivity. The technology transformations described in the previous sections are the essential foundation for this business model revolution. We believe that it will enable the emergence of a new type of operator: the "in-building operator" (IBO). The IBO will provide a unified

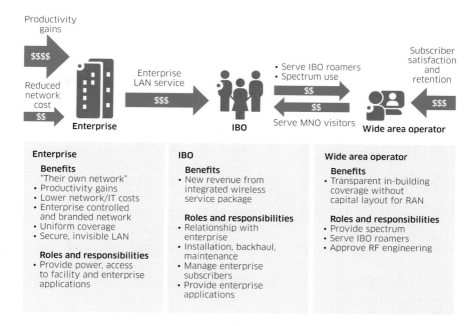

FIGURE 7: New business model for wireless enterprise LAN service

wireless enterprise service. The resulting realignment we envision is illustrated in figure 7, with different responsibilities (and benefits) for the different parties.

IBOs may be independent operators, or IT providers that offer in-building connectivity as part of an overall IT package. In this future reality, IBOs will become the single point of contact for managing enterprise subscribers, installing and managing in-building access and providing value-added business applications that can be amortized over multiple enterprises. This abstraction from today's multi-provider model, which typically involves enterprises, IT providers and multiple wide area operators, simplifies the relationships, eliminates multi-party negotiations for roaming, and provides public access in the enterprise and support for BYOD. Using in-building, multi-operator CECs, as described earlier, the deployment of mobile-operator-specific nodes is also eliminated, resulting in a massively simplified enterprise wireless LAN solution.

We foresee the IBO relationship with the enterprise taking two different forms:

1. The IBO is paid to install and manage the in-building wireless network on behalf of the enterprise and manage the roaming arrangements with the wide area operators whereas the enterprise pays for the in-building

In the RAN, vendors are starting to support converged, multi-technology small cells, with combined Wi-Fi and 3GPP cellular access.

network and retains ownership of the wireless network infrastructure.

2. The IBO funds the capital cost of the in-building network, operates the network for the enterprise, manages the roaming agreements with the wide area operators, and is compensated with a long-term contract for the in-building service by the enterprise.

In either form, the existing wide area operators benefit by gaining in-building coverage for their subscribers, without the capital outlay and management costs for the in-building network. In exchange for this benefit, they allow use of their spectrum in the building and provide connectivity for IBO subscribers in the wide area cellular network. This symbiotic relationship is illustrated in figure 7.

In figure 8, we show the potential business value of all-wireless access. The yearly value to a large enterprise, per employee, is estimated at $1,900, with 85% of that coming from a 6% improvement in employee productivity. A 34% reduction in infrastructure cost contributes 4% per employee in yearly savings, with the balance of the benefit coming from IT staff and other productivity gains. In addition to employee efficiency gains, there will be substantial benefits obtained by seamlessly connecting IoT devices, enabling much improved enterprise business automation.

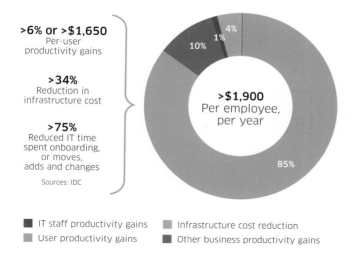

>6% or >$1,650
Per-user
productivity gains

>34%
Reduction in
infrastructure cost

>75%
Reduced IT time
spent onboarding,
or moves,
adds and changes

Sources: IDC

>$1,900
Per employee,
per year

4%
1%
10%
85%

■ IT staff productivity gains   ▨ Infrastructure cost reduction
▨ User productivity gains   ■ Other business productivity gains

FIGURE 8: Enterprise benefits from new all-wireless enterprise (Greene 2014)

The future begins now

The future we have outlined in this chapter is already beginning. Since 2012, network equipment vendors have been transforming their portfolios from dedicated hardware platforms to virtualized functions linked together and programmed using SDN. This has become one of the most important initiatives in the wireless infrastructure industry.

In the RAN, vendors are starting to support converged, multi-technology small cells, with combined Wi-Fi and 3GPP cellular access. For several years, networks have supported un-trusted access via Wi-Fi, as well as trusted Wi-Fi access (as defined in 3GPP specifications), providing IP session continuity as a user moves between Wi-Fi and 3GPP coverage. Integration of Wi-Fi and LTE is being further developed by 3GPP in Release 12, using a mechanism similar to dual connectivity, which allows LTE resources at different base stations to simultaneously serve a user device. One such solution sends packets over both access layers, but off-loads the Wi-Fi uplink traffic to the LTE interface. This improves signal strength and significantly reduces the contention associated with multi-user Wi-Fi due to unmanaged competition for resources over the uplink. A similar approach will be needed to efficiently allocate air-interface resources in the all-wireless enterprise LAN.

Network operators have also been adopting key technologies that will support the new virtual private wireless LAN service. For example, network sharing

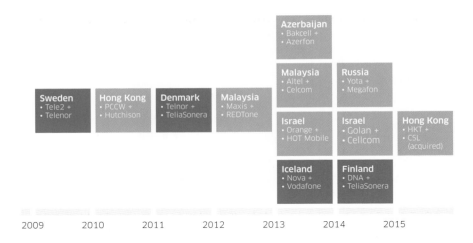

FIGURE 9: Recent operator deployments of RAN sharing using 3GPP multi-operator core networks (MOCN) (Coleago 2015, ZTE 2015).

of RAN resources is used to overcome capital investment constraints. The 3GPP-defined, MOCN functionality enables the sharing of a single LTE-RAN base station among subscribers from up to six operators by directing traffic from the RAN to the subscriber-specific, operator core network. As illustrated in figure 9, operators, particularly in Asia and Europe, have been using this MOCN-based sharing for some time. In some cases, it is the simplest way to combine networks of two companies that have merged, but more frequently, MOCN is used to reduce deployment costs of two or more separate operators.

This MOCN-based RAN sharing among operators is similar to the RAN sharing between IBOs and wide area operators that we envision for the enterprise. Today, wide area operators use MOCN to share a RAN but compete to provide wide area mobile coverage for the same subscribers. But the enterprise customer base is distinct, so that IBOs can target enterprise customers and partner with wide area operators who merely target individual subscribers. Of course, wide area operators can also choose to become IBOs as well.

Summary
We envision a future where the enterprise undergoes a radical infrastructure change with the creation of an invisible all-wireless enterprise LAN that provides always-present, untethered connectivity. Tablets, smartphones, wearables and IoT devices will all be continuously connected to the applications and services they need via a network that tailors itself to provide the required service, with

optimum bandwidth, latency, security and reliability, and requiring no human intervention – or *zero-touch*.

The new wireless LAN will use converged, multi-operator cells, provided by an IBO in collaboration with the enterprise and the mobile operators. Devices will be automatically connected through the right combination of access technologies to meet user-centric requirements. The supporting network functions will be virtualized and linked using SDN, allowing deployment of dedicated, secure enterprise networks with improved economics.

Enterprise employees will experience zero-touch access on an always-available network, with security provided by software-defined VPNs that connect enterprises and data centers, and extend all the way to the device. Security will be an aspect of network connectivity, rather than an overlay application requiring client support and periodic user intervention. Similarly, virtualized network slices will automatically marshal the resources required to provide a service tailored to the device and application requesting it.

Finally, we see a new IBO business model emerging, one that eliminates the historical division between IT providers and wide area operators, as well as a similar division between wide area operators. As part of an expanded IT service, IBOs will partner with wide area operators to provide integrated, in-building wireless access. The net effect will be that a *new deal* economy is created, leading to enhanced productivity and efficiency for enterprises, greatly improved in-building coverage for wide area operators and their subscribers, and a new revenue stream for in-building operators.

References

Alcatel-Lucent, 2015. "Wireless Unified Networks," *Alcatel-Lucent web site* (https://www.alcatel-lucent.com/solutions/wireless-unified-networks).

Bell Labs Consulting 2015a. "Mobile Demand Consumption Model – NAR forecast 2015-2025," internal tool and presentation, March 27.

Bell Labs Consulting 2015b. "Wireline Demand Model – NAR forecast 2015-2025," internal tool and presentation, May 11.

Bell Labs Consulting 2015c. "IoT/M2M Forecasting Model: Global Forecast 2015-2025", internal tool and presentation.

Canalys 2008. "Smart mobile device shipments hit 118 million in 2007, up 53% on 2006", Canalys Web Site, February 5 (http://www.canalys.com/newsroom/smart-mobile-device-shipments-hit-118-million-2007-53-2006).

Cisco Systems 2008. "Cisco Visual Networking Index: Forecast and Methodology, 2007–2012," White Paper, May.

Cisco Systems 2008. "WLAN Radio Frequency Design Considerations," *Enterprise Mobility 4.1 Design Guide*, 9 December (http://www.cisco.com/c/en/us/td/ docs/solutions/Enterprise/Mobility/emob41dg/emob41dg-wrapper.pdf).

Cisco Systems 2009. "Cisco Visual Networking Index: Forecast and Methodology, 2008–2013," White Paper, May.

Cisco Systems 2010. "Cisco Visual Networking Index: Forecast and Methodology, 2009–2014," White Paper, May.

Cisco Systems 2011. "Cisco Visual Networking Index: Forecast and Methodology, 2010–2015," White Paper, May.

Cisco Systems 2012. "Cisco Visual Networking Index: Forecast and Methodology, 2011–2016," White Paper, May.

Cisco Systems 2013. "Cisco Visual Networking Index: Forecast and Methodology, 2012–2017," White Paper, May.

Coleago Consulting 2015. "Mobile Network Infrastructure Sharing," *Presentation*, February.

Custom Cable 2011. "CAT3 vs. CAT5 vs. CAT5e vs. CAT6 vs. CAT6e vs. CAT6a vs. CAT7," *Custom Cable Blog*, December 24 (http://customcable.ca/cat5-vs-cat6).

Dell'Oro Group 2015. "Ethernet Switch Layer 2+3 Report Five Year Forecast 2015 – 2019", July 17.

Entner, R., 2011. "International Comparisons: The Handset Replacement Cycle," *Recon Analytics*, p. 2 (http://mobilefuture.org/wp-content/uploads/2013/02/mobile-future.publications.handset-replacement-cycle.pdf).

Entner, R., 2013. "Handset replacement cycles haven't changed in two years, but why?" *Fierce Wireless*, March 18 (http://www.fiercewireless.com/story/entner-handset-replacement-cycles-havent-changed-two-years-why/2013-03-18).

Entner, R., 2015. "2014 US Mobile Phone sales fall by 15% and handset replacement cycle lengthens to historic high," *Recon Analytics*, February 10 (http://reconanalytics.com/2015/02/2014-us-mobile-phone-sales-fall-by-15-and-handset-replacement-cycle-lengthens-to-historic-high).

FCC 2012. "In the Matter of Improving Spectrum Efficiency Through Flexible Channel Spacing and Bandwidth Utilization for Economic Area-based 800 MHz Specialized Mobile Radio Licensees," *Commission Document*, May 24 (https://www.fcc.gov/document/800-mhz-smr-band-order).

FCC 2014a. "Annual Report and Analysis of Competitive Market Conditions with Respect to Mobile Wireless, Including Commercial Mobile Services," *Commission Document*, December 18 (https://www.fcc.gov/document/17th-annual-competition-report).

FCC 2014b. "Auction 97: Advanced Wireless Services (AWS-3)," *FCC Auction Site* (http://wireless.fcc.gov/auctions/default.htm?job=auction_factsheet&id=97).

FCC 2015. "FCC Releases Rules for Innovative Spectrum Sharing in 3.5 GHz Band," *Commission Document*, April 21 (https://www.fcc.gov/document/citizens-broadband-radio-service-ro).

Frost & Sullivan 2013. "Global Mobile VPN Products Market: Meeting the Wireless Security Challenge with an Emerging VPN for the Remote Worker," *Slideshare*, May 14 (http://www.slideshare.net/NetMotionWireless/frost-sullivan-global-mobile-vpn-products-market).

Goldstein, P., 2011. "AT&T to launch LTE Sunday, Sept. 18," *Fierce Wireless*, September 15.

Goldstein, P., 2015. "AT&T expects to start deploying 2.3 GHz WCS spectrum for LTE this summer," *Fierce Wireless*, March 30.

Greene, N., and Perry, R., 2014. "The Tipping Point Is Here – All-Wireless Workplaces Show Benefit over Traditional Wired Technology," *IDC White Paper*, October.

Hoelzel, M., Ballve M., 2013. "People Are Taking Longer To Upgrade Their Smartphones, And That Spells Trouble For The Mobile Industry," *Business Insider*, September 12 (http://www.businessinsider.com/the-smartphone-upgrade-cycle-gets-longer-2013-9#ixzz3hnoTkF9k).

Kauffman 2013. "Wi-Fi security: history of insecurities in WEP, WPA and WPA2," *Stack Exchange*, IT Security Blog, August 28 (http://security.blogoverflow.com/ 2013/08/Wi-Fi-security-history-of-insecurities-in-wep-wpa-and-wpa2/).

Korotky, S. K., 2013. "Semi-Empirical Description and Projections of Internet Traffic Trends Using a Hyperbolic Compound Annual Growth Rate," *Bell Labs Technical Journal* 18 (3), 5–21.

Metro Research 2011, "Summary Report: Small business telecoms survey," *Slideshare*, March 29, slide 7 (http://www.slideshare.net/smorantz/mobile-telecoms-for-small-business).

Neville, R. 2014a. "The Un-carrier Network: Designed Data-Strong," *T-Mobile blog post*, June 18 (http://newsroom.t-mobile.com/issues-insights-blog/the-un-carrier-network-designed-data-strong.htm).

Neville, R. 2014b. "T-Mobile LTE Now Reaches 250 Million Americans - Months Ahead of Schedule," *T-Mobile blog post*, October 27 (http://newsroom.t-mobile.com/issues-insights-blog/t-mobile-lte-now-reaches-250-million-americans-months-ahead-of-schedule.htm).

NTIA 2015a. "Fifth Interim Progress Report on the Ten-Year Plan and Timetable," US Dept. of Commerce Document, April.

NTIA 2015b. "Technical Studies to Support Examination of Sharing Options in the 2900–3100 MHz Band," NTIA Blog, April (http://www.ntia.doc.gov/blog/ 2015/technical-studies-support-examination-sharing-options-2900-3100-mhz-band).

Ovum 2012. "Corporate-liable mobile connections and revenue forecast: 2012–17", October.

Ovum 2014, "Corporate-Liable Mobile Connections and Revenue Forecast: 2013–18", January.

PCAST 2012. "US Intellectual Property Report to the President: Realizing the Full Potential of Government-held Spectrum to Spur Economic Growth," *Executive Office of the President of the United States Document*, July (https://www.whitehouse.gov/sites/default/files/microsites/ostp/pcast_spectrum_report_final_july_20_2012.pdf).

Pyramid Research 2012. "Global Smartphone Forecast," November.

Pyramid Research 2013. "Global Smartphone Forecast," February.

Pyramid Research 2015a. "Global Fixed Forecast Pack" and "Fixed Communications: Global Forecasts," June.

Pyramid Research 2015b. "Global Smartphone Forecast" and "North America Smartphone Forecast", July.

Rayment, S. G., 2014. "The Role of Wi-Fi in 5G Networks," *Presentation to Johannesburg Summit*, May.

Security Uncorked 2008. "A Brief History of Wireless Security," Aug. 8; updated January 31, 2012 (http://securityuncorked.com/2008/08/history-of-wireless-security).

Sprint 2012. "Sprint 4G LTE Launch Extends to 15 Cities Throughout Portions of Georgia, Kansas, Missouri and Texas," *Press Release*, July 16 (http://newsroom.sprint.com/news-releases/sprint-4g-lte-launch-extends-to-15-cities-throughout-portions-of-georgia-kansas-missouri-and-texas.htm).

T-Mobile 2014. "Mobile Celebrates 1st Anniversary of LTE Rollout By Launching Major Network Upgrade Program," *Press Release*, March 13 (http://newsroom.t-mobile.com/news/t-mobile-celebrates-1st-anniversary-of-lte-rollout-by-launching-major-network-upgrade-program.htm).

US Cellular 2012. "U.S. Cellular Announces Launch Of 4G Lte Network Next Month Along With Upcoming Devices," *Press Release*, February 1 (http://www.uscellular.com/about/press-room/2012/USCELLULAR-ANNOUNCES-LAUNCH-OF-4G-LTE-NETWORK-NEXT-MONTH-ALONG-WITH-UPCOMING-DEVICES.html)

Verizon 2010. "Verizon Wireless Launches The World's Largest 4G LTE Wireless Network On Dec. 5," *Press Release*, November 30 (http://www. verizonwireless.com/news/article/2010/12/pr2010-11-30a.html).

Wikipedia 2015a. "Twisted Pair," updated June 6 (https://en.wikipedia.org/ wiki/Twisted_pair).

Wikipedia 2015b. "Ethernet over twisted pair," updated July 17 https://en.wikipedia.org/wiki/Ethernet_over_twisted_pair).

Wikipedia 2015c. "IEEE 802.11," updated July 22 (https://en.wikipedia.org/ wiki/IEEE_802.11).

World Bank 2015a. "Employment to population ratio, 15+, total (%) (modeled ILO estimate)," *World Bank Web Site* (http://data.worldbank.org/indicator/ SL.EMP.TOTL.SP.ZS/countries).

World Bank 2015b. "Total Population (in number of people)," World Bank Web Site (http://data.worldbank.org/indicator/SP.POP.TOTL).

ZTE 2015. "ZTE Enables World Leading MOCN/CA/VoLTE-based Converged 4G Network for HKT," *Business Wire: ZTE Press Release*, March 2 (http://www. businesswire.com/news/home/20150302007025/en/ZTE-Enables-World-Leading-MOCNCAVoLTE-based-Converged-4G#.Vctkb1wk_wx).

CHAPTER 9

THE FUTURE of COMMUNICATIONS:

The emergence of a unified interaction paradigm

Markus Hofmann, Anne Lee
and Bo Olofsson

The essential vision

The essence of communication is "the imparting or exchanging of information" (Collins 2015). We can broadly define three primary types of communication – personal, media and control communications, each of which results in the exchange of information between people or systems. Personal communication can be regarded as the exchange of *information* or *perspective* with one or more people; media-type communication occurs when a system exchanges *content* with a person or another system; and control-type communication occurs when a person exchanges *commands* or *information* to/ from a device or system. A single communication session can be comprised of more than one of these types.

The goal of *unified communications*, a single system of communication that enables all three types of communication, has remained elusive. Different, often proprietary, protocols and encapsulations of information and content have been, and continue to be, used. Sometimes they persist due to their ubiquity. Others provide security or protection. But increasingly, the boundaries between the three modes of interaction – personal, media and control – are blurring as new IP-based protocols and message handling schemes allow communication of content, information or commands by and with people, systems and devices.

We believe that a new, unified communications paradigm will emerge in the future, taking us into an era of truly unified communication, with anyone or anything, and accessible by anyone or anything. This will profoundly change how we interact. It will enable a new digital existence, one in which we are always connected and can communicate with any system, device or person, allowing continuous and perpetual augmentation of our lives.

Past and present

In order to understand future potential, it is always instructive to look at the evolution of human communications to date. As Father Culkin once described the central idea of Marshall McLuhan's work, as humans we "become what we behold, we shape our tools and our tools shape us" (Culkin 1967). From the invention of the phonetic alphabet to the printing press, radio, telephone, television and the World Wide Web, we have responded to major technical innovations in communications by changing the ways we interact with each other and with our environment. Before electronic networks, long-distance communications were focused on protracted exchanges of personal information and media in the form of written letters and printed images, which were transported by mail using whatever transportation mechanism was available. With the advent of electrical networks and related systems, electronic communications became possible and the telegraph, telefax machine and telex

We believe that a new, unified communications paradigm will emerge in the future, taking us into an era of truly unified communications with anyone or anything, and accessible by anyone or anything.

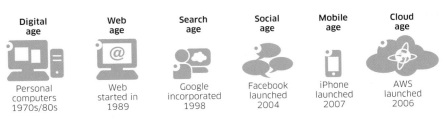

Digital age	Web age	Search age	Social age	Mobile age	Cloud age
Personal computers 1970s/80s	Web started in 1989	Google incorporated 1998	Facebook launched 2004	iPhone launched 2007	AWS launched 2006

FIGURE 1: Different ages of digital communications

network enabled nearly instantaneous global transmission of text and images. The invention of telephony then led to the emergence of real-time (analog) verbal communication between people. The telephony network evolved over decades, replacing human-assisted call routing with automated switches, from vacuum tubes to transistors, from wired to wireless, and from analog to digital.

The emergence of personal computers in the 1970s and 1980s created a foundation for the widespread move to multimedia, digital communications. People used computers to create and exchange significant bodies of information and media with other people and systems, in what could be considered to be the beginning of the *Digital age* of communications (figure 1).

The phenomenal growth of the World Wide Web beginning in 1989 led to the *Web age* of communications in which people also began to interact by publishing media and content to systems for global retrieval by anyone. It quickly became apparent that uniform resource locators (URLs) were too difficult to recall, and the massive amount of publication that occurred was creating a corpus of data too large to be managed without automation of discovery, which led to the creation of search engines, and the *Search age*.

As the Internet became a globally popular repository for all kinds of information and media content, it naturally became a means of sharing media content. We saw the emergence and proliferation of peer-to-peer file sharing applications. The majority of the content shared did not have the requisite permissions due to restricted content rights, but this illicit content sharing directly led to the desire to share personal content on the Web by communicating with groups on social networking sites, creating the foundation for the *Social age* in the early 2000s.

At around the same time (circa 2000+), mobile telephony started to experience significant uptake due to the creation of a legitimate content sharing framework (the Apple iTunes store) and the coincident launch of intuitive new multimedia communication devices (also pioneered by Apple, with the launch of the iconic iPhone in 2007), marking the start of the *Mobile age*. Communication devices had now taken a profound shift to evolve from the initial fixed-line human personal communication paradigm to a mobile one. Or, in other words, communication was no longer with places, but truly with people, wherever they were. And what was communicated was no longer just verbal information, but digital files and media content.

So, by the year 2000, people-centric information and media/content sharing communications services were beginning to co-evolve in time based on separate infrastructures and protocols, typically within geographic boundaries and primarily in developed economies. The primary communication problem then became one of seamless global scaling. The Internet made obsolete the previous paradigm, where traditional communications systems existed within national boundaries and scaling only involved the interconnection of these systems via international exchanges. This global scaling was achieved in the mid-2000s by the rise of a global cloud infrastructure of computing and storage resources, which could be utilized to store media and information economically at massive scale, and with acceptable latency and throughput. This was the beginning of the *Cloud age*.

Interestingly, this era was based on an evolution of the processor technology that had begun the *Digital age* in the 1970s and that ultimately became the basis of cloud servers. The same processor technology continued to be used in personal computers, which ultimately evolved to the more portable form of laptops. The related development of low-power processor technologies formed the foundation of smart mobile devices, which arrived on the scene at approximately the same time as the cloud. This was not coincidental, as a major component of the success of the iPhone – after prior smartphones had failed to impress consumers – was the connection to the cloud: the availability of content (via iTunes) and novel applications (via the App Store). The global cloud thus first appeared as a global store with which to request and exchange these new forms of media.

As a consequence, personal and media communications services between people, and people and servers or systems, were beginning to coalesce, leveraging the same device (smartphones) and the same network (broadband access and Internet/web infrastructure). But the protocols and software systems remained mostly separate, with personal communications leveraging one set of protocols (for example, SIP, XMPP, or proprietary equivalents), and media communications using another set (for example, HTTP and RTSP). In the last couple of years, a bridge has begun to form between these two communications worlds with the rise of WebRTC protocols coupled to Session Initiation Protocol (SIP)-based frameworks; they both integrate personal communications into web browsers, and are the primary interfaces to the Hypertext Transfer Protocol (HTTP) (web server) domain.

The brief history above shows a clear convergence beginning to occur between media and personal communications. As we enter the era of the digitization and connection and communication with any*thing*, it is reasonable to conjecture that a further convergence could occur with machines to allow seamless communication with the Internet of Things (IoT). We explore this idea further in this chapter.

Personal communications

Personal communications is anchored by two foundational modes – the spoken and the written word, both with and without images. The ability to convey information using either mode among multiple parties over any distance has been the focus of the telecommunications industry for more than 100 years.

As early as the mid-1700s, experiments were conducted to transmit information over some distance using electricity (Marland 1964). By the 1830s, the commercial telegraph and Morse code had been invented enabling a very basic form of written-word, electronic communications. Then, in the 1840s, the Scottish inventor Alexander Bain invented the first chemical-mechanical telefax machine, allowing transmission of simple graphical images. Bakewell and Caselli made multiple further refinements, with commercial service becoming available by the 1860s (Wikipedia 2015b). And in the 1870s, the first practical telephone was invented by Alexander Graham Bell, making spoken-word, electronic communications possible over long distances.

Systems for long-distance communications of both spoken and written word evolved separately for a time. In the 1840s, the teleprinter, an electro-mechanical typewriter, was invented to replace the need for Morse code operators in the telegraph system. In order to make this form of communication ubiquitous, the telex network was created in the 1930s. It connected teleprinters together using telex exchanges developed in compliance with International Telecommunication Union (ITU) recommendations, which allowed for interoperability between telex operators around the world. This text-based network and system can be regarded as the earliest forerunner of modern email and text messaging systems and infrastructure discussed below.

Spoken word systems

The telephony system matured rapidly and eventually grew beyond voice to subsume and surpass both the telex network capabilities and the telefax services. The electromechanical switch was invented in the late 1800s with the patent issued for the Strowger switch 1891 (Tollfreenumber 2012). It eliminated the need for a fully interconnected mesh of physical cables between the telephones of every user, household, and business and ultimately eliminated the manual telephone switchboard. These switches were also built in compliance with ITU recommendations, enabling interoperability between telephony operators around the world – forming what we know now as the public switched telephone network (PSTN).

Over time, ITU recommendations for the PSTN evolved from in-band signaling to out-of-band signaling. The in-band approach used multi-frequency tones for signaling, which greatly restricted new service innovation. Signaling System 7 (SS7) introduced out-of-band signaling in the 1970s, which enabled many new features, including multi-party calling and call forwarding. SS7 standards were extended further in the 1980s to cover mobile systems enabling a ubiquitous, global wireless ecosystem of mobile operators called the public land mobile network (PLMN), which is only now being replaced with SIP/IMS and VoLTE.[1]

Although, the hexagonal cell concept was proposed for a wireless system in 1947 (the same year as the invention of the transistor) and vocoders were invented even earlier, it took decades before supporting cellular technologies, infrastructure and regulations enabled mass mobile communications service. Around the world, large-scale commercialization, deployment and adoption of mobile systems supporting handheld devices began in earnest in the 1980s – just over 100 years from the start of fixed line voice service. As with the PSTN, the PLMN initially used analog circuit technology and evolved along with computing technology into digital systems with packetized interconnect.

Over the century between the introduction of fixed to mobile telephony, major technology breakthroughs enabled the global build-out and advancements in both the PSTN and later the PLMN. Manual, human-assisted switching gave way to automated, electro-mechanical switches. Vacuum tubes were replaced by transistors, which also spawned the modern

---

1 - SIP is Session Initiation Protocol, IMS is IP Multimedia Subsystem and VoLTE is Voice over LTE (Long-Term Evolution). We will discuss them in greater depth below.

computing industry. And, in turn, the use of computers as a basis for telecom systems enabled digital switching, improved operations for higher reliability, as well as new end-user features, such as caller ID, call waiting and voicemail. Fixed, residential phones evolved slowly from hand-cranked operator-assisted candlesticks to rotary dial, to touch tone, to even include small screens for information such as caller ID. In contrast, mobile phones advanced much more rapidly, becoming lighter, thinner and much more functional, in part because they evolved in parallel with the development of the personal computer. Both trends converged in the introduction of the iPhone, the first successful smartphone and the beginning of truly mobile computing, largely driven by the vision of one man – Steve Jobs – and his focus on the creation of transformative, consumer devices and interfaces.

Since the 1990s, there have been efforts to develop and deploy IP communications solutions. SIP is one of the enduring standardized protocols for IP communications. It had its beginning in 1996 and was first standardized by the IETF in 1999. SIP began as a signaling protocol for real-time communications and was quickly extended to include messaging and presence capabilities. In 1999, 3GPP began work to transform the existing wireless circuit-based telecom network into an all-IP network based on a framework known as the IP multimedia subsystem (IMS), using SIP signaling to replace SS7 and the cumbersome new H.323 ITU-T recommendation. SIP also incorporated many of the elements from HTTP and SMTP[2], which was also seen as another important advantage.

From the beginning, IMS was designed to be an end-to-end, rich communication solution (RCS) that supported voice, video, messaging, file transfer, address books and presence, all of which were seen as key table stakes features. It also included the concept of a services factory – a platform enabling the creation or enhancement of new and existing features and applications. IMS was also designed to be used with any access technology and to ultimately replace both the PLMN and PSTN. This goal placed additional requirements on the system, including support for interworking with all legacy systems, plus support for circuit switched voice services of all types (for example, payphones and ISDN lines), as well as fax services and regulatory requirements.

Consequently, as fully envisioned, the development of IMS took time – too much time – to be realized. The transition has begun, however, first for

2 - HTTP or hypertext transfer protocol is the key protocol enabling the World Wide Web, and SMTP or simple mail transfer protocol, enables email.

wireline-based TDM voice replacement, and more recently with the deployment of the first VoLTE systems that will allow retirement of the dedicated circuit switched wireless infrastructure and refarming of the associated spectrum for use in the more efficient IP-based LTE networks.

With the advent of network functions virtualization (NFV), further evolution of these IMS systems will leverage cloud technologies and pave the way for a more flexible, scalable solution and the adoption of new business models. In particular, the scale of the IMS infrastructure can be adapted for different deployments and even be offered as software-as-a-service (SaaS) to various vertical industry sectors and enterprises generally. The transition to the cloud also provides the opportunity for the use of cognitive technologies, such as machine learning, which may be based on support vector machines or neural networks. These technologies may make it possible for these platforms to be self-aware and intelligent, leading to autonomous solutions capable of self-management, self-organization and self-healing, as described in chapter 5, *The future of the cloud.*

The evolution of written word systems

The earliest form of electronic mail emerged in the 1960s when the compatible time sharing systems (CTSS) team at MIT created a way for their members to send messages to each other's terminals and/or electronic mail boxes. Other electronic mail systems were also

The transition to the cloud also provides the opportunity for the use of cognitive technologies, such as machine learning, which may be based on support vector machines or neural networks.

developed in that time period but none of them were interoperable. It was in the 1970s that ARPANET introduced a standardized email system and IETF specifications for electronic mail were first proposed. In the 1980s, today's specification for SMTP was adopted worldwide, allowing for interoperability between email providers around the globe. This adoption modernized written word electronic communications from the early interoperable telex networks and telefax services to interoperable email.

Also in the 1980s, work began on a new form of written word electronic communications – text messaging. Led by Friedhelm Hillebrand, the GSM standards body worked on defining specifications for short message service (SMS) that culminated in 1992 with the first SMS text message delivery to a mobile device. Using the SS7 signaling infrastructure to relay these messages limited the maximum message length to 160 characters (Milian 2009). In the 1990s, SMS on mobile devices along with PC-based IP messaging apps such as ICQ, AOL Instant Messenger, Yahoo Messenger and MSN Messenger were the start of the genre of messaging services with which everyone is familiar today.

It is interesting to consider why such an apparently constrained communication type has become so essential to modern human existence. Perhaps the most obvious characteristics are the asynchronous nature of the communication, combined with near-real-time response that is available but not required (at the human level) in the exchange. In combination, these attributes allow:

- *Multi-tasking* – the ability to send a (short) message while participating in another spoken conversation and/or several simultaneous SMS-based conversations
- *Non-intrusiveness* – the ability to choose when and where to respond
- *Instantaneousness* – the ability to respond immediately when necessary
- *Unobtrusiveness* – the ability to have a private exchange in a public setting
- *Persistence* – the ability to have a continuous or persistent exchange over a prolonged period

This combination of attributes has proven so attractive to modern existence that messaging has arguably become the dominant form of communications today. Indeed, North American usage data supports this assertion. In 1995, the average American user sent 0.4 texts per month. In 1999, interoperability allowed users of any telephony provider to send text messages to users of any other telephony provider so that by 2000, the average American user sent 35 texts per month. Usage continued to rise and by 2007, Americans sent more text messages than made phone calls (Erickson 2012).

With the near ubiquitous deployment of wireless communications, and the relatively high costs of messaging services by network operators, the emergence of new third-party messaging applications followed. Because these applications were not tied to the underlying transport network but were delivered over the top of telecom broadband data networks from a cloud, they were often "free" and spread rapidly across the globe. Moreover, the users of these platforms began using messaging apps instead of SMS not only to avoid messaging limits and usage charges, but also because they preferred the asynchronous and less scheduled nature of messaging over live, spoken communication in the form of voice calls. Thus, while messaging grew dramatically, voice calls in general were flat to declining, as shown in figure 2. Also key in the rapid growth of messaging was the innovation in messaging platforms, which the app-developer ecosystem has driven.

It is interesting to consider the associated economics. WeChat with 549 million monthly average users (MAUs) is not the largest new messaging platform but it brought in the most revenue at $1.1 billion in 2014 (Millward 2015, Paypers 2015). Although not insignificant, this is small compared with the revenues of the largest telecom service providers who make in excess of $100 billion annually (AT&T 2015). According to Pyramid Research, telecom service providers still make roughly half of their revenue from voice and SMS services. This is the classic innovation scenario, with the incumbent provider focusing on sustaining innovation and revenue streams, and new entrants disrupting the market by entering at the low end, with lower revenue expectations (and needs), and serving an unmet user demand. In the following we will argue that these worlds can and will further coalesce around the concept of a new global unified communications paradigm.

A portent of this future is provided by the recent evolution of these new messaging applications to become genuine communications platforms — adding voice calling, voice messaging, video calling, video messaging, file sharing and collaboration — including the use of the emerging set of WebRTC-type capabilities, in addition to proprietary extensions. And they have started experimenting and branching out beyond traditional communications to offer the following (Meeker 2015):

- Connecting to mobile commerce by creating a conversational or "social commerce" platform

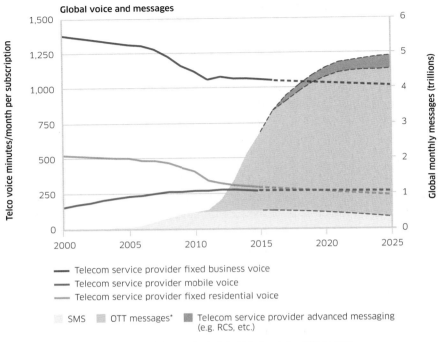

FIGURE 2: Illustration of the symbiotic nature of voice and text-based communications (Pyramid 2015a-c, Ovum 2015, Statista)

- Offering peer-to-peer payment schemes
- Partnering with content providers (for example, offering integrated streaming video)
- Partnering with gaming companies (for example, giving users access to games within their apps)
- Building applications within the messaging application in order for users to further multi-task

Media communications

Storytelling has been an important and changing part of human communication and social behavior since the beginning of time – examples include prehistoric paintings of tribes shown gathering around campfires listening to the elders, the Greek oral tradition of epic poems, of which Homer's *Odyssey* and *Illiad* are the written record, all the way to the last 100 years, with the first silent movies in the early twentieth century, to today's major motion picture movies and TV series. How this tradition is changing, and will continue to change, is largely driven by new technologies; thus it is important to understand storytelling in the context of the new digital era.

As storytelling evolved from oral, face-to-face communication, to print (books and periodicals), to images and moving images, it became more expensive and complex. It evolved from something anyone could do to something reserved for professionals and large corporations. In the early days of the printing press, it was much less expensive and was used widely, playing a major role in the Reformation with the printing of the Bible, and in both the French and American revolutions through the printing of pamphlets. But as it was industrialized in the nineteenth century, large publishing companies came to dominate the production of newspapers, magazines and books. From the early days of movie production using nitrate-based film and movie theaters for playback, movie-making was an activity that was reserved for movie studios with big budgets or a few committed and well-funded amateurs.

Although the invention of television increased the distribution by putting a "movie theater" in every home, production was still controlled by corporations (movie studios and TV studios/networks). Not only did they control what was produced, they also controlled where and when people could watch their content. It was a one-to-many, broadcast distribution where viewers had to be at a specific place (for example, a cinema or in front of a television) at a specific time to see a presentation. And the production companies leveraged this system to maximize their revenue and profits by scheduling different viewing windows, each of which they could monetize, in the case of movies, through box office receipts, and television, through advertising. Remarkably, this same system is essentially still in use today. However, it has recently been threatened by a number of technology developments, resulting in the return to an age where everybody can tell their story to anyone who wants to listen.

The full digitization of video production and distribution in the mid-to-late 1990s[3] was driven primarily by the aim to increase TV picture quality to HDTV and to make satellite and over-the-air broadcasting more efficient. However, digitization also had a huge impact on the production and distribution models. With the plummeting cost of computing, digital production cost and tools fell within the budget of small companies and even individuals. Digital files

---

3 - Many professional production tools went digital in the 1970s and 1980s, but distribution was still analog through physical film or analog terrestrial broadcasting.

Consumers are no longer willing to be tied into a linear schedule as the only way to watch content and they expect to be able to choose not only when, but also where and on which device they watch it, and to control the playout as they watch.

being capable of being copied and viewed endlessly, without loss of quality, meant that consumers and applications developers such as Napster (music) and BitTorrent (video) used the Internet to illegally share media files. This new form of distribution massively perturbed some media industries, music initially, until Apple legitimized digital music distribution with iTunes – although this changed the dominant music format from the album to the song, with significant loss of revenue for the music studios. Netflix has had a similar effect on digital video distribution and consumption habits, changing viewing from serial viewing (an episode per week) to "video binging" watching an entire TV season as if it were a book. This has led to a second "golden age" of TV and a related decline in the popularity of the feature movie format (Nielsen 2013, Wikipedia 2015b).

The new digital video capture and editing tools allowed any individual to create their own digital media. The introduction of QuickTime in 1991, a multimedia framework by Apple, and RealPlayer in 1995, the first media player for streaming over the Internet by RealNetworks, showed a glimpse of what was coming, although by today's standards the quality was very low. The personal video era started with the introduction of consumer digital video cameras in the 1990s, and accelerated with the large-scale adoption of smartphones with image and video capture capabilities around 2010. The amount of *user-generated content* (UGC) consequently exploded, creating new UGC media sites such as YouTube, founded in

2005 and bought by Google in 2007 for what was considered, at the time, the staggering sum of $1.65 billion. Today, in 2015, YouTube has over 1 billion users and 300 hours of video are uploaded every minute. We are truly back to an era where everyone can tell their story, but now they can record and distribute it for global consumption with no direct cost to them.

Another important development that has affected media consumption is the explosion of digital viewing devices in consumers' homes and in their pockets. The very rapid transition to digital flat-panel HDTVs in the home during the 2000s, driven by HD adoption and extreme price competition in the TV set market, made a second (and many times even a third) TV set common in many homes, even in less developed countries. The increase in home computers and personal laptops also increased the number of potential viewing screens. But most of all, the availability of smartphones (iPhone in 2007) and tablets (iPad in 2010) also extended the reach of digital content in time and space. This capability has drastically changed the viewing behavior and even introduced an entirely new way of watching media content – "snacking" – in which video is consumed in small amounts continuously throughout the day at any available moment. This new consumption model is particularly suitable for UGC content, as it tends to be shorter in length, but has also given rise to a completely new form of professional short form (2–10 minutes) content, for example the popular series "Comedians in cars getting coffee" by Jerry Seinfeld. Interestingly, the rise of UGC watching minutes and snacking has not – until now – meant a corresponding drop in minutes watching "traditional" TV/entertainment content (figure 3).

These developments have permanently changed user expectations for TV and video services. Consumers are no longer willing to be tied into a linear schedule as the only way to watch content and they expect to be able to choose not only when, but also where and on which device they watch it, and to control the playout as they watch. This makes unicast delivery (one stream per user) mandatory, negating the efficiency of broadcast/multicast (one stream for many users), which drives the need for higher capacity in today's networks and will continue to do so in the Future X Network. It has been reported that up to 36.5% of the traffic in US networks at peak TV viewing time (7–10 pm) consists of on-demand video watching from Netflix only (Sandvine 2015). The effect on the mobile networks has been even more dramatic with a reported 69% traffic increase in 2014 reaching a staggering 2.5 exabytes per month; it is predicted to continue to grow with a CAGR of 57% until 2019 (Cisco 2015).

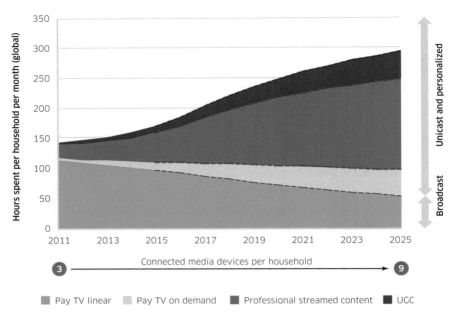

FIGURE 3: Viewing hours and connected devices increase, showing the increase in overall consumption and the surge in popularity of unicast delivery methods[4]

Fortunately, the evolution of encoding technology provides some help in managing the increase of video traffic in the networks. Every 10 years, encoding algorithms and methods have evolved, from MPEG-2 to H.264 (MPEG-4 Part 10) to High Efficiency Video Coding (HEVC), H.265 (figure 4). Each new codec standard provides ~50% reduction in bit rate for similar picture quality, measured quantitatively as peak signal-to-noise ratio (PSNR) or qualitatively as viewer mean opinion score (MOS). This has been achieved by a combination of algorithmic enhancements and the effect of Moore's Law allowing more sophisticated processor technologies processing higher bit and color resolution. These advanced encoding formats are supported on a growing number of chip sets, at ever lower cost, making digital video available on an increasing number of screens and devices.

The big challenge for real-world deployment of a new codec is the wide availability of low-cost decoders. Even today (2015), 12 years after H.264 was ratified by the ITU, there are many systems still using MPEG-2 streaming in order to support older set-top boxes (STBs) and other devices. And the adoption of HEVC/H.265 (ratified in Jan 2013) has only just begun, but will

4 - Analysis and modeling by Bell Labs Consulting using various sources (Pyramid 2014, SNL Kagan 2014).

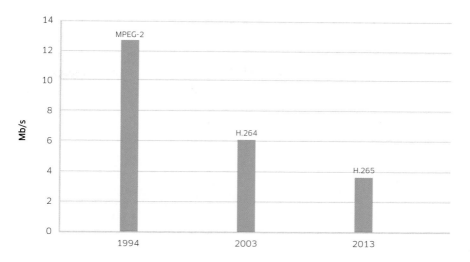

FIGURE 4: Evolution of compression performance of MPEG-2, H.264 and H.265 over time.

become an important tool for all video providers in the coming decade as more and more devices have native, on-chip support for the new codec.

The distribution of digital content (music, video, etc.) to viewing devices has evolved over time from using a constant bit rate (CBR) encoding and the associated Real-Time Streaming Protocol (RTSP) with reliable delivery over TCP to variable bit rate encoding and unreliable delivery over the HTTP (web) protocol. In the RTSP world, a constant bandwidth on the order of 2 Mb/s for SDTV and 8 Mb/s for HDTV (MPEG-2 level of encoding) was required, and if the required capacity was unavailable the playout would freeze and/or have to resort to progressive download mode, where successive pieces of the file were pre-loaded into the device buffer before playing out. Clearly, this delivery model was premised on having well-defined demand patterns. For instance, initially the known demand was for professional, long-form video-on-demand movies, but that has changed; the new demand patterns are quite different, driven by snacking and binge-watching and catch-up TV viewing. And the available network capacity simply did not support such fixed bandwidth regimes for this new demand. Consequently, there has been a wholesale migration to HTTP-based adaptive streaming (HAS), which uses an *adaptive bit rate* (ABR) approach where it encodes and stores the video content at a variety of different bit rates. The client video player senses the current network bandwidth availability by looking at the packet arrival rate and requests lower or higher bit rate "chunks" of the video – every 2 seconds or so. This allows continuous playout of the video independent of the available capacity, albeit at variable quality.

The network has also been optimized to support higher quality video delivery, even at the new (and predicted) unicast demand levels. There are essentially three primary approaches taken:

- Increase the capacity of the network by adding fiber-like capacity in access and massively expanding and optimizing the wide area network. This is discussed in chapter 4, *The future of wide area networks*; chapter 6, *The future of wireless access*; and chapter 7, *The future of broadband access.*

- Locate the video delivery server close to the end user to circumvent core network capacity constraints.

- Optimize the delivery of different video streams over access links – most importantly the mobile access link – using network optimization techniques as described below.

In order to optimize video delivery, companies such as Akamai, Limelight and Level3 Networks built content delivery networks (CDNs) starting in 1999. CDNs use a distributed cloud approach, which moves and caches content at the optimal location based on demand and available network capacity. This was essentially the precursor to the evolution to the new edge cloud architecture, as described in the chapter 5, *The future of the cloud*.

The optimization of the delivery of mobile access networks is also an area of active research, with feedback from the mobile access network increasingly being used to inform the video server or client of the instantaneous state of the network and either provide an adapted set of target encoding chunks to be requested or by request use of a higher priority delivery path over the radio access link (figure 5). Such approaches will become commonplace as we enter the new communications era, where media and content become an integral part of the new unified communications paradigm described below.

Today the highest quality video is most often an HD picture with 1080p resolution, but soon it will be with auto-stereoscopic 3D and ultra-HD with 4K or 8K pixel resolution, high dynamic range, high frame rate, increased color space, requiring large unicast data streams to be transported through the network with low-latency and high-quality.

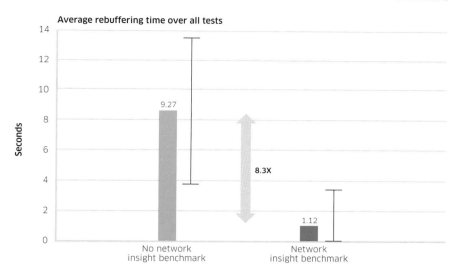

FIGURE 5: Example of QoE improvement (less re-buffering time) with intelligent video delivery based on network insight, Bell Labs

It is instructive to consider the nature of the communications between the end device and the video server. In the past, this was primarily via a handheld remote control to an STB using simple commands communicated over an infrared wireless interface. These commands (such as play, pause, fast forward, rewind) were then interpreted and forwarded to the designated video server using the requisite protocol (RTSP or HTTP). Although this media control communications is still widely in use today for broadcast media, it has been replaced by touch-screen equivalent controls on smart devices for the vast majority of unicast media. We believe that this will be a dominant method for media communications in future, and will be integrated into a unified messaging and communications platform.

In addition, there will be the increasing number of "sensory" devices capable of accessing and displaying digital media content, many of which will be increasingly personal (for example, smartwatches and virtual reality devices like the Oculus Rift or Microsoft HoloLens) with virtual reality now becoming an accepted media format supported on media sharing sites like YouTube. Therefore, in the future sensory or "perspective-based" media communications will be required, but again we believe that this will be mapped into text-based commands, either through voice-to-text conversion or other direct mappings of movement or sensing into messages (for example, a head nod to approve an action, or an eye movement to select an object)

Given the human nature of control communications and the preference for humans to communicate using their native language, rather than machine or system-specific languages, natural language processing is a central focus for the evolution of control solutions.

which is then translated into the text-based action.

Control communications

Humans communicate with a variety of different control systems today, whether home automation and facilities surveillance, or interactive voice response (IVR) systems. In the earliest incarnation (1877), home automation focused on centralized alarm systems that could be used to monitor intrusion over the emerging telephone system. As shown in figure 6, fast forward 100 years and these simple alarm systems were augmented with new monitoring and surveillance systems to protect against fire, carbon monoxide poisoning or theft in the home. These systems continued to leverage the telephone line as a reliable, fail-safe connection back to a control center in order to convey changes in alarm and other state information, as broadband data services and the Internet did not yet exist. Another type of control communications system that became widely utilized during the 1980s was an IVR solution focused on optimizing calls (again over the existing telephony infrastructure) to customer care call centers by attempting to route calls to the right agent, as well as allowing automated information retrieval via the phone.

The growth of the Internet had a profound impact on these control solutions, with the exponential growth of IP-connected devices making it possible to leverage broadband IP networks and integrate multimedia, multisensory communication clients directly into these devices or into network proxies that aggregate for the

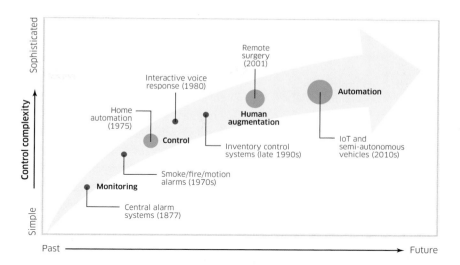

FIGURE 6: The evolution of control communications solutions over time

endpoints. Now, not only could simple data commands and information (on/off, open/closed) be transferred but a rich set of video and audio (and now multi-spectrum) measurements and captured content as well.

This massive facilitation of remote human communication with machines and systems will be the foundation to the new digital era. The utilization of new WebRTC methods now provides a unified, standardized approach to support VoIP, video and data channel services (for example, messaging, file transfer and data information sharing), on any device either natively or through a browser across public and private networks. Today, many solutions and software developer kits (SDKs) allow developers to add communications interfaces to their Internet-connected devices (Harris 2015), allowing users to control these devices using voice or text commands from anywhere in the world.

Given the human nature of control communications and the preference for humans to communicate using their native language, rather than machine or system-specific languages, natural language processing is a central focus for the evolution of control solutions. Due to the pioneering work of Vapnik and LeCun at Bell Labs in the 1990s, fundamental technologies such as support vector machines and neural networks now enable intelligent services including the ability for device users to simply talk into a microphone on their smartphone and be understood by an intelligent virtual assistant (IVA) without extensive training – something that was impossible only a decade or so ago. Such IVAs became widely popular with the introduction of Siri on the Apple iPhone in 2011, allowing users to retrieve information based on simple queries.

Today, work is progressing to realize a second generation of IVAs that leverage contextual information to augment their intelligence, and to become a genuine personal assistant to coordinate relevant information gathering and presentation, calendar and schedule management, and other mundane or time-consuming tasks (Levy 2014). The next generation of IVAs will also include specialized IVAs created for various vertical industries or enterprise operations and processes such as event-based alerts for specific groups of employees with relevant expertise (for example, security or emergency event handling, supply chain management and field force deployment). A new class of control system is also emerging – remote controlled and partially autonomous vehicles, such as cars and drones – which require ultra-low-latency communications as described in described in Chapter 11, *The future of the Internet of Things*.

The ultimate goal of all these control systems is to augment human existence as described, whether it be in a consumer or enterprise context, as discussed in chapter 10, *The future of information* and chapter 2, *The future of the enterprise*.

It is important to recognize that in nearly all IVA and human control communications systems, voice commands are converted into text – text that can be consumed and acted upon by digital systems and software platforms. So, in many ways, text and text messaging is the fundamental or primary mechanism of control communications, just as it has become for personal communications and media communications as described above. We explore this convergence or unification of communications around new multi-dimensional messaging systems and platforms in the next section.

The future of communications

The landscape of past and current communication experiences and enabling technologies outlined above clearly points to the expanding and converging frontiers of communication among people and systems, as illustrated in figure 7. We clearly identify the three types of existing communications services: personal, media and control communications as possessing different attributes in terms of interactivity and realism, with interactivity ranging from days or weeks (mail) to milliseconds (tactile and immersive communications), and realism ranging from simple digital messages and commands (telegraph and simple machine control) to full multiparty human representation (face-to-face meetings).

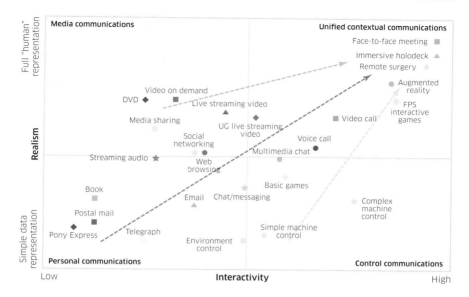

FIGURE 7: Depiction of the evolution of communications and trend to future convergent communications

While human communications span the entire space outlined in figure 7 and will continue to do so, the future of communications lies in the domain of highly realistic, highly interactive digital experiences that augment human existence in all dimensions:

- Personal communications have moved from episodic, asynchronous exchanges (postal mail) to the global interactive exchange of media over the Internet (e.g. multimedia video chat and conferencing). We are now on the verge of creating augmented realities and synthesizing virtual environments using cloud based processing and platforms.

- Media communications have moved from asymmetric and asynchronous broadcast analog forms (newspapers and analog broadcast TV) to ad hoc viewing of digital multimedia from any source (for example, online music and VoD). We are now on the verge of using this ability to generate, augment and orchestrate augmented realities by creating personalized multisensory media composites.

- Control communications have moved from infrequent transmission of basic machine commands (alarms and resets) to simple end devices to interactive, time-sensitive exchange of critical alerts and near-real-time control of remote devices, robots and vehicles. We are now on the verge

Increasingly messaging clients are comparable to contemporary web browsers; they can process messages according to dynamic formatting guidelines, which can be controlled by content providers, and they can execute software applications.

of being able to dynamically control remote objects anywhere around the globe, with interactivity only limited by the speed of light.

In each case, the interactivity is becoming increasingly real-time and the realism of the representation becoming increasingly "real". This suggests a convergent future for all types of communications, which in turn suggests that unifying communications platforms will emerge as a critical technology enabler that seamlessly and conveniently unifies all communication modes. In the remainder of this chapter, we will describe the essential attributes of such platforms essential for convergence and for meeting the needs of all the systems connected by the Future X Network.

We believe there will be three key technology elements in a convergent communications platform in 2020:

- *Augmented messaging platform* – The massive increase in popularity of chat/messaging systems indicates that a message-based communications framework – one that treats every interaction as a message exchange – will be the common base for future digital communications. Messages will support personal interactions, access to digital content and control of "things" within a common framework.

- *Stream-handling platform* – Streams of data will become far more numerous as multi-component sensors, cameras, vehicles, robots and drones are more widely deployed. A new real-time data

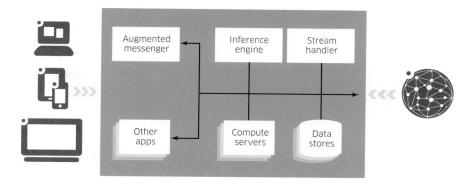

FIGURE 8: The essential components of a future communications platform

stream-handling framework will be required that will provide mechanisms for naming, finding, opening, manipulating and terminating these flows and the data being transported within them. This will require new stream-level functions (for example, procedures to identify streams by their semantic properties rather than by name, to identify properties of data within streams, and to reconfigure stream sources and destinations). It will also require new content-level functions (for example, procedures to package content, combine content from multiple streams, and filter content according to control guidelines).

• *Contextual inference platform* – In order to limit big data overload it will be vital to intelligently forward only relevant information in the form that most closely approximates actionable knowledge. Such a contextual inference engine will organize and curate all known information and be able to provide rapid summary perspectives for every current context.

A high-level architecture is shown in figure 8. An *augmented messenger* will handle all primary communications among people, devices and systems and will share summary data with a contextual inference engine, which stores summary perspectives about the exchange. A ubiquitous stream-handling framework will identify and correlate relevant data streams, extract relevant information and make it available to the same inference engine, based on the user's specified preferences and control.

The following sections will describe in more detail the essential functions, their roles in shaping future communications and the transformational potential of these three, synergistic communication elements.

The augmented messaging platform

Due to the pivotal nature of communications in human existence, major communications technologies generally do not appear frequently, and when they do their impact is gradual. For example, email was initially a simple electronic version of mail, and was therefore treated as a relatively formal type of correspondence. But over time, the higher level of responsivity it offered resulted in its use as a preferable alternative to phone calls, in many instances. Similarly, the World Wide Web has moved from its initial use as a large distributed system of (relatively static) information sites with which people communicated (searched), to a dynamic system for performing a wide range of transactions, with unprecedented access to information and processes. When combined with the emergence of broadband mobile communications, it has enabled people worldwide to communicate with computers and systems that respond to queries and support of essential services for businesses, governments and every kind of organization.

Most recently, a new communications paradigm based on messaging has emerged, as discussed previously. It has the potential to become as important as the existing Web, although merely an overlay on top of that basic infrastructure. Most people view chat/message systems as a natural and convenient way to communicate quickly, with immediacy, brevity and unobtrusiveness that make it preferable to email. It retains the same optionality in response time as email and so allows more user control than traditional synchronous, telephony-based communications.

As more people adopt such chat/messaging technology as their preferred communications method, enterprises are beginning to respond. For example, Salesforce Chatter supports live chat interactions between customer support staff and people seeking online help (Salesforce 2015), and the mobile messaging app, Line, allows people to pay for their product purchases through chat messages (Line 2015). Similarly, Walmart and AliExpress customers can search for and purchase products within the Tango messaging application (Jaekal 2015). A "WeChat City Service" tab allows WeChat users to access several municipal services (for example, to pay utilities or search for library books) within the messaging application (Chen 2015).

The history of the World Wide Web provides hints as to the requirements for a convergent augmented messaging platform. The Web has evolved to become an active architecture with more dynamic code creation and execution (JavaScript), to support access to increasingly personalized and up-to-date information and services. Web browsers also now process multiple data streams to update displays continuously according to dynamic formatting guidelines and content, which can be controlled by content providers. In essence, the World Wide Web has become an information marketplace, where content producers and consumers exchange data and content "goods". Moreover, anyone can establish a web presence by creating and publishing a web site using a variety of software tools; the site can then be promoted via well-established marketing channels and social media; and it can then scale based on pay-as-you-grow IaaS cloud platforms to achieve global reach and scale within days or even hours.

A similar evolution is taking place with messaging platforms and services. Originally text messages contained static information as they were used to exchange text, photos and videos. However, increasingly messaging clients are comparable to contemporary web browsers; they can process messages according to dynamic formatting guidelines, which can be controlled by content providers, and they can execute software applications. For example, ChatOps allows third-party software modules to be associated with messages (Daityari 2014). Similarly, the recently announced Facebook Messenger Platform allows third-party developers to build applications that integrate Messenger (Franklin 2015).

These third-party modules allow people to interact with one another and with other modules. Using this new messaging software environment, modules called messaging agents can be built to act as participants in chat sessions. These agents will contain interactive functions to represent devices, organizations and act as intelligent virtual assistants – essentially an evolution of the "chatbots" that are used for online customer care today. But, these new messaging agents will also allow direct access to information and to applications, to control and query, and discover and program the web.

These new control functions will be pushed to the user as executable code inside the messaging client whenever the appropriate request is made, or based on a user context. Examples of such controls include:

- Device controls for TVs, media players, video servers, printers, projectors, consumer electrical goods, home and building controls
- Decision controls to individuals or groups to determine opinions and preferences, such as surveys, voting, decision-making
- Direction and function control of robots, vehicles, drones and instruments

- Discovery controls for contextual search and exploration of data, images, maps, preferences and intents

These control functions will be embedded inside augmented messages, so they can be forwarded, shared or re-assigned as readily as forwarding a text message in current messaging applications.

As an example of a future, augmented messaging control scenario, consider the following: a user at work will be able to unlock doors, query a group of colleagues to arrange a meeting time, locate local colleagues, invite remote colleagues to participate via immersive video session, or via a "presence robot", create a dynamic digital workspace, book a room to make a presentation, be guided to the room and have the workspace connected to the room, control the screens and projector and share the controls as needed, capture audio and video into the shared workspace, and then share or print any required physical documents. At home, a user will be able to lock/unlock doors, control lighting, heating and cooling, control and query appliances, search from a personalized menu of media content, control the play-out of the content, share the controls as needed, and much more, and all from within the same platform as they use to communicate with people, and the same platform they use at work for enterprise tasks. Although some of these capabilities are either already available or emerging, they have been developed haphazardly with no unified approach to security, generality, or multiple platform support.

The messaging agents described above will play the analogous role of the web page in the World Wide Web; as entities have a web presence today, they will have a messaging agent tomorrow with augmented messages continuously updating content and executing (small) software applications as part of message processing. And just as the web growth has depended upon search tools to find the pages hosting content and services, there will be a need to develop tools to find messaging agents that represent particular content and services and to create "agent mash-ups," allowing messaging agents to use the services of other agents.

The stream-handling platform

The rise of the Internet and the web was fueled by people's desire to share and exchange static or "non-real-time" content – text documents, emails, photos, and images – produced once and then made available for access by uploading and storing it on servers, with relatively long "shelf-life" (measured by continuing interest over time).

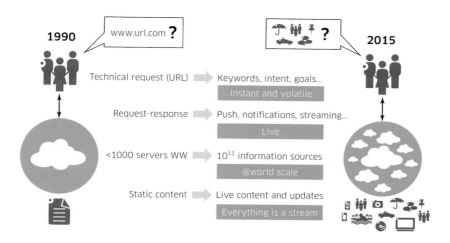

FIGURE 9: Differences between the World Wide Web (WWW) and World Wide Streams (WWS)

With the emergence of massive numbers of sensors, smart devices and connected objects in the new digital era, a different type of content will gain importance; this content is produced and made available on a continuous basis in the form of long-lived data streams. Any ephemeral data or information contained within these live streams is most valuable at the time it is produced and quickly loses value thereafter. For example, streaming sensor data from cars is of maximum value immediately – to optimize coordinated driving of cars in platoons or to stream current situational awareness and images for traffic and safety optimization – but less relevant thereafter (although it may prove useful as aggregate data for forensic studies after an incident). Likewise, sensory information can be extremely valuable for steering and controlling other machines, and also requires instantaneous, real-time updates. Figure 9 explains the differences between the World Wide Web and World Wide Streams (WWS).

Today the web clearly provides a powerful platform for publishing, finding, accessing and connecting non-live content and it supports easy creation of content mash-ups – new content and services created by connecting or overlaying various available content pieces. For example, real estate applications overlay house market information with geographic maps and current interest, offering a convenient way to find relevant properties for sale and assess potential demand. Similar capabilities are needed for the emerging world of live data streams. The required platform will allow creators of data streams to name and describe those streams and to make them available to others. At the same time, mechanisms are needed to allow users and applications to find relevant data streams and to connect them into data stream mash-ups. For example, if an automobile could, in real time, find the acceleration and directional movement data from nearby cars, accidents could potentially be prevented, or driving optimized. But how can data streams be made available in a manner that they could be discovered and accessed by others in real time, while maintaining the desired privacy and security? Just as the web provides such capabilities for non-live content, a WWS platform must provide similar capabilities for data streams. This new WWS experience will be essentially about publishing, finding, filtering, contextualizing and mashing-up ephemeral data streams to detect relevant live phenomena and to make them actionable. This will be enabled through a few key functional concepts:

- *Hyper-personal contextualization:* In the early days of the web, users had to remember the URLs of web sites they wanted to visit. This syntactic addressing scheme was cumbersome and inconvenient. Soon search engines were implemented that allow users to express their interests rather than having to specify syntactic strings. Likewise, users cannot be required to remember the names of billions of data streams; they should be able to simply express their interest, and the system automatically identifies relevant data streams, finds and accesses them, and combines their data to provide the most relevant response to a user's inquiry. A language will need to be defined that allows users to express their interests in clear form, and methods need to be developed that map such interests into suitable combinations of available data streams. Then logic will need to be implemented that finds the answer to users' interests by extracting and combining the relevant data pieces from the data streams, as described in the next section.

- *Hyper-volatile:* As the number of data streams to and from devices increases into the billions, most flows will be of interest to only a small number of consumers and the applications they use, with little re-use potential for others. These highly personal streams will contain personal mash-ups of content, which will be created by small, localized processing units. At the same time, user interests will be dynamic and change frequently, requiring a data stream processing framework that is highly agile and can easily adapt to changing user interest – depending on the application – even on the timescale of seconds. At the same time, data streams can themselves be volatile intrinsically – appearing in an instant

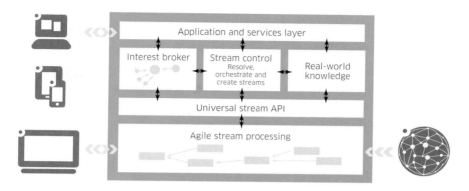

FIGURE 10: High-level architecture of future real-time data stream processing platform

and disappearing a moment later. This makes real-time matchmaking between users' dynamic interests and available data streams extremely challenging, as both can experience significant churn. Methods will need to be developed that enable the system to identify instantaneously the relevant data streams in a flood of mostly irrelevant big data streams. This search also needs to support semantically rich inquiries, from structured queries to natural language questions.

- *Hyper-scale:* With the explosion of data-generating devices, any solution must be able to handle an extremely large number of data streams, at any time. The hyper-personal and hyper-volatile nature of these streams magnifies the resource consumption problem. Therefore, methods will need to be developed that ensure a close-to-optimal usage of the underlying network resources – data streams need to be processed and relevant insights need to be extracted as close to the source as possible, eliminating the need to transport data all around the world. Prior analysis of data streams should be re-used wherever possible, irrelevant streams should be discarded and stream processing should be pushed to the network edge into the *edge cloud*, closer to the data sources.

As illustrated in figure 10, these concepts will be brought together into a modular, scalable and robust real-time stream-processing platform, which takes advantage of distributed cloud resources to enable fine-grained resource management.

Sensors, devices, and application data streams will be processed by an agile stream-processing component to be named and made available to other elements through a universal stream API. Applications will use a "discovery"

We are already moving into a future that is built upon messaging solutions, ones that go far beyond simple exchanges of text, photos or videos, to become a universal communications method also for business transactions and control functions.

augmented message communication to express the interests of their users within a certain context through an interest broker function, which leverages different inputs to identify, access or create the data streams needed for satisfying user interests. The interest-broker request forwarding mechanism is described more fully in chapter 4, *The future of wide area networks.*

Consider the following use case: vehicles such as drones or cars or freight delivery vehicles will provide continuous speed, direction and on-board sensor information to the agile stream processing component, which will be complemented with video feeds from internal cameras and any other relevant external video feeds. Interested (and authorized) parties will be able to express an interest in such local vehicular data streams without any knowledge of the specific streams, or vehicles and the relevant streams will be discovered and forwarded based on the inferred probability of relevance to the specific user as described below. Although some of these capabilities are available in a piecemeal fashion today, across multiple proprietary platforms with varying degrees of security, we suggest that a unified approach will democratize the sensor information.

In order to deliver the optimum economics for the scale required (billions of objects), this data stream processing platform will be deployed as a massively distributed system leveraging the edge cloud architecture that is a fundamental element of the Future X Network, and which is described in chapter 5, *The future of the cloud.*

The contextual inference platform

In chapter 10, *The future of information*, we describe how in the future it will be possible to create a massive graph construct containing all the relevant attributes of people, things and their inter-relationship. The ability to rapidly read this structure and infer user preferences will be critical to enable automated control of tasks, contextualization of all communications and meet the fundamental goal of saving time. The challenge is to move beyond conventional user data management, toward a deeper understanding of the user's needs and constraints, that is, to move toward user intent management. New approaches will have to be developed −algorithmic as well as systems-related innovations are needed. For example:

- *Enriched user definition:* Typical data management systems record user information such as age, race, gender or contact information in relatively straightforward data schema. But in the future, more broadly defined indicators for user behavior and intent will need to be captured, stored and retrieved based on large graph analysis and new categorization schema.

- *Enlarged user database:* While today's solutions interpret the term "user" as a human, this will have to be extended to include non-human entities such as cars, aircrafts, sensors, robots, devices and appliances, thus generating billions or even trillions of entries within the next few years.

- *High-frequency updates:* New services will require very different data update intervals. For example, location data of a moveable object like a drone or car might have to be updated every second, while more static user information such as home address or credit card information will need to be updated only every few years.

The future begins now

Elements of the vision we have outlined above are already beginning to become a reality. Hundreds of companies − both large and small − are developing the next generation of messaging applications and they are rapidly evolving toward more unified communications platforms, as described by Mary Meeker in her "Internet Trends 2015" report (Meeker 2015). For example:

- WeChat has aggressively pursued new user experiences while bringing in new revenue. They are pioneers in social or conversational commerce, enabling their users to include third-party products that can be purchased during a chat session. WeChat acts as a video distribution channel by letting users watch videos within the app. WeChat users sent over a billion "red envelopes" filled with e-money during the Chinese New Year celebration. Lastly, users can order taxis or food delivery by simply sending instant messages to businesses.

- The Line mobile messaging application offers similar enhanced services – they provide users with games playable within the app and Line's Kawaii characters have their own cartoon series in Japan. Line has said that they see themselves as an entertainment – not a technology – company.

- The Tango messaging app announced a partnership with Walmart and Alibaba to let their users chat and purchase products at the same time.

- Facebook Messenger has rolled out free, peer-to-peer payments and will be allowing users the ability to chat with businesses.

- Telegram is a new chat/messaging app that provides a platform SDK enabling third-party developers to create Telegram "Bots", which their users can chat with using the Telegram app communications services.

- WebRTC has been adopted by Facebook Messenger, Google Hangouts, Slack and other high-profile services integrating their messaging platform and communications services with the web.

These examples – and their broad adoption – show that we are already moving into a future that is built upon messaging solutions, ones that go far beyond simple exchanges of text, photos or videos, to become a universal communications method also for business transactions and control functions. And the emergence of open interfaces for integrating third-party software into these platforms supports this vision of messaging being the nexus of the future of communications.

In terms of new media communications and consumption:

- In February 2015 Meerkat launched in the App store. Meerkat gives anyone with a smartphone and Internet connectivity the ability to broadcast content so that anyone on the Internet can find and watch, based on Twitter hashtags. Twitter launched Periscope in March and in early August Facebook launched their Mentions functionality (currently reserved for "celebrities"). Statistics already show the impact; Periscope announced in August 2015 (4 months after launch) that they have 10 million active accounts watching the equivalent of 40 years of content every day! (Periscope 2015).

- In the Ofcom 2015 Communications Market Report for the UK, it was found that adults aged 16 to 24 spend only 50% of their viewing time watching traditional television, and 72% of people claimed to watch short-form video (such as clips and music videos on services such as YouTube), with 32% viewing either daily or at least weekly. In the same age group, computers (57%) or smartphones (45%) were used more often than an STB (40%) for viewing on-demand and catch-up services on a monthly basis (Ofcom 2015).

- In May of 2014, Netflix was the first to commercially make available content in ultra-high definition (UHD/4K). By combining their cloud-based over-the-top (OTT) streaming infrastructure, the availability of UHD-capable TV sets (with HEVC decoders), and their own content ("House of Cards") Netflix was able to launch "4K/UHD from Netflix".

- Facebook with Oculus Rift and Microsoft with HoloLens have demonstrated virtual and mixed reality immersive services. Mozilla has published a WebVR API for Firefox allowing web developers to plug in an Oculus Rift and create a virtual reality experience with some JavaScript.

Control communications are quickly evolving toward systems that are built on top of a popular and/or standard messaging platform. Many of these control systems support chat-based, convenient, easy-to-use interactions among people and devices. Others focus more on real-time responsiveness rather than language processing tasks. For example:

- Select appliances from LG Electronics (LG) support chat control interfaces. Their HomeChat service uses the Line mobile messaging application to transmit recommendations and control settings and uses natural language processing to interpret message content. The resulting interface lets people communicate with LG's smart refrigerator, washing machine or oven through convenient language-based chat sessions (LG 2014).

- In a similar cooperation between a device control specialist and a provider of a popular chat system, Ayla has worked with the WeChat messaging application to create a "smart" conference room. It allows multiple devices within a room, from the lights to the curtains, to be controlled through a WeChat public account. Room users can sign up for the account by scanning a QRcode (Xiang 2014).

- Device-to-device communications are gaining more acceptance and attention. One example is particularly noteworthy – vehicle-to-vehicle (V2V) communications. These communications let cars talk to each other by automatically sharing reports of their position, speed, steering-wheel position, brake status and other data. They can even send alerts to each other if a crash seems imminent. The U.S. Department of Transportation has begun taking steps to enable V2V communication (SaferCar 2015).

- Facebook's wit.ai, Apple's Homekit, and Google's Brillo/Weave coupled with Now are solutions that enable developers and IoT vendors to add a natural language interface through text and/or voice to their products. Some technologies also give developers the capability to utilize their preferred communications app for remote control through instant messaging or a voice call.

Last, with regard to IVAs:

- Apple's Siri, Microsoft's Cortana and Google Now are all high-profile examples of IVAs that are pre-packaged with smartdevices. Viv, a new assistant by the inventors of Siri, is expected to include cognitive capabilities beyond the more basic, initial information retrieval or question-and-answer function of today, thus making it a true personal assistant.

- Specialized intelligent virtual personal assistants are also emerging. Genee and x.ai are examples of meeting scheduler assistants that will integrate with any chat/messaging app that has an open interface.

All these examples show that the future of communications is beginning now. Technologies for the next generation of convergent messaging platforms are being developed. Hyper-personalized content is being created, distributed and consumed in large amounts, and new methods are being deployed for controlling objects and the environment.

Summary

As we enter the era of digitizing anything, we are at the brink of experiencing a true convergence between personal, media and control communications. The fundamental goal is the desire to save time, by integrating and contextualizing all human interactions or communications. Synchronous, intrusive and obtrusive communications by voice is coming to an end, to be replaced by asynchronous, unobtrusive and convenient communications, with messaging at the core.

Current messaging applications will further evolve into augmented messaging services that bring messages to life, just as the World Wide Web evolved from its rather static nature at inception into the highly dynamic and customizable form that is so valued today. The addition of stream-handling capabilities to the web infrastructure to create a World Wide Stream extension will make it possible to fully leverage the emerging flood of data streams, generated not only by smartphones, but by every kind of device, sensor and wearable. Contextual inference engines will help navigate through this data "tsunami" with ease and provide a converged, contextualized communications experience – integrating not only personal and media communications, but also providing powerful control to the user. In combination, this new communications reality will allow "life automation" and savings of our most scarce resource – time.

References

AT&T 2015. "AT&T Reports Strong Subscriber Gains and Solid Revenue Growth in Fourth Quarter". AT&T Press Release, January 27 (http://about.att.com/story/att_fourth_quarter_earnings_2014.html).

Chen, T., 2015. "Check out 14 new features in WeChat made specifically for Shanghai users," *Shanghaiist* August 6.

Cisco 2015. "Cisco Visual Networking Index: Global Mobile Data Traffic Forecast Update 2014–2019," February 3.

Collins, 2015. "Definitions of communications," *Collins Online Dictionary* (http://www.collinsdictionary.com/dictionary/english/communications?showCookiePolicy=true).

Comedians In Cars Getting Coffee 2015. *Single Shot web site* (http://comediansincarsgettingcoffee.com).

Culkin, J., 1967. "A Schoolman's Guide to Marshall McLuhan." *Saturday Review* March 18 (1967): 51-53, 71-72.

Daityari, S., 2014. "An Introduction to ChatOps: Devops Meets IM," *Sitepoint*, October 30 (http://www.sitepoint.com/introduction-chatops-devops-meets-im).

Erickson, C. 2012. "A Brief History of Text Messaging," *Mashable*, September 21 (http://mashable.com/2012/09/21/text-messaging-history).

FCC 2015. "What You Need to Know About Text-to-911," *Federal Communications Commission*, April 20, (https://www.fcc.gov/text-to-911).

Franklin, L., 2015. "Introducing Messenger Platform and Businesses on Messenger," *Facebook Developers web site*, March 25 (https://developers.facebook.com/blog/post/2015/03/25/introducing-messenger-platform-and-businesses-on-messenger).

Harris, D. 2015. "Facebook acquires speech-recognition IoT startup Wit.AI," *Gigaom*, January 5 (https://gigaom.com/2015/01/05/facebook-acquires-speech-recognition-iot-startup-wit-ai).

Jaekel, B. 2015. "Walmart, Tango and Alibaba bring conversational commerce to fruition," *Mobile Commerce Daily*, May 13 (http://www.mobilecommercedaily.com/walmart-tango-and-alibaba-bring-conversational-commerce-into-fruition).

Kelly, S. M., 2015. "YouTube's 360-degree video could mean big things for VR," *Mashable*, March 16 (http://mashable.com/2015/03/16/youtube-adds-360-video-support).

Levy, S. 2014. "Siri's Inventors Are Building a Radical New AI That Does Anything You Ask" *Wired*, September 12 (http://www.wired.com/2014/08/viv).

LG Electronics 2014. "LG Rolls Out Premium Smart Appliances that Chat," *LG Press Release*, May 6 (http://www.lg.com/ae/press-release/lg-rolls-out-premium-smart-appliances-that-chat).

Line 2015. "Line Pay, a secure and easy payment platform," *LINE web site*, (http://line.me/en/pay).

Marland, E. A., 1964. *Early Electrical Communication*. London: Abelard-Schuman, pp. 17-19.

Meeker, M., 2015. "Internet Trends 2015," *KPCB web site*, May 27 (http://www.kpcb.com/internet-trends).

Milian, M., 2009. "The Business and Culture of Our Digital Lives," *The Los Angeles Times*, May 3 (http://latimesblogs.latimes.com/technology/2009/05/invented-text-messaging.html).

Millward, S., 2015. "WeChat grows to 549M monthly active users," *TechinAsia*, May 13 (https://www.techinasia.com/wechat-549-million-active-users-q1-2015).

Nielsen, "'Binging is the New Viewing for Over-the-Top Streamers," *Nielsen* September 18 (www.nielsen.com/us/en/insights/news/2013/binging-is-the-new-viewing-for-over-the-top-streamers.html

Ofcom 2015. "The Communications Market Report: United Kingdom – The UK is now a 'smartphone society'," *Ofcom web site* August 6 (http://stakeholders.ofcom.org.uk/market-data-research/market-data/communications-market-reports/cmr15/uk).

Ovum 2015. "OTT Communications Tracker - 4Q14," *Ovum*, March.

Paypers 2015. "Messaging apps to become commerce platforms," *The Paypers*, March 31 (http://www.thepaypers.com/mobile-payments/messaging-apps-to-become-commerce-platforms/759322-16).

Periscope 2015. "Periscope, by the Numbers," *Periscope web site*, August 12 (https://medium.com/@periscope/periscope-by-the-numbers-6b23dc6a1704).

Pyramid Research 2014. *Fixed Communications*, Pyramid Research.

Pyramid Research 2015a. *Fixed Communications Demand* – Residential, Pyramid Research, June.

Pyramid Research 2015b. *Global Voice Usage*, Pyramid Research, July.

Pyramid Research 2015c. *Mobile Data*, Pyramid Research, June.

Portio Research 2014. "Mobile Messaging Futures 2014–2018," *Portio Research*, September 15 (http://www.portioresearch.com/en/messaging-reports/mobile-messaging-research/mobile-messaging-futures-2014-2018.aspx), quoted in Wikipedia, "Short Message Service," (https://en.wikipedia.org/wiki/Short_Message_Service#cite_note-4).

Portio Research 2015. *Analysis and forecasts for Mobile Messaging worldwide.* Portio Research.

SaferCar 2015. "Vehicle-to-Vehicle Communications," *SaferCar.gov web site* (http://www.safercar.gov/v2v/index.html).

Salesforce 2015. "Chatter: Take action at the speed of social," *Salesforce web site* (http://www.salesforce.com/chatter/overview).

Sandvine 2015. Sandvine Global Internet Phenomena Report, Sandvine.

Shu, C., 2013. "Messaging App Line's Kawaii Characters Get Their Own Cartoon Series in Japan," *TechCrunch*, April 7 (http://techcrunch.com/2013/04/07/line-offline).

SNL Kagan 2014. *US connected device outlook - a smorgasbord for video delivery*, SNL Kagan.

Statista. (http://www.statista.com).

Tollfreenumber 2012. "Who Is Almon Strowger?" *Tollfreenumber.ORG* (http://www.almonstrowger.com).

Wikipedia 2015a. "Fax," Wikipedia, last updated July 27 (https://en.wikipedia. org/wiki/Fax).

Wikipedia 2015b "Golden Age of Television (2000s–present), Wikipedia, last updated August 15 (https://en.wikipedia.org/wiki/Golden_Age_of_Television_ (2000s%E2%80%93present)).

Xiang, T., 2014. "Riding the Internet-of-Things Trends, Solution Providers Emerge in China," *TechNode* October 27 (http://technode.com/2014/10/27/ internet-of-things-solution-providers-emerge-in-china).

YouTube 2015. "Statistics," *YouTube web site*, August 7 (http://www.youtube. com/yt/press/en-GB/statistics.html).

CHAPTER 10

THE FUTURE of INFORMATION

Christopher A. White

The essential vision

We sit on the cusp of a revolution in information driven by the intersection of a massive increase in the sources of data and the ability to augment our intelligence and use critical thinking to turn this data into knowledge.

Information can be defined as the "communication or reception of knowledge or intelligence; knowledge obtained from investigation, study, or instruction" (Merriam-Webster 2015). The acquisition of information requires the ability to capture data, compute something based on the entirety of data available and then communicate the resulting insight. These "3Cs" are the essential elements that are inherent to the synthesis of information.

This revolution in information we see will be defined by three primary trends:

1. *Data and information will become "free":* The cost of data sets and the relationships between them will tend to zero as data becomes ubiquitous and commoditized.

2. *Big data will become "small":* The hyperbole of the *big data* phenomenon will transition into new, *small data* applications that provide real knowledge (and value).

3. *Intelligence will be "augmented":* New *augmented intelligence* tools will change the way we explore and interact with information and our world to massively enhance a fourth "C" – cognition – the acquisition of knowledge.

The cost of information can be attributed to the cost of the "3Cs". The rise of the Internet has dramatically reduced the cost of communications of data and the widespread deployment of cloud computing has similarly reduced the cost of computation. Capturing data has remained the primary barrier to obtaining information, but as we enter the era of the digitization of everything and the Internet of Things (IoT), this key constraint will be removed, resulting in increasingly large volumes of data that are both free and available.

Over the last decade, the promise of big data has captured the imagination of corporations around the globe (Manyika 2011). The expectation that this data will have significant value in the future has driven the purchase and deployment of scalable storage and processing systems with the intent to preserve every byte of corporate and customer data. In reality, much of this data does not contribute any insight, as it is either a replica of existing data or noise that has no real information content. As storage costs mount, the hype around big data will rapidly evolve into a new focus on applications of small data where the value arises not from the scale of the data set, but from the ability to extract useful information and to make decisions based upon the smallest data set. The goal will not be to measure and store every byte; it will be to measure and store "just the right amount" of data.

This new small data-based information trend will both be continuously enabled by the network today and provide critical insights into the requirements of the network of tomorrow. The network will provide a critical role in providing the near-real-time access to data from a multitude of sensors and a multitude of augmented intelligence tools running on a massive distributed set of computational resources. The new knowledge that results will transform human lives, creating new businesses and industries. This will drive a virtuous cycle, where the knowledge gained will be used to continuously optimize the underlying communications network and its ability to connect global *capture* and *compute* resources tomorrow.

In essence, as trends surrounding the 3Cs coalesce to provide a low-cost data acquisition, storage and processing architecture with global scale, the value of information will no longer be related to its cost, but rather to the new knowledge that results. Our enhanced cognition (the fourth "C") will be enabled with unprecedented facility by the application of new augmented intelligence tools. This will result in new understanding and control of the global phenomena and processes underlying the human and natural environments, forever changing the way we interact with the things we own, with the world and with each other.

The past and present

When we consider the definition of information given at the beginning of the chapter, as well as other writings on the distinction between information and knowledge (Kivumbi 2011, OED 2015), we identify three constituent components that will prove valuable to the discussion, as follows:

- Data: the actual captured *observations*
- Information: the determination of *relationships* between data
- Knowledge: the determination of models that describe the *meaning* of Information

These will be the definitions we use throughout this chapter. Using these definitions, it is clear that data is essential to the creation of information and that, in turn, information is required for the creation of knowledge. The history of critical human thinking is almost entirely a story about the disruptions that impact (lower) the barriers to achieving scale in the ability to capture, compute, and communicate (and store) information from data and then, where possible, to build models that encapsulate knowledge – the act of *cognition*.

In the earliest human eras, the capture, computation and communication of information was all by humans through observation, mental analysis and talking to other humans, usually within a few meters – the normal carrying power of the human voice. Over time, the primary changes have been in the scale of communications and computation resources available, as shown in figure 1.

In each information era identified in figure 1, there is a first disruption that causes a profound change but only with local impact. Then follows a period of *scaling*, where decreased cost and increased access to data and resources enable an individual or group with lesser means, perspective or analytical ability, to become more informed. In the following, we discuss this in the context of the "3Cs" that all evolved to a greater or lesser extent to move from one era to the next, and the enablement of the fourth "C": cognition.

Communication of data

The creation of information begins with communication of observations (the captured data) that potentially address questions of interest. Early communication (8000 BCE) was limited to spoken language and thus required all parties to be present in the same location at the same time. This locality requirement was a huge barrier to the creation of information; given the existence of a limited number of experts on any given topic, the physical movement of these experts to understand the data and find connections between data sets was largely impossible. So, during this time period, the creation of information arose entirely from direct, local observation and data analysis with no larger cumulative or global learning.

The development of written language (5000 BCE) dramatically changed human existence by addressing this "localized" constraint for the first time. Questions, observations and answers could now be communicated without the need for direct interaction or physical presence. A single document could be shared with many individuals — amplifying the ability to communicate broadly to a diverse set of people who could add to, edit or correct the information and re-communicate the new perspective. Communication was unconstrained by a *location or time*, as the information could be accurately reproduced and augmented over generations and epochs. This led to disproportionally greater value as the result of a *network effect*, often described as Metcalfe's Law.[1] This disruption in scale was driven by the mail and transportation network, but was also catalyzed by related technological inventions such as movable type and the printing press, which dramatically lowered the cost of producing "data records" so that multiple copies could be made and distributed *in parallel* (Eisenstein 1983).

The next major disruption in data communication addressed both greater interactivity and scale. The invention of the telegraph and the telephone removed the locality constraint for *interactive communication*. For the first time, conversations could take place in real-time over enormous distances, allowing the facile and rapid sharing of data and perspectives and the accumulation of aggregated information on the timescale of days rather than months. At the same time, broadcast radio and television enabled large-scale, multimedia communication for the first time, further enhancing the ability to share information with millions of people in parallel (Sarkissian 2001). The variety of recording technologies developed in this period of rapid growth also allowed the widespread use of time–shifting and replay of transmissions, which further extended the global reach, both to different times and places (often where live transmission infrastructure did not exist).

1 - Metcalfe's Law states that the value of the network (information) rises as the square of the number of the connected nodes (data sources). Theodore Vail, President of Bell Telephone, first presented the idea in Bell's 1908 annual report. Metcalfe later popularized it to promote 3Com's Ethernet products (from Wikipedia, "Network effect").

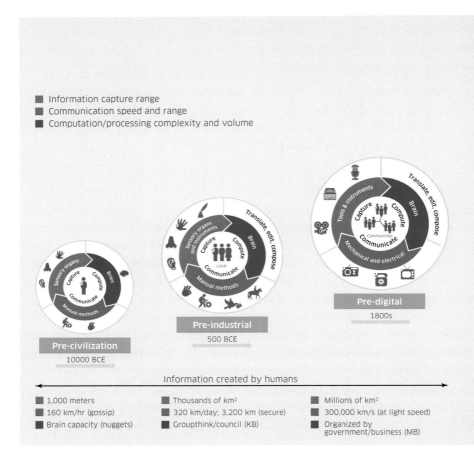

■ Information capture range
■ Communication speed and range
■ Computation/processing complexity and volume

Pre-digital
1800s

Pre-industrial
500 BCE

Pre-civilization
10000 BCE

Information created by humans

■ 1,000 meters
■ 160 km/hr (gossip)
■ Brain capacity (nuggets)

■ Thousands of km²
■ 320 km/day; 3,200 km (secure)
■ Groupthink/council (KB)

■ Millions of km²
■ 300,000 km/s (at light speed)
■ Organized by government/business (MB)

FIGURE 1: Evolution of the capture, computation and communication of information (Bell Labs Consulting)

Figure 2 shows the growth in scale that resulted with the subsequent invention of IP networking and the Internet, aided by broadband access, mobile data and communications services and devices, as well as the digitization of existing media and information. This has accelerated the disruption, allowing communication of data to achieve global scale, in real time, at any location, and with massive parallelism. Everything is now accessible almost everywhere, by almost everyone, at little or no cost. The ability of enterprises and consumers to create their own web presence and to expose data to the world marks a significant change. Data that was once privately held and sold at a premium is now exposed to everyone: financial market data, weather data, map and location data, housing and travel data,

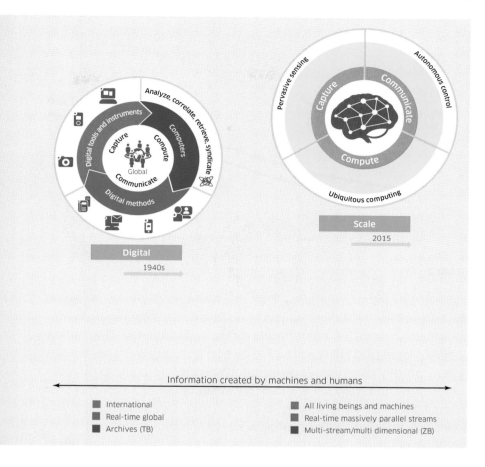

Information created by machines and humans

- ■ International
- ■ Real-time global
- ■ Archives (TB)

- ■ All living beings and machines
- ■ Real-time massively parallel streams
- ■ Multi-stream/multi dimensional (ZB)

consumer goods pricing data and personal preference data are all freely available at little or no cost. This has essentially been driven by the new economic deal offered by the web: so-called free web services in exchange for personal data and behavioral tracking (through clickstreams and cookies). When this is combined with the ability to communicate data from anywhere to anywhere at low-cost, one of the fundamental barriers to the creation of information has been removed.

Computation of data

The second component of information creation is the computation and analysis of data to perform the "critical thinking" to address the problem of interest and creation of information by finding relationships and correlations in the data. Looking back at pre-history, an acknowledged expert typically accomplished this "computation" by using analysis and intuition derived from

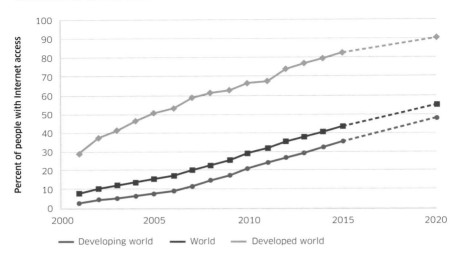

FIGURE 2: Growth in the proportion of people with Internet access (ITU 2015).

a limited data set to create simple heuristics that were enshrined as customs and folklore. The first true disruption in computation was the development of methods to formalize, automate and communicate expertise. The creation of mathematics (2000 BCE) and logic (500 BCE) enabled experts for the first time to formally share their expertise with others (figure 1) and to develop models of existence and the first real "knowledge" from information. But the "computation" during this period was an exclusively local process, performed by individual humans, with larger (parallel) computation scale requiring the physical movement of experts to the same location. In the absence of this, computation was inherently sequential, with individual local perspectives shared in non-real-time by mail and published documents.

Notably, the level of knowledge required to apply a formula or model is significantly less than the level of expertise required to devise such a model. Much like written language, mathematics allows critical expertise to be shifted and expanded in both time and space. A mathematical or computational model can be communicated and then used to facilitate understanding at another location and time. The same model can be used over and over (the Pythagorean theorem, dating from around 500 BCE is still in use today), greatly amplifying the ability to solve multiple similar problems and thereby accelerating understanding and knowledge creation.

Written language, mathematics, logic and elementary science and engineering, together with the spread of formal education, helped reduce the cost of critical thinking. They allowed models to be shared and used for problem solving in parallel, which led, in turn, to broader-based cognition. The most important advance came with the development of the new mechanical computing machines in the 1830s. These new computing devices invented by Babbage and others could tirelessly execute logic and model-based programs to produce new results at super-human speed, achieving an increase in *speed* of computation impossible with human effort alone (Computer 2015). The development of digital mainframe computers in the 1940s and the PC in the 1980s ushered in a new era that allowed every person with access to a network or a personal computing device to apply sophisticated models and algorithms to data sets, for any need they may have, whether personal or professional; the world was becoming cognitive.

Computing machines made global scale a possibility, but it took advances in computer operating systems, computer languages and the training of computer programmers to create scale in the implementation of software systems and applications that encoded these models into software that could be distributed and used by the global populace (Open 2012, Kernighan 1978). In essence, programming encoded expertise in a form consumable and potentially extendable by anyone with a basic education. Indeed, we now live in a time where "apps" seem to exist for every task (figure 3). This was only made possible by the power and simplicity

Data that was once privately held and sold at a premium is now exposed to everyone ... driven by the new economic deal offered by the web: so-called free web services in exchange for personal data and behavioral tracking.

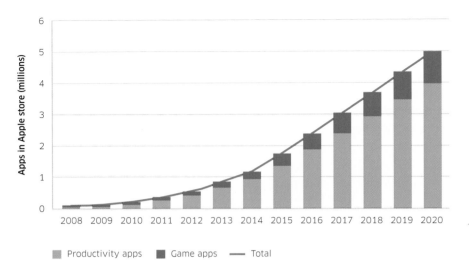

FIGURE 3: Growth in the number of mobile computing "apps" in the Apple app store[2]

of modern programming languages, the increase in computer literacy and higher level languages (such as HTML and JavaScript) and user interface and graphical design utilities. Children can now create web pages; teenagers create sophisticated web mash-ups; and young adults can create billion dollar software-based companies that disrupt legacy business models.

The most recent disruption in computing is the growth of cloud computing combined with smart devices and broadband access. This computing revolution has again removed the locality constraint on computing, due to the global distribution of cloud resources, and access to those resources from anywhere, by anyone with an Internet-connected device.

Given this availability of computing resources, and the ability to create applications to utilize these resources, models are required that form the basis for the computation. In some cases, the model is well known and easy to apply; for example, predicting hours of sunlight based upon location and date is a straightforward computation based upon historical data. Similarly, predicting the driving time between two cities based upon historical traffic data is easy to compute with a few assumptions and simple mathematics. In such cases as spam detection, handwriting analysis and natural language processing, our current understanding of these problems has not yet provided simple deterministic algorithms or models, although as humans, we flawlessly

2 - Modeling by Bell Labs Consulting using data from PocketGamer (PocketGamer 2015).

and effortlessly perform these tasks every day. This has led to a recent focus on sophisticated techniques such as *machine learning*.

Machine learning aims to develop flexible, parametric computational models to address these tasks without deterministic models. A limited set of training data is used to determine the values of the parameters and through this training process, the system "learns" the model. Once trained, a machine-learning model can be applied to new data for classification and prediction. One well-known application of machine learning, spam detection, uses models that have been created with a set of adjustable parameters. A set of known spam messages can be fed into the model and the parameters estimated; if the model is flexible enough, and the training data covers the messages of interest, then a highly effective classification model will result. Further, feedback from users designating certain messages as "junk" allows the model to adapt and improve over time.

In recent years, *deep learning* techniques (figure 4) have shown great promise in extending the reach and applicability of machine learning. Deep learning can be distinguished from machine learning based upon the number of parameters computed (millions) and the flexibility in the model in terms of the ability to adjust both parameters and the interconnectivity of the machine learning layers; a typical large-scale model might include millions of parameters expressed as adaptive sub-units arranged in a nearly limitless number of potential configurations (Schmidhuber 2015). Deep learning techniques based on artificial neural networks (ANN) use analogies with the human brain to define the sub-units and connectivity and consequently excel at applications such as speech recognition, image recognition or natural language processing. The widespread availability and low-cost of cloud computational resources has made such deep learning models economically possible.

One clear limitation in both machine learning and deep learning is the loss of insight into the operation of the trained model being applied. The model may excel at classification or prediction, but it may be nearly impossible to understand exactly how this success is achieved. A facial recognition system may learn to recognize an individual from a set of training images with a high success rate, but exposing exactly which image features enable that recognition is beyond the abilities of most deep learning systems. The recognition process is encoded in the interactions of millions of "neurons" in a manner that is not suitable for human understanding, so there is little capability to understand, optimize or modify any model and therefore acquire "knowledge".

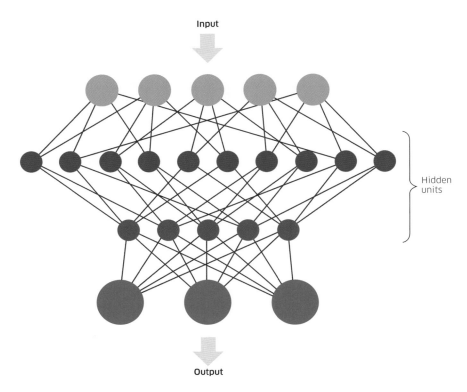

Input

Output

FIGURE 4: Artificial neural networks (ANN) provide the basis for many deep-learning techniques. Deep learning is distinguished from shallow learning by the number of processing steps or linkages between input and output quantities (Bell Labs).

It is tempting to view the recent progress in deep learning as a precursor to the development of true artificial intelligence. Indeed these models attempt to "think like humans". But machine learning arose from the artificial intelligence (AI) community as an application of AI techniques to more focused problems of practical interest, and this is where the majority of the effort will be – to solve specific problems – rather than generalizing the approach to solve nearly any problem, as a human brain can.

Capture of data

Experimental observation of the natural world dominated early science with scientists creating extensive tables of data and mechanical or physical tools to allow use of these observations to navigate or design or understand the physical world (Kosso 2011). Moreover, books containing tables of these of physical and elemental phenomena were published and distributed for widespread propagation of data in the pre-digital world.

Beyond scientific endeavors, enterprises historically dominated data capture, investing heavily to collect, manage and sell large volumes of data such as geographic, scientific, weather and financial data, which were all highly valued. But the data was captured locally, by manual recording or analog image capture, which limited the ability to distribute the data at scale and in parallel.

With the advent of the digital era in the 1940s and new digital capture technologies in the 1990s, such as photography, video and specialized sensors, there was a proliferation of data, as it no longer had to be manually reproduced and distributed in physical form. Again, the locality constraint was removed, and the ability to copy became limitless, instantaneous and with no loss of fidelity. This created an industry dedicated to the purchase and re-synthesis of data by companies such as Informa and Dun & Bradstreet for re-distribution to enterprises. However, this model is increasingly being challenged and/ or complemented with the rise of publicly accessible user-generated and government-sponsored data records (for example, in the US, www.data.gov), as well as the rise of user-generated social and behavioral data, which take us into the present era of big data, as discussed previously.

The distribution of information and knowledge via sites like Wikipedia (figure 5) has also changed how people access information and knowledge and critically think; what once required a reference library with reference tools such as the *Encyclopedia Britannica* in printed

Big data projects that are focused on a specific application are significantly more successful than big data efforts that warehouse data with the anticipation that an application will arise in the future.

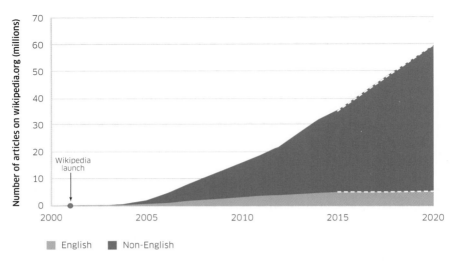

FIGURE 5: Growth of Wikipedia articles over time[3]

form to be purchased by every household at considerable cost, has now become freely available and constantly updated to reflect new knowledge.

The most recent, large-scale disruption in the area of capture is the proliferation of *user-generated content*. People and systems everywhere now have ready access to smart devices that allow the capture of pictures, video and sensory data with the ability to instantaneously upload it to a cloud-based store or sharing web service (for example, YouTube, Instagram). Indeed, the social media and networking platforms especially favored by Millennials actively encourage the widespread creation of data and shareable content. Furthermore, we are at or near the point where the reduction in the cost of sensors means that they can be incorporated into nearly every device, object or thing (vehicles, homes, wearables, infrastructure and environmental sensors), which is giving rise to the IoT. It will soon be possible to continuously capture and share data about seemingly everything, in every context.

The application of big data collection has led to several interesting applications in recent years. The most surprising of these applications typically arise from the use of big data in one domain to make predictions in an entirely different domain. One such example is the widely cited Google Flu project. The Centers for Disease Control monitored trends in flu-related search words and were able in this way to anticipate flu outbreaks (Mayer-Schönberger 2013). This is part of a larger trend in big data to use human-generated data, such as social media posts and hash

3 - Modeling by Bell Labs Consulting using data from Wikipedia.

tags, to monitor customer satisfaction, predict commercial success in marketing and quantify and map natural disasters. This trend leverages humans as sensors to map the world wherever we are active, connected and communicating.

In the enterprise context, similar applications exist on a much larger scale. The Weather Company and IBM have combined to provide nearly real-time microforecasts (Morgan 2015), allowing optimized decision making for a variety of applications impacted by weather from retailers, to utilities, to emergency management. Healthcare is another important area where big data is just beginning to bear fruit. Companies like 23andMe offer genetic testing services, and they combine these results with customer surveys to predict correlations between known genes and human characteristics. They have collected over one million tests with 80% of customers allowing these test results to be included in research studies, and as a result have contributed to hundreds of different studies in the last five years (23andMe 2015).

While long enthusiastic about the promise of big data, enterprises are beginning to recognize the limitations as well. Big data projects that are focused on a specific application are significantly more successful than big data efforts that warehouse data with the anticipation that an application will arise in the future. Indeed, a recent Cap Gemini report showed that 60% of executives felt that big data was important to their future business, but only 27% would indicate that their efforts to date were successful (Consultancy 2015).

The success of these analyses relies on three critical components: a data source, a relationship between this data and the quantity of interest, and an application of interest. The data is typically the easiest component to capture or obtain. In many cases the relationship is determined through a statistical search for correlated quantities. This can be powerful, requiring little actual knowledge of the data, but can equally be quite dangerous because correlation does not imply causation (merely because two properties change in the same pattern does not mean the changes in one property actually cause the changes in the other). Consequently, the determination of applications for which the correlation is meaningful will be the primary determinant of success for big data approaches.

The future

Data and information will become free

As discussed above, the cost of information derives from the cost of its constituent parts: communication, computation and capture. The Internet,

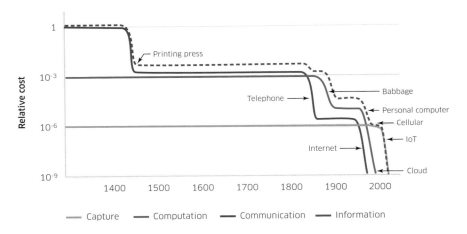

FIGURE 6: Calculation of the reduction in the cost of capture, computation and communication over time leading to the reduction in cost of information (Bell Labs)

the growth of cloud computing and the widespread sensing of the world (IoT) have all contributed to reducing the cost of the 3Cs and thus disrupting the cost of information, as illustrated in figure 6. We have modeled the relative cost of each dimension by estimating the capture, computation and communication costs of roughly one Gutenberg Bible worth of data. We estimate that there have been several thousand-fold impacts in relative cost in all three dimensions, with the result that the cost of obtaining information will soon decline to a point where it is vanishingly small per byte – in effect, free. This disruption represents a billion-fold decrease in the cost of information over 8,000 years and nearly a million-fold decrease just in the last 200 years.

Given the massive increase in the volume of data generated, however, we will still need even lower-cost capture, computation and communication than we enjoy today. In particular, given the constrained nature of the network highlighted in chapter 4, *The future of wide area networks*, and chapter 7, *The future of broadband access*, the connectivity required for computation and capture will depend heavily on the Future X Network, which will need to support greater responsiveness (lower latencies) as well as high volumes of small payload transmissions such as video sensing applications and services. In order to achieve the bandwidth scale and low-latency required, the network and cloud will become integrated and distributed to form the cloud-integrated network (CIN), described in chapter 5, *The future of the cloud*. As described in chapter 9, *The future of communications*, this new CIN architecture will make it possible to create and process new streaming data in real time. It will also make it possible, as described below, to discern the essential small-data set that can be further distributed through the local and global network. This distinction is critical: although the cost of big data collection, computation and communication will tend toward zero, the value of small data in the form of actionable knowledge will be immense, with price determined by the cost of the problem being addressed and the efficiencies that are created. We discuss this further below.

Big data will become small

Big data is an often used – though rarely defined – term, which typically refers to the collection, storage and manipulation of data sets too large for conventional data analysis tools. As conventional tools evolve, so too does the distinction between data and big data. One kilobyte was considered big data in the 1800s, one megabyte was big data in the 1960s, and today systems exist to manage multiple petabytes of data. Given this exponential growth rate, it is no surprise that the primary focus in this space has been on new scalable hardware and software technologies suitable for managing the rising flood of data from sensors, corporate operations and consumers. However, this big data quest overlooks the critical point that the *value of data does not grow proportionally with the size of the data*.

Humans have an intuitive feel for when they have "just enough" data to make an informed decision or to understand the nature of a situation. They intuitively understand that additional data does not always add additional information; the value of additional data depends strongly on the intended use of the data. We illustrate this in figure 7 with examples that show how the required data size is naturally limited by the application.

The first limitation is based upon mathematical fact; in a noise-free system, fitting a model that has N parameters requires only N independent observations. Similarly when sampling a bandwidth-limited function (of highest frequency f hertz), the Nyquist theorem states that only 2f measurements/second are required for a perfect reconstruction and that further measurement therefore has no value (Wikipedia 2015a). The introduction of noise in both situations can increase the number of required observations for a reliable model, but even in this case, the marginal value of new data decreases with increasing numbers of observations.

Statistical limits are similar; the sample size needed to achieve 95% certainty in a survey is relatively small – in the hundreds of respondents (Wikipedia 2015b).

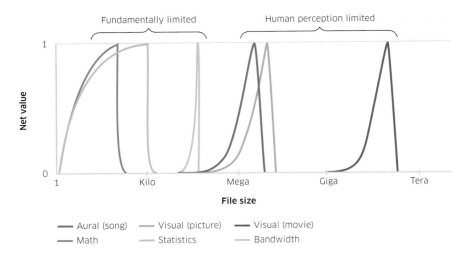

FIGURE 7: Illustration of the net value (computed by intrinsic value/opportunity cost) for increasing sizes of different data sets (Bell Labs)

This holds true independent of the number of people in the survey set. Certainly additional sampling will increase the accuracy of a result, but in most situations 95% accuracy is statistically "just enough".

Beyond the fundamentally limited data sets shown in figure 7, sets of perceptually limited data also exist, as humans have a finite ability to perceive the world. High-definition video, MP3 music, and JPEG images all exploit this fact to achieve the optimum balance between data size and human perception. Increasing the pixel density in a photo or the number of digital frequency samples in a song provides little additional value. In fact storing additional data beyond "just enough" quality actually incurs an opportunity cost relative to storing additional content – using twice as much data than required reduces the number of files stored by a factor of two.

In a general sense, the value of data derives from the knowledge or cognitive experience extracted from the data. This clearly suggests a future in which there will be a shift in focus from development of tools and technologies to manage large amounts of data, to the development of algorithms and methods that focus instead on creating and leveraging small data, by the creation of new models and the use of inference analyses (White 2012).

The rise of models for decisions

One way to formalize the idea of "just enough" expressed in the previous section is the development of a representational model for the data that allows both anticipation and prediction of future outcomes and data. As an example, rather than store the coordinates of all of the points that make up a line, it is optimal to just store the equation for that line. This provides two important advantages: first, it dramatically shrinks the amount of storage required; second, it allows for the prediction of data points not included in the original data set. This clear value in efficiency of cognition will drive the shift toward the development of representational models over big data storage, mining and correlation.

By analogy with the models developed in the physical sciences, which have enabled tremendous advances in understanding and optimization, in the future we will develop specific models tailored to the system being observed by using knowledge of the system to achieve high accuracy and fidelity. In particular, we will see the creation of quantitative models about how people and things communicate, how they interact and even how they think. These models will be used to optimize both personal and professional lives and to achieve the ultimate goal of "saving time".

The true value of a model derives from its application in decision-making, control and optimization, and the amplification of the utility and value of data. Consider that a *single bit* of data that contradicts a known model or expected result might be highly valuable as the effective trigger for a decision. Obtaining that single bit in time to make a critical decision is infinitely

> Consider that a single bit of data that contradicts a known model or expected result might be highly valuable as the effective trigger for a decision. Obtaining that single bit in time to make a critical decision is infinitely more important than the number of data bits collected.

more important than the number of data bits collected. This is the essence of the small data trend outlined earlier, which we believe will become the focus of future information gathering. It will employ sophisticated representational models that create new frameworks for control and optimization.

The rise of models for prediction and inference

As discussed in the previous section, the appropriate model paired with just the right small data set provides the ability to make decisions, but perhaps more importantly also provides the ability to make predictions about the future, often with higher precision than one can experimentally measure. This is certainly true of the models devised in the physical sciences. Given the cost in both time and resources to measure, transport and store observations, model-based inference will become a primary method for predicting new values or outcomes based on the smallest number of actual observations possible. This is the second application of small data – the capture and retention of the smallest amount of data necessary. This suggests a pre-processing application that resides in the edge cloud, as described in chapter 9, *The future of communications*, whose purpose is to analyze real-time streaming data and decide on the data set that should be retained and stored for usage by, and refinement of, existing and new models.

In fact, even limited models can often produce accurate predictions. People tend to believe that they are unique; however, only a few pieces of personal information can often reliably predict a multitude of facts about an individual. For example, tracking just the location (GPS coordinates) and emotional state (heart rate, galvanic skin response, etc.) of an individual might be enough to predict their socio-economic status, their job, their friends, the location of their home, their favorite content, when and where they eat, and so on. All of these things are challenging to measure, but can potentially be inferred from a small set of assumptions and a limited set of small data.

Intelligence will be augmented

In the preceding sections we describe a future based on free information and sophisticated representative models for prediction and decision-making. Enterprises have begun to leverage the current (nearly) free information to more effectively market to consumers, using a consumer's own information to optimize, control and personalize marketing of products and services. But the ultimate goal must be larger than simple marketing and advertising – it must be to save time, by creating knowledge that can be used to optimize any process, system or schedule.

From our definition of knowledge as discovering the meaning in information, we can define the five basic knowledge acquisition (or cognition) questions: who, what, when, where and why? These questions are an ideal starting point for assessing the relative value of information. Indeed, there has been a lot of recent attention given to the creation of automated assistants to provide answers to these basic questions. These assistants attempt to recognize spoken, natural language and perform relevant searches of freely available information. But there is a clear intent to extend the deep learning methods they use for natural language processing to become artificially intelligent personal assistants, such as Apple's "Siri" (Apple 2015).

But one of these questions – why? – is fundamentally different from the other four because the answer to a "why" question typically is not simply a database query; it implies some understanding or "knowledge" of a situation. "Who was at a location?" is easy to answer with a location-based search, but "Why were they at that location?" is a much more difficult question to answer without first understanding or analyzing the context as well as perhaps the history of known behaviors or models that describe the scenario – it requires cognitive reasoning.

Bloom's taxonomy provides an instructive mechanism to better understand the nature of cognition (figure 8). This taxonomy was created in 1956 to encourage the development of educational methods that focus on the development of higher-level, critical thinking skills in contrast with more traditional, rote learning methods that concentrated primarily on just remembering and recalling information (Bloom 1956, Anderson 2001). The hierarchy of skills described in the taxonomy can also be viewed as levels of difficulty – it is necessary to master the first levels before acquiring the higher levels. This simple diagram is also useful in describing the cognitive goals for augmenting intelligence.

The first level of Bloom's taxonomy, *remembering*, involves simply recalling information, such as an individual's ability to repeat facts or data. Each layer builds on this foundation adding increasingly sophisticated manipulation of data, ideas and information. The taxonomy progresses through *understanding* and *applying*, which seek to interpret and use data to create information that can be used in other contexts. The next levels of *analyzing* and *evaluating* concern an individual's ability to support informed, contextual judgments about information. Finally at the highest level of the taxonomy, *creating*, is the ability to synthesize new ideas and thoughts.

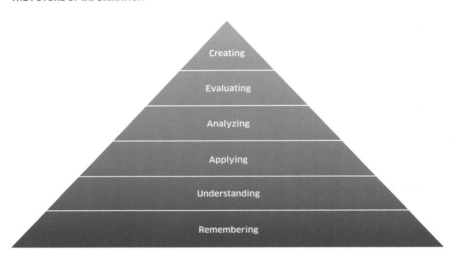

FIGURE 8: Bloom's taxonomy of higher-level thinking (Bloom, et al. 1956)

Using this taxonomy of higher level thinking and applying it to systems whose goal is to mimic, assist or enhance human capabilities, we see that existing automated assistants and information services focus almost entirely on the lowest level of thinking: recall or remembering. In a large part web search has become an extended memory providing real-time access to up-to-date answers to simple factual questions. For example, Apple's Siri performs very well when asked a question such as "Who is Benjamin Franklin?" or "Who invented the telephone?" but fails at answering questions such as "Why did Alexander Graham Bell invent the telephone?" As such, current search and personal assistant functions provide a firm *foundation* for thinking, but do not themselves support the higher levels of critical thinking.

Bloom's taxonomy can therefore provide a virtual roadmap for the future development of increasingly sophisticated thinking support tools. The ultimate assistant will be able to not only recall data but to actually synthesize or create knowledge and apply or learn models that provide cognitive assistance and support knowledge creation via critical thinking. However, the next logical evolution in intelligent assistant functions will be to focus on allowing people to better understand and apply information and answer questions.

It is instructive to come back to the nature of the question "why?" in the context of cognition. "Why?" is distinct from "who, what, when and where?" because forming an answer to "why?" begins at the second level of Bloom's taxonomy. It requires some understanding of the situation

	Big data	Machine learning	Artificial intelligence	Augmented intelligence
Focus	Large-scale data collection, manipulation and storage	Algorithms that can learn and make predictions from data	Thinking machines that can perceive the world and react	Enhance the ability of humans to think
Method	Scalable, distributed databases and data correlation techniques	Parameterized linear and nonlinear data models	Analogy: neural networks Mathematical: logic Knowledge-based	Computed similarity and relevancy provide connections
Key aspects	Scale	Flexibility and adaptability	Autonomy	Augmentation, personalization and interaction
Challenge	Volume, velocity, and veracity of data	Algorithm development	Fundamental advances in how we think	Algorithms and methods of interaction
Output	Pattern, correlation	Learning models	Automated intelligent agents	Agents participate as partner in user's thought process
Application	Business intelligence	Control and optimization	Robotics, control, automated agents	Personal assistants, "thinking" tools

TABLE 1: Comparison of data and information processing fields (Bell Labs)

rather than simply a response with a remembered fact. Such questions are relatively straightforward to answer for well-defined quantitative systems, but require real contextual understanding to answer for less well-defined, subjective systems, which suggests that the answer lies in augmenting human intelligence, rather than attempting to replicate or replace it. We explore this further in the next section.

Augmented not artificial

Most humans have experienced having seemingly all-knowing intelligent assistants that planned our schedules, attended to our needs, provided relevant context for decision-making and salient guidance at key moments. This is the vaunted role of parents, who provide this function during the early years of our lives. Their goal is to teach us to think for ourselves and become independent so that we can expand beyond our familial – and familiar – environment and knowledge base.

When we look at the history of humankind, we have developed many tools that provide mechanical, sensory, communications, memory or computational advantages (figure 9), but we have created very few tools that enable us to

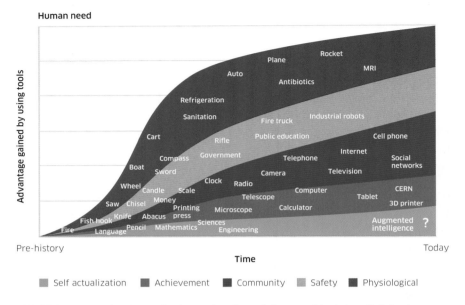

FIGURE 9: Human advantage gained over time through the use of tools to satisfy basic human needs (Bell Labs)

think better — just language, mathematics, logic, scientific and engineering concepts and written language.

In fact, many would say that the current thinking tool of choice is old-fashioned paper and pencil, which is still preferable to any electronic medium for quick exploration of ideas with mathematical, textual and/or visual elements. In a time period where it is possible to mechanically move things that are thousands of times larger, travel thousands of times faster and further, and see things millions of times smaller, we still use the same writing and drawing we have used since nearly the beginning of existence to improve our ability to think.

However, we believe that this is about to change with the advent of a new set of *augmented intelligence* (AugI) tools that enhance the thought process. The intent of these tools will be to present potential answers or possible paths for exploration, not to simply suggest an ordered list of answers. They will function as a personalized partner in the thinking process presenting relevant suggestions guided by access to external information and data, but more importantly guided by the user who will dynamically assess the relevance through interaction with the tool.

Such tools will use deterministic algorithms to compute and visualize the probability associated with different potential answers and help explain why an answer was suggested (as discussed above, answering the "why?"). This could be considered "meta-cognition", insofar as it supports the understanding of both an answer selected by the user and the thought process that results in that answer, which is a unique element of the augmented cognition approach. In addition, the supervised learning will enable an AugI tool to become more valuable with use, which is the ideal behavior of any such system.

These tools will draw on private and publicly available data – data about people, objects and the environment. Figure 10 is a representation of a data structure that encodes this information as a large-scale interconnected graph. Given the amount of data to represent (potentially trillions of nodes) the creation of such a "life graph" provides several advantages in terms of representation, computation and distribution of data.

Such a graph encodes information not only in the nodes of the graph, but also in the relationships between nodes as expressed by the edges of the graph, making it an ideal structure for encoding qualitative relationships that are difficult to express in other ways. Adding a new piece of information only requires an understanding of the relationship of that new information with a single other node on the graph; this single connection then provides adequate information to infer other potential (inferred) relationships on the graph.

A graph also provides ready access to a number of highly scalable graph algorithms for computing both local and global functions of the graph structure and contents. The approximation techniques developed over the last two decades provide speed computation (Kennedy 2013, Saniee 2015). But perhaps more important than the speed and efficiency of computation, these algorithms provide a visual intuition that maps local and global results directly onto human-understandable associations. The Google PageRank algorithm is probably the most famous large-scale graph algorithm. The page rank of a node is a mathematical operation on the graph (of web pages), and maps well to the user's intuitive view of most important or most relevant nodes (web page), so that the user is "impressed" by the relevance of the search results presented.

Last, graphs can be readily sub-divided without impacting the scaling of the underlying graph algorithms. This allows information relevant to a specific person or context to be managed and processed locally. A combined "life

...the methods for fully personalized search results require order N^2 operations or order N^2 storage where N is the number of objects in the database. This results in massive scale, which is cost-prohibitive when we anticipate having billions of such objects in the database.

graph" for everyone can be constructed and stored in the cloud, but for the vast majority of questions access to only a small fraction of that total graph is required, so hierarchical graph algorithms can be constructed that allow:

1. A first answer to be obtained from a local graph (stored on a smart device)

2. An improved answer based on connections to a larger graph (stored in a local edge cloud node)

3. A further improved answer using connections to the full graph (stored in a national data center)

Such progressive processing techniques allow fast, responsive tools to be created that leverage the power of enormous volumes of data and the future distributed cloud architecture.

Personalization

It should be apparent that when creating new critical thinking or augmented intelligence tools, personalization is a key goal. Most current web sites, such as Wikipedia, provide essentially the same information to all users. Google's search and mail applications employ rudimentary personalization by way of targeted advertising – based on a personal history of terms searched or content of personal email messages – but the actual results are still generic. The current generic search model places the effort to find the most relevant results in the hands of the user. With augmented intelligence, the tool will determine the relevant content for the specific user, not merely advertising. In other

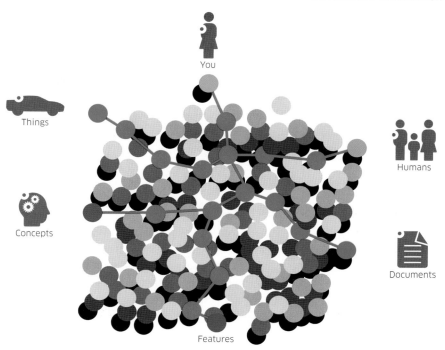

You

Things

Humans

Concepts

Documents

Features

FIGURE 10: Large-scale graph representation of the connections between humans, objects, documents, features and concepts. The large-scale graph stores relationships in the linkages between nodes. Scalable algorithms can be built that operate on this graph to provide valuable insights into these connections (Bell Labs).

words, one individual's search results will be different from the search results of an individual with a different background or searching in a different context.

Google's inability to do more in the way of personalization is related to scale; the methods for fully personalized search results require order N^2 operations or order N^2 storage where N is the number of objects in the database. This results in massive scale, which is cost-prohibitive when we anticipate having billions of such objects in the database. However, using a large-scale graph representation of the entire data set and novel algorithms, it will be possible to use inferred similarity to provide personalized search results with sub-linear complexity (figure 10).

Last, in addition to personalized results, it is also possible to use the same approach to explore the same data space but with the perspective of another – a "borrowed" – perspective. This will allow users to assess how other individuals understand the world and learn based on these borrowed perspectives, which are different to their own.

Interactivity

Since augmentation is by definition about adding to an existing process, it requires a degree of interactivity. To enhance and not distract from the thinking process, this interaction will need to be carefully designed. It will need to be intuitive and allow navigation through large volumes of data while still providing detailed focus on just the critical data. We illustrate one such approach in figure 11. In this representation of the "answer", the object of interest is in the center, and the most similar objects are shown by size and distance from this origin. The user then visually inspects this set of organized "sub-answers" to discern the answer of the most relevance to them at a given point in time.

Using such a visualization schema, the user is in control of the query but the AugI layer augments their understanding and also allows the exploration of other answers. This is achieved by moving another item, or person to the focus in the center, and then the presentation is recomputed from that perspective (Lipsky 2015). In all cases, the "reasoning" is shown by visual "parsing" of the related items, which can also be selected (tapping or clicking) to show more information about a given item and its relationship with the focus item.

Many other visual representations are possible and will be developed over time. In particular, as neural networks and deep learning models are developed based upon the workings of the human brain, we expect similar advances in understanding of how to visually assist critical thinking that will further enhance the efficacy of AugI tools.

The future begins now

We are at the very beginning of this transformation in information processing and augmented cognition, so there is very little apparent in terms of commercial deployments of constituent elements of systems. But we have identified key technological enablers for this new reality that will be required for this vision to be realized:

Network – Information requires the capture, computation and communication of data and the cloud-integrated network will play a critical role in all three of these functions. The following networking trends are foundational:

- The embedded edge-cloud computing infrastructure created from a repurposing of existing infrastructure where possible (for example, hub office or central office) will allow high-scale processing of streaming data, as described in the chapter 9, *The future of communications*. Early indicators

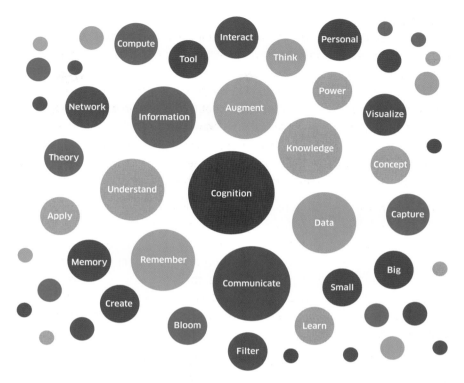

FIGURE 11: Interactive multi-dimensional display of information. Relationships between data elements can be encoded in position, size, color and proximity (Bell Labs).

of this trend are current network operator explorations of edge-cloud infrastructure and mobile edge-computing initiatives (MEC 2015).

- The access network will also be optimized for IoT devices that will be deployed en masse, allowing massive scale capture. The evolution of LTE to support so-called LTE-M (or Cat-M device) enhancements will be a key first step in 2016/7, but the full enablement of massive scale IoT will depend on the subsequent 5G standards and deployment, as discussed in chapter 6, *The future of wireless access*.

- The network will also be the source of critical location data that can be used to discern the co-location (and therefore similarity) of all connected people and things.

Data – The continued evolution of distributed cloud storage infrastructure, combined with the evolution of smart personal devices will allow creation of a distributed graphical representation of data, with support for progressive/partial download of relevant graph structure for the user's current context.

Tools – The recent rise of personal assistants such as Apple's Siri, Microsoft's Cortana and services like Google Now all provide a level of assistance in either generic queries (Siri), contextual queries (Cortana) and predictive schedule and travel alerts (Google Now). In addition, the set of partnerships that IBM is forging based on its Watson analytics research program is leading to new recommendation "assistant" applications, for example, for new recipes (IBM 2014). In addition, there is research ongoing into new user interfaces for information and knowledge visualization approaches (Lipsky 2015).

Summary

We have described a future in which data and information will become free and freely available, big data will be reduced to the smallest amount possible in order to discern the meaning of the information or "knowledge", and new augmented intelligence tools will be developed that assist in the acquisition of knowledge (cognition) by enabling critical thinking from multiple perspectives. This could represent the first breakthrough in human thinking methodologies since the discovery of elementary mathematical and scientific or engineering principles. Furthermore, new related visualization technologies may finally provide augmentation to the classical pen and paper techniques that have dominated idea exploration to date.

References

23andMe 2015. "23andMe fact sheet," *23andMe web site* (http://mediacenter.23andme.com/en-ca/fact-sheet).

Anderson, L.W., Krathwohl, D.R., Airasian, P.W., Cruikshank, K.A., Mayer, R.E., Pintrich, P.R., Raths, J., Wittrock, M.C., 2001. *A Taxonomy for Learning, Teaching, and Assessing: A revision of Bloom's Taxonomy of Educational Objectives*. Pearson, Allyn & Bacon.

Apple 2015. "Apple Siri," *Apple web site* (http://www.apple.com/ios/siri).

Bloom, B.S. (Ed.), Engelhart, M.D., Furst, E.J., Hill, W.H., Krathwohl, D.R., 1956. *Taxonomy of Educational Objectives, Handbook I: The Cognitive Domain*. David McKay Co Inc.

Computer History Museum 2015. "The Babbage Engine," *Computer History Museum web site* (http://www.computerhistory.org/babbage).

Consultancy 2015. "Capgemini Consulting: Big Data trends & challenges," *Consultancy.UK web site*, February 26 (http://www.consultancy.uk/news/1516/capgemini-consulting-big data-trends-challenges).

Eisenstein, E. L., 1983. *The Printing Revolution in Early Modern Europe*, Cambridge University Press.

IBM 2014. "IBM Cognitive Cooking," *IBM web site* (http://www.ibm.com/smarterplanet/us/ en/cognitivecooking).

ITU 2015. *World Telecommunication/ICT Indicators database 2015*, 19th edtion, June (http://www.itu.int/en/ITU-D/Statistics/Pages/publications/wtid.aspx).

Kennedy, W.S., Narayan, O., Saniee, I., 2013. *On the Hyperbolicity of Large-Scale Networks* July (http://arxiv.org/abs/1307.0031).

Kernighan, B. W., Ritchie, D. M., 1978. *The C Programming Language*, 1st ed., Prentice Hall.

Kivumbi 2011. "Difference Between Knowledge and Information," *DifferenceBetween.net*, September 6 (http://www.differencebetween.net/language/difference-between-knowledge-and-information).

Kosso, P., 2011. *A Summary of Scientific Method*, Springer.

Lipsky, J., 2015. "Bell Labs Bucks Short-Term Research Trend," *EE|Times*,

July 15 (http://www.eetimes.com/document.asp?doc_id=1327109&page_number=3).

Manyika, J., Chui, M., Brown, B., Bughin, J., Dobbs, R., Roxburgh, C., Hung Byers, A., 2011. "Big data: The next frontier for innovation, competition, and productivity," *McKinsey Global Institute Report*, May (http://www.mckinsey.com/insights/business_technology/ big_data_the_next_frontier_for_innovation).

Mayer-Schönberger, V. and Cukier, K., 2013. *Big Data: A Revolution That Will Transform How We Live, Work, and Think*, John Murray.

MEC 2015. *Mobile Edge Computing Congress web site* (http://meccongress.com).

Merriam-Webster 2015. "Information" (http://www.merriam-webster.com).

Morgan, L., 2015. "Big Data: 6 Real-Life Business Cases," *Information Week – Slideshows*, May 27 (http://www.informationweek.com/software/enterprise-applications/big data-6-real-life-business-cases/d/d-id/1320590?image_number=4).

Open Group 2012. "Unix past: history and timeline," *The Open Group web site* (http://www.unix.org/what_is_unix/history_timeline.html).

Oxford 2010. "Encapsulate," *The Oxford Dictionary of English*, Stevenson, A. (Ed.), Oxford University Press.

Oxford 2015. "Information contrasted with data," *Oxford English Dictionary* (www.oed.com).

PocketGamer 2015. "App Store Metrics," *PocketGamer.biz*, (http://www.pocketgamer.biz/metrics/app-store).

Saniee, I., 2015. "Scalable Algorithms for Large and Dynamic Networks: Reducing Big Data for Small Computations, *Bell Labs Technical Journal*, vol. 20, pp. 23-33, Wiley, July (http://ieeexplore.ieee.org/stamp/stamp.jsp?tp=&arnumber=7137721&tp=?ALU=LU1007233).

Sarkissian, J., 2001. "On Eagle's Wings: The Parkes Observatory's Support of the Apollo 11 Mission," *Publications of the Astronomical Society of Australia*, 18:3, February 1 (http://www.parkes.atnf.csiro.au/news_events/apollo11).

Schmidhuber, J., 2015. "Deep Learning in Neural Networks: An Overview", *Neural Networks*, vol. 61, January, pp. 85–117 (arxiv/pdf/1404.7828v4.pdf).

White, C. A., 2012. "The Promise of Small Data: Communications, Control, Context, Cognition," *Bell Labs Presentation*, September (http://ect.bell.labs.com/who/whitec/ ChristopherAWhite_SmallData.pdf).

Wikipedia 2014. "Metcalfe's Law," updated March 13 (https://en.wikipedia.org/wiki/Metcalfe's_law).

Wikipedia 2015a. "Nyquist-Shannon sampling Theorm," *Wikipedia*, updated July 13 (https://en.wikipedia.org/wiki/Nyquist%E2%80%93Shannon_sampling_theorem).

Wikipedia 2015b."Sample Size determination," *Wikipedia*, updated August 16 (https://en.wikipedia.org/wiki/Sample_size_determination

CHAPTER 11

THE FUTURE of
THE INTERNET of THINGS

Harish Viswanathan and Francis Mullany

The essential vision

Machine communication for monitoring and remote control predates the Internet, with supervisory control and data acquisition systems (SCADA) having been used to monitor and control machines since the 1960s. The deployment of devices for monitoring and control has accelerated in recent years, aided by ubiquitous wireless connectivity and declining communication costs. Nevertheless, until recently machines still communicated only with purpose-built applications and with limited data acquisition or analysis capabilities.

Connecting things to the Internet is a completely different proposition that is occurring on a massive scale. By bringing together a wide variety of devices and applications, there is the potential for a major transformation in how humans interact with the physical world, both in a personal and professional context. The future of the Internet of Things (IoT) can be viewed as bringing the physical world into the digital realm, "animating" physical objects to augment our existence and "saving time" by increasing knowledge and awareness. The essential ingredients of this vision are as follows:

1. *The dawn of the era of pervasive digital automation* – The world will begin to witness a revolution in human life comparable in scope to the Industrial Revolution and computerization. Sensing and control with global reach and

with real-time communication and analysis will enable the physical world to become a seamless part of a sentient digital whole.

2. *The network as a digital nervous system* – The connectivity network will become the nervous system upon which the automation is built. The network interfaces and architecture will be reimagined to provide the control plane and data plane connectivity and processing at the required cost, energy autonomy, responsiveness and reliability. The network will have the following characteristics:

- A massive increase in control-plane capacity to handle sporadic transmissions from a large number of new devices

- A significant change in the traffic characteristics with the proliferation of "short-burst" type communications, in addition to large-scale upstream video streaming

- Minimal consumption of device energy for sporadic communication so that devices with infrequent communication have battery lives in excess of 10 years

3. *The rise of augmented IoT solutions* – Future IoT solutions will include key functions such as augmented contextual analytics, device security verification and data privacy, as well as automated device management and control to enable the rapid deployment and easy operation of myriad devices and device capabilities in the new digital era.

In this chapter we explore these elements in detail in order to help provide a prescription for what promises to be a transformative economic reality – one in which our relationship with the physical world will become at once more global and more local, more remote and more interactive.

The past and present

The evolution of automation

Control of our environment, both natural and man-made, has been a human imperative from our earliest history. For much of that time, however, control over distances was severely limited, due primarily to the need for line-of-sight monitoring and to be within arm's reach for actuation and control. Mechanical automation in the nineteenth century enabled men and women to reach much further control processes, but the remote-control scale was still within a building or vehicle and monitoring was still mostly limited to direct visual observation. A good example of this was the railway network;

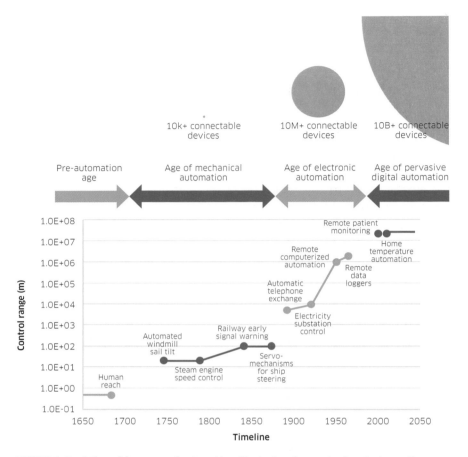

FIGURE 1: Depiction of four ages of automation, illustrating the reach of control over time (bottom) and the increase in the global number of connectable devices (top)

train track points were visually monitored and changed from a signal box that was always within sight.

The dawn of electrical communications technology vastly increased the scope of monitoring and control. The solutions were, however, bespoke and mostly confined to controlled environments with dedicated personnel. For example, the "engineer order telegraph" was the throttle signaling system between a ship's bridge and the engine room. True industrial automation had to wait until the post-World-War II era, where the combination of control theory, rudimentary computing and reliable digital communications networks enabled the first long-distance monitoring and actuation. Technologies like SCADA, proportional-integral-derivative (PID) controllers, embedded electronics and long-distance wired communication opened up a range of applications from vehicular traffic control and robotics in manufacturing, to electronically automated telephony exchanges.

However, these solutions were still tailored to a specific purpose, scaled poorly and expensive to install and run. Consequently, commercial implementations were limited to only those applications with the best return on investment (for example, control and automation of large-scale industrial plants or, at the other end of the scale, the use of smart meters for better managing electricity consumption).

The first wave of IoT

With advances in technology driving down device and communication costs, the vision of pervasive digital automation began to be realized, starting with those applications with high value to end users in specific industries. Inevitably, the initial applications were those in which critical resources needed to be managed much more efficiently. These can be broadly summarized as the following:

- *Energy costs in large buildings and homes* have increased dramatically over the last decade. The average annual cost of direct energy consumption for a UK household in 2012 was £1,250, up from £552 in 2004 (Telegraph 2012). Optimized comfort control and energy consumption to use off-peak power can minimize the excess costs.

... predictions abound on the number of IoT devices that will be deployed ... they all point in the same direction: an explosion in the number of devices to multiples of tens of billions of devices by 2020 and beyond.

- *Chronic (long-term) illnesses* are becoming a major economic burden, with 75% of US healthcare costs being consumed by the treatment and management of chronic diseases. This equates to roughly $8,000 per unhealthy individual per annum in 2007, with the key expense being the cost of highly skilled medical personnel and hospitalized care (GSK 2015). Automated management of chronic illnesses can enable the patient to reside at home for longer periods, thereby minimizing the use of premium and scarce hospital resources.

- *Water conservation* is now a fact of life in most developed economies, with the average water and wastewater bill for a UK household in 2010 being £340 (Offwat 2009). This is only being exacerbated by climate change events, such as the 2015 drought in California, which has reached unprecedented severity (Guzman 2015). Effective monitoring and control of all points in the water collection, delivery and consumption chain (and the associated trend analytics) would allow superior management of this increasingly precious resource.

The common need in each of these cases is to manage distributed infrastructure, connecting and controlling scarce resources, valuable processes and assets in order to increase productivity and save time.

The growth of IoT is the subject of much speculation, and predictions abound on the number of IoT devices that will be deployed. While the projected numbers vary, they all point in the same direction: an explosion in the number of devices to multiples of tens of billions of devices by 2020 and beyond. Figure 2 shows the total number of IoT devices across all economic sectors and communications technologies as predicted by Bell Labs Consulting. It also compares with other industry estimates, with both a conservative and an aggressive view proposed.. In either case, the exponential growth in IoT is immediately apparent, with more than 25 billion IoT devices expected by 2023. In the lower panel in figure 2, the number of devices connected via cellular technologies is projected. This is expected to grow rapidly to more than 3 billion devices by 2023, driven by the deployment simplicity offered by wide area connectivity – the absence of a need for a local access point as is the case for short-range radio technologies, the relative ease of configuration and management (SIM-based authentication, no passwords or user configuration required) and the intrinsic support for mobility.

The scaling of networks

Various IoT applications and devices have different connectivity requirements and thus a wide variety of networking technologies serve

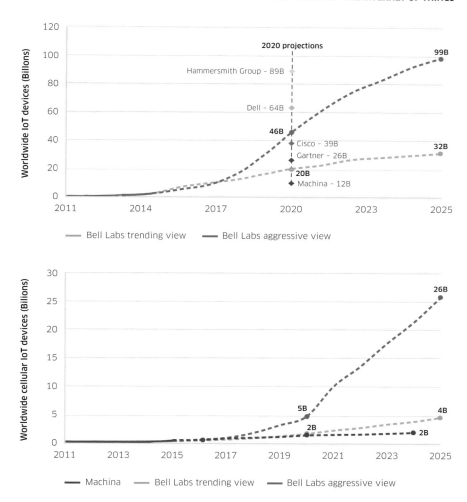

FIGURE 2: Projected worldwide growth in IoT: top chart is all devices; bottom, cellular only[1]

the needs of IoT. From a networking perspective, IoT applications can be categorized along two major dimensions: geographical spread and mobility. Geographical spread refers to whether the devices deployed are concentrated in a small area – within a few hundred feet of each other – or dispersed over a wide area. Mobility refers to whether the devices move around, and if so, whether they need to communicate while on the move. For localized devices, a short-range local area network such as Wi-Fi or Bluetooth is most appropriate as it allows the use of unlicensed spectrum and maximizes

1 - Data analysis by Bell Labs Consulting from various sources (Beecham 2008, Machina 2011a-e, Pyramid 2011, 2014).

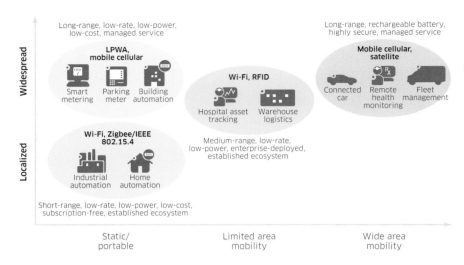

FIGURE 3: IoT application categories and network technologies they use

battery life while meeting the connectivity requirements. For applications such as connected cars, or freight tracking, the device moves over a wide area for which the mobile cellular network is the most suitable. In addition, for applications such as remote metering where the devices are widespread but there is little need for mobility, a wide area network is required but it does not have to support seamless mobility. Consequently, low-power wide area (LPWA) networks, which are specifically designed for low data rates and extended coverage with deep reach into buildings, have been deployed since 2013 with technologies that support long battery lives for the devices. Increasingly, mobile cellular networks that were designed for broadband traffic are also being optimized to address these non-mobile IoT applications by reducing device cost and extending coverage beyond the traditional cellular baseline. 5G systems are intended to make significant further improvements, as discussed in the next section.

Figure 3 roughly summarizes the application of different network technologies to IoT use cases in the market environment in 2015. Cellular networks are expanding from the purple category (widespread, wide area mobility) to the blue category (widespread, no mobility) and are also applied in some applications in the green and yellow categories because of the relative simplicity of deployment and management, as described above. However, there is renewed interest in the deployment of LPWA (blue bubble) due to the extended battery life and the low deployment costs. (LPWA radio towers use low data rates and

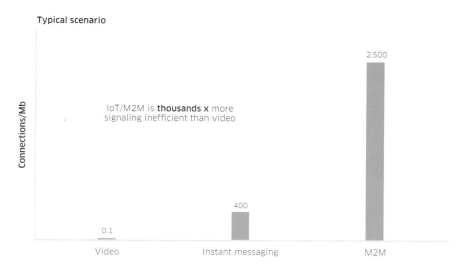

Typical scenario

FIGURE 4: Comparison of the number of connections required to transfer 1 MB of data. Each set up generates the same amount of signaling in the network.

can be situated as much as tens of kilometers away from the end device and therefore serve a very wide area with little infrastructure investment.)

A key additional attribute of many (non-video enabled) IoT devices is the substantially higher volume of signaling traffic that they generate, relative to data traffic. This is because such IoT devices typically transfer only a small amount of data in a given transaction, with each communication requiring signaling to set up the radio connection. Figure 4 shows the number of transactions or connections that are needed to consume 1 MB of data for video, messaging, and short-burst IoT transactions; three orders of magnitude more transactions are required compared to video applications, thus the network must be able to efficiently support significantly more control traffic than the number of attached devices.

The control software
Today the IoT ecosystem is characterized by significant market fragmentation. The wide diversity of end devices, coupled with the system complexity needed to deploy and operate end-to-end solutions, has resulted in purpose-built solutions, optimized for each application or service. Each solution typically uses a proprietary communication protocol and incorporates a customized approach to device discovery, communication, security, management, diagnostics, analytics and enterprise back-office integration.

Figure 5 shows the number of different companies across the value chain in North America providing solutions for the same function, illustrating the vertical market specialization even though the functions in these domains are largely common across the segments. The result is not one, but an ensemble of competing ecosystems with non-interoperable technologies and limited economies of scale. This must be addressed in the future if we are to achieve the optimum price/performance metrics and support the rapid adoption of IoT solutions.

Since the beginning of this decade (2010), the industry has begun to address this fragmentation issue, recognizing that many of these functions are common across multiple IoT solutions in different vertical industries. Consequently, horizontal software platforms that perform such functions and expose common application development interfaces (APIs) for applications to be built using standard web technology (such as REST[2]) have recently emerged in the market. Figure 6 shows the increasing level of venture capital investment in IoT and in particular, such software platforms. In the future, we expect that software platforms will consolidate around a connectivity platform, a device and service management platform and a services creation support platform, as described in the next section.

The future
The future of IoT will be driven by technology advances in wireless networking, software platforms and device technologies. In this section, we capture some key innovations that we believe will be instrumental in shaping the future of IoT.

The future of IoT networking
Wireless communications has advanced significantly in the last few decades turning the Internet into a mobile Internet. We envision a future where multiple wireless technologies further evolve to efficiently and cost effectively connect the billions of things to the Internet. In this section we discuss how cellular, short-range, and low-power wireless technologies are being optimized to create massive-scale connectivity in an IoT world.

The evolution of LTE
Traditionally, wide-area wireless networks, such as LTE, have been focused on enabling wireless broadband connectivity with technology enhancements introduced over multiple releases targeting increased spectral efficiency, user throughput and system capacity. Once it became clear that IoT devices would form the next wave of wireless network growth, the 3GPP standards body began focusing on optimizing LTE standards for narrowband IoT devices, as well.

2 - Representational state transfer, sometimes REsT, an architecture for designing network applications.

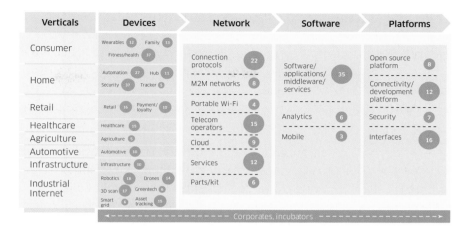

FIGURE 5: IoT industry fragmentation across vertical industry segments. Bubbles indicate the number of companies providing a specific function in the value chain

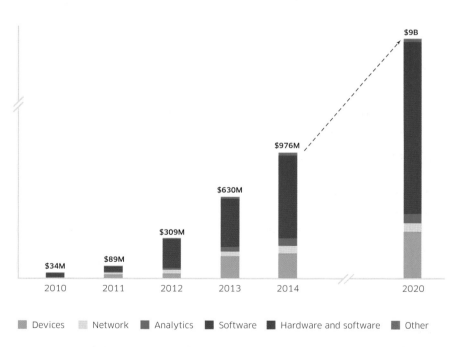

FIGURE 6: Cumulative venture capital investments in various IoT solution domains 2010–2014, with extrapolation to 2020[3]

3 - Data analysis by Bell Labs Consulting from various sources (Magee 2014, Postscapes 2015)

Enterprise-specific services such as virtual private networks, static IP addressing and more detailed call-data records are also more easily provisioned on a separate core network.

Consequently, since 2013, LTE has included features specific to IoT that:

- Significantly reduce device costs by defining limited capability device categories. These new devices operate in a narrow bandwidth of 1.4 MHz within the LTE carrier, support only low-peak rates compared to broadband devices, are single antenna without multiple input-multiple output (MIMO) support, and operate in half-duplex (only transmit or receive at any given time) mode.

- Extend coverage by 20 dB to reach devices deployed in the basements of buildings. This is primarily achieved by introducing repetition in physical signaling and data channels to increase the link budget.

- Improve battery life by introducing a new power-saving state in which the devices are in sleep mode for an extended period without monitoring the paging channel.

- Prevent congestion in the network because of IoT devices. This is achieved by creating a new "low-priority" indicator that can be sent by IoT devices or set in the device subscription to indicate to the network that scheduling these devices can be delayed under overload situations, and the network can selectively bar access to such low-priority devices.

- Include APIs for applications to trigger or send notifications to the devices. Short message service (SMS) is currently used to transport these triggers; alternate mechanisms are under consideration for standardization.

We expect deployment of LTE-M (containing additional IoT features) to begin in 2017 and continue for the next decade into the 5G releases.

In parallel to the evolution of radio access, we expect network functions virtualization (NFV) of the core network to play a major role in supporting the rapid expansion of IoT. Virtualization of mobile core network elements – the mobility management entity (MME), serving gateway (SGW) and packet network gateway (PGW) – allows operators to create dedicated networks for IoT cost effectively. These dedicated core networks for IoT have several advantages over a single network for all traffic types:

- Virtualization allows on-demand network scalability using available data center computing resources. This is particularly useful for IoT as the bulk provisioning of devices for a service and the correlated behavior of devices (in time and location) can result in a rapid increase in the number of devices in the network.

- Different IoT applications have different traffic profiles, especially with respect to the proportion of signaling to data traffic. As noted earlier, many IoT applications have a disproportionately large amount of signaling compared to consumer broadband devices. A virtualized core network element such as PGW or gateway GPRS support node (GGSN) can be optimized for IoT by allowing for more bearer contexts, more signaling traffic to be processed and less packet throughput (bearer radio resources) for the same amount of computing and memory resources compared to the corresponding instance for consumer traffic.

- Enterprise service level agreements (SLAs) for IoT may be quite different from those for consumer services. So, network operators will need to expose parts of their evolved packet core (EPC) for automating on-boarding and diagnostics through enterprise portals. Having a dedicated network or a network slice restricts this exposure to allow enterprise control to enterprise devices only. Other enterprise-specific services such as virtual private networks, static IP addressing and more detailed call-data records are also more easily provisioned on a separate core network.

- A dedicated core or network slice provides isolation across different sets of devices, thus preventing failure of network services to all devices when certain network functions fail. In particular, with IoT devices initiating connections in an automated fashion, anomalous device behavior could result in unexpected traffic scenarios that can impair the network; separation of this traffic onto a separate core network helps to isolate the larger network from such occurrences.

vEPC for IoT

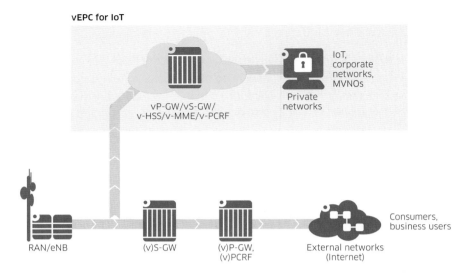

FIGURE 7: Network with separate virtual evolved packet core functions for IoT

Figure 7 illustrates the concept of a virtualized core network for IoT operating in parallel with the physical "consumer" network and sharing the same radio access layer. The set of separate core network functions deployed for IoT could range from only the virtualized packet network gateway (vP-GW) to all functions of the EPC.

The evolution to 5G

The 5G system is expected to be further optimized for IoT devices and improve significantly on battery life and network efficiency compared to LTE, as discussed in detail in chapter 6, *The future of wireless access*. The basic framework, such as the signaling and radio waveform in LTE, was designed for long-lived data flows, such as video streaming and web services, and is not optimized for small packet transfer between IoT devices and their applications. Thus the extent to which LTE can be optimized for IoT while being backwards compatible is limited. Without the constraint of complete backwards compatibility, 5G will be designed from the beginning not only for smartphones but also for IoT devices and thus will be better optimized for IoT. In particular, the fact that IoT involves significant number of short duration short-burst sessions will be taken into account to ensure small packet transfers do not generate significant amounts of network signaling.

We expect the following technologies to be part of 5G IoT:

- Asynchronous uplink communications using some filtered versions of orthogonal frequency division mutiplexing (OFDM) waveform such as the universal filtered OFDM (UF-OFDM) to enable devices to avoid the closed loop synchronization signaling that consumes control plane processing resources and reduces device battery life

- A "connectionless" mode of transmission in which devices can send user packets without establishing radio bearers, thereby minimizing excess signaling

- Radio resource allocation mechanisms designed to handle the transfer of a small number of packets from a very large number of devices

- Multiple symbol and frame durations to support different latency requirements efficiently – in particular, we expect support for very low-latency that will enable mission-critical IoT in vertical industry segments, such as public safety, smart grids and vehicle safety

The role of low-power wireless access (LPWA)

An alternative to mobile cellular networks for wide-area IoT are LPWA networks that are designed specifically to meet the communication requirements of low-throughput, stationary or nomadic IoT devices.[4] These networks are designed to cost little to deploy. They are typically based on proprietary technologies and are deployed in (small) slices of unlicensed spectrum in the low-frequency (<1 GHz) bands. As a result, they deliver only very low data rates in the order of a few 100 b/s. This enables low-power devices (for example, 14 dBm transmit power, which is about 9 dB less than that of typical cellular devices) to communicate over a range of tens of kilometers in suburban environments. In addition, the low-signaling overheads make it possible to achieve a 10-year battery life for devices and support a few transactions per day.

Given the suitability of these networks for a subset of IoT applications, we expect some adoption of LPWA networks by network operators and enterprises/industries, despite the following limitations of such approaches:

- Remote device management solutions, such as firmware upgrades, are not practical because of limited downlink capacity.

- Only a narrow range of IoT devices and applications can be addressed due to the very low data rate, insufficient downlink communication and no mobility support.

4 - For example, the Sigfox network is an LPWA network. Details can be found at www.sigfox.com.

- The inability to provide guaranteed service availability using unlicensed spectrum bands. Flexible sharing of licensed bandwidth between IoT and consumer devices is not possible and spectrum acquisition would be uneconomical.

- Some LPWA networks have poor support for localization of devices, so locale-specific analysis or spatial mapping of a measured value are not possible.

So we believe LWPA will remain a niche networking solution for IoT, especially if LTE-M and 5G can meet the promise of low-cost and long battery life for IoT devices.

The role of short-range local area networks

A large number of IoT applications involve devices that are deployed within the confines of a building or home, and involve sensors that are within a short range of a concentrator or a gateway device connected to the Internet. Therefore, such sensor devices only need a short-range (~100 feet) wireless connection to connect to this gateway. A number of short-range wireless technologies such as IEEE 802.11 (Wi-Fi), IEEE 802.15.4 (the MAC and PHY for technologies, such as ZiGigEe) and Bluetooth are being used to address the short-range communication needs for such devices. New technologies being standardized in several IEEE working groups are also expected to improve each of these existing short-range solutions in the future:

- An ultra-low-power physical layer with peak total device power consumption of less than 15 mW, allowing energy-harvesting devices to be used (IEEE 802.15.4 – discussed later in this chapter)

- Peer-to-peer direct device communication providing features such as discovery of peer information, with a scalable number of devices (up to 10) during discovery, and scalable data transmission rates of ~10 Mb/s

- A new version of Wi-Fi, 802.11h, for sensor devices operating in unlicensed bands below 1 GHz

The future of IoT software platforms

IoT solutions have a number of common functions across different applications and verticals. First, the devices have to be provisioned and activated in the network, which involves provisioning in systems such as the home subscriber server (HSS) in the case of a mobile network. Once the device is on the network, device management and diagnostics become critical for smooth operation of the device and associated service. In the data plane, protocol translation proxies that perform protocol conversion and perform routing to appropriate application servers may be needed to collect data and send commands to the device. Analytics and additional processing functions are also required in almost all applications to derive useful information from streams of data. The need for such common functions across different IoT use cases has led to the development of horizontal software functions that we believe will become increasingly important in the future of IoT, even if they do not achieve the "holy grail" of complete unification across all verticals. Some key emerging software platforms are described below.

The connectivity management platform

The primary function of connectivity management platforms is automation of the provisioning life cycle of the IoT devices in the mobile network, including SIM order management, activation, suspension, deactivation in the network and over-the-air provisioning. Specialized rating and billing functions for IoT are also sometimes included in the connectivity management platform.

An emerging new functionality is over-the-air personalization of so-called "white label" SIMs, also known as machine identification modules (MIMs) when used in a machine. Currently, device or machine manufacturers have to employ operator-specific embedded SIMs in the machines at the time of manufacturing or install removable SIMs at the time of shipping. This is cumbersome for multinationals such as car manufacturers that do not want their machines to be tied to a particular operator before it is sold to the end customer. They prefer to have the flexibility to choose the operator after the machine is deployed in the field, and also want the option of changing operators at a later time. In the solution standardized by the GSM Alliance (GSMA), the SIM cards can be provisioned with security credentials of a subscription manager that can then install operator credentials when the operator is selected. This approach is being called subscription management and will likely become a significant element of future IoT connectivity platforms.

The device and service management platform

To scale IoT applications to a large number of devices, it is essential to be able to remotely manage the devices, to troubleshoot and fix issues as they occur. This is particularly important for IoT deployments, since IoT devices are deployed by vertical enterprises, often with infrequent or no human involvement. Since truck rolls to the device location would make IoT services cost prohibitive, there must be a set of remote-management functions, including device configuration for new services or reconfiguration, remote device diagnostics, remote device reboots and firmware upgrades in the device. Such device and service management platforms will therefore be an integral part of future IoT solutions.

IoT has a dramatic effect on the volume, variety and rate of data processing as billions of devices sense and push information back at an ever-increasing rate. As video sensing takes hold, the volume of data will increase by several orders of magnitude.

The services creation support platform

The third emerging software platform cutting across several IoT verticals is a platform that enables the device data to be accessible to the Internet, permitting web application APIs to expose data, and allowing web services to enable rapid new service creation and usability on multiple end-user device platforms. These platforms will typically include the following:

- A common communications platform, abstracting different device types, and the protocols supported by different devices, and providing normalization of data so it can be used by multiple applications

- Enablement of cross-vertical and multi-tenant rapid deployment of new services, by providing a common services layer with catalogs of resource models and templates based on device, data and service primitives that are made available to applications through APIs

- Common application creation and execution capabilities that include tools for application creation and debugging; a set of simple, base applications and application templates; simulators and sand-boxes for testing; and utilities for data generation with real-time data visualization

A streaming analytics and edge cloud platform

One of the primary goals of IoT is the generation of business intelligence that

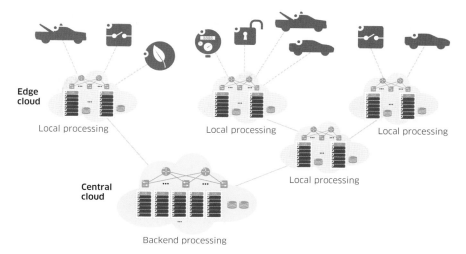

FIGURE 8: Future distributed streaming analytics architecture

can make business processes more efficient and save time. Big data analytics technologies are increasingly being used to extract useful information from myriad data sources accumulated as part of business operations. IoT has a dramatic effect on the volume, variety and rate of data processing as billions of devices sense and push information back at an ever-increasing rate. As video sensing takes hold, the volume of data will increase by several orders of magnitude. Increasingly, a combination of historical and real-time streaming analytics will be required for organizations to achieve resource optimization, to improve operational efficiency and to achieve real-time situational and contextual awareness.

The current practice of IoT data being transported to a central location for processing does not scale well and will not meet the real-time latency requirements of some key use cases. Processing of data in an *edge cloud* (small, highly distributed data centers) will be key to meeting the new requirements. The emerging stream processing frameworks will process the data as it is ingested and typically split analysis into independent modules that run on separate servers. This will allow parallel nodes to be added on the fly, scaling up processing resources as and where required. A future architecture is presented in figure 8; a hierarchical aggregation network is composed of edge clouds responsible for processing events originating from sensors in a specific geographical area. These edge clouds are physically located between the event sources (sensors) and the back-office (centralized) data center.

In this architecture, each edge cloud handles raw data streaming from a limited number of sensors to which they are directly connected. They pre-process events into a reduced data set that is forwarded to the central cloud – as much as possible, without losing information. In cases where real-time analytics can be performed by the edge cloud, avoiding the time delay of a round trip to the core network significantly reduces latencies.

The future of device technology

Pervasive digital automation will also be enabled by major leaps forward in device technology. Deployments of tens of billions of devices will only happen with devices that deliver new information and control functionality at a cost point significantly lower than that for low-end mobile handsets. We see six key technologies that will drive this revolution:

- *Integrated circuitry* – Moore's Law continues to play a critical role in increasing sophistication of control equipment and objects. Declining computing costs will enable sensing and automation over a broader range of processes or use cases, with low-power electronics that support extended battery lifetimes.

- *Micromechanical devices (MEMS)* – The rapid maturing of electronically controlled mechanics at the micrometer scale will allow for the control of a much wider range of small objects and components and the conversion of mechanical actions into an electronic signature. This increases both the scope of IoT (new objects connected) and the sophistication of the control enabled, such as in vehicular systems.

- *Miniaturized biological/chemical platforms* – The modes of interaction with the analog, physical world will proliferate beyond today's limited range of transducers and mechanical devices. Biological and chemical "lab-on-a-chip" devices, using technologies such as microfluidics and nanotechnology-based engineering, open up direct sensing and manipulation of biological systems, including the human body, in situ and with complete freedom of movement.

- *Device packaging* – Rapid miniaturization of these electronic and sensing components will result in the device size depending more on integration and packaging technologies. SiP (system-in-a-package) solutions and ultimately SoC (system-on-a-chip) solutions will be increasingly important, with the movement from 2D-chip integration to more flexible and compact 3D integration within the packaging giving rise to a significant reduction in device size. On a larger scale, flexible and stretchable printed circuits (Kim 2008)

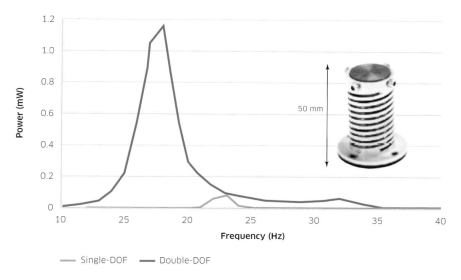

FIGURE 9: Compact, dual degrees-of-freedom vibration energy-harvesting technology from Bell Labs, resulting in an order of magnitude more energy harvested compared to state-of-the-art single degree-of-freedom solutions (Bell Labs)

open up new applications that will play an increasingly important role (for example, conformal devices for instrumentation of biological organs or 3D printing of simple devices and electronic/mechanical substrates).

- *Energy harvesting and storage* – Energy autonomy can be defined as the elimination of battery replacement over a device lifetime. This could result in a major reduction in the per-device operational costs of an IoT device. Ongoing advances in energy storage and harvesting (solar, vibrational or electromagnetic) will extend the effective lifetime of the energy supply beyond the device's lifetime. This will help drive the massive growth of IoT devices for which recharging is unfeasible or uneconomical. Figure 9 shows an example of a mechanical energy-harvesting device that could be used to power any device that has the requisite size and motion.

- *Virtualized devices* – For many objects, adding processing, memory or displays will be cost prohibitive. In such cases, only the essential sensing functionality and a wireless interface will be supported, and the remainder of the device logic will be virtualized and run in the cloud. Such device virtualization is described further in chapter 12, *The future of the home.*

Future devices and applications

While the pervasive digital automation brought about by IoT will be transformational in all walks of life, a few key applications and devices

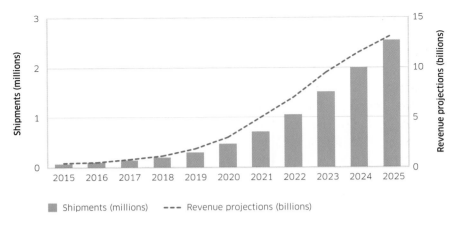

FIGURE 10: Bell Labs projection of growth of commercial drones. (Tractica, 2015)

warrant a separate discussion. By far the largest segment of the wide-area wireless IoT – the connected vehicle – is set to expand further with self-driving cars and drones; these two will likely dominate IoT in economic value and introduce substantially new requirements on communications networks.

The use of drones for commercial activities is growing rapidly. Mary Meeker in her "Internet Trends 2015" report (Meeker 2015) shows more than a doubling of consumer drones each year for the last two years, and a Bell Labs projection in figure 10 shows rapid growth of commercial drones over the next decade. Drones are being used in a variety of industries such as agriculture, oil and gas, utilities and for disaster relief. In many cases, the drones are used for aerial photography, and the data is then analyzed to extract information about faults or other abnormal situations. As drones become more common and more sophisticated, the communications requirements will also increase. We expect real-time streaming of high-resolution video from the drones, rather than the capture and store approach that is typical today. We also expect the need for fast, real-time drone-to-drone communication, as well drone-to-ground station communication. These will ensure optimization of flight paths and dynamic control from ground stations to modify mission objectives based on real-time processing of gathered data. The latter will be especially relevant for public-safety applications such as fire fighting and disaster recovery after major incidents.

Self-driving cars rely on a large number of onboard sensors and cameras to automate the process of driving. While today's autonomous cars do not assume any special infrastructure or vehicle-to-vehicle communication, major benefits will be realized from ultra-low-latency vehicle-to-vehicle and vehicle-to-roadside communication. The benefits will come from better use of roads and improved accident avoidance through "platooning" – organizing vehicles to be closely spaced with respect to each other like a platoon as they travel at high-speeds. Platooning requires low-latency, direct vehicle-to-vehicle communication and broadcast communication among vehicles to maintain appropriate vehicle speeds under various road conditions. Low-latency vehicle-to-vehicle communication also eliminates the need for street signals, thereby minimizing stopped times, and localized broadcast of emergency notifications between vehicles, which can help to significantly reduce accidents by warning drivers or control systems of the occurrence of an incident in their proximity.

Both drones and self-driving cars are among the new applications that call for low-latency and highly reliable mission-critical communication that will dominate our lives in the future, to allow the real-time remote control, warning or reconfiguration of these "digital things" to respond to their environment or address software, system or security issues, as discussed further in chapter 5, *The future of the cloud*. Similarly, remote robotic or industrial machine control and some critical infrastructure sensing applications (such as for an impending critical event)

The use of drones for commercial activities is growing rapidly. Mary Meeker in her "Internet Trends 2015" report shows more than a doubling of consumer drones each year for the last two years

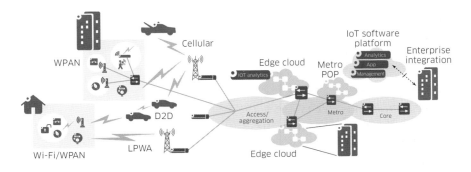

FIGURE 11: Future end-to-end architecture for IoT solutions.

are other examples of the new class of applications that will be enabled by ultra-low-latency and high-reliability networks.

The net effect will be the realization of a broad and affordable range of options for connecting and interfacing with the physical world. This will result in the proliferation of ubiquitous digital automation in a multitude of areas where we are only able to scratch the surface today, as imagined in figure 11.

Sensors in the home and office will be connected to gateway nodes through short-range wireless personal-area networks or Wi-Fi access points. Some low-mobility sensors will also be connected through long-range LPWA to remove the dependence on local Internet connectivity. High-mobility sensors and other outdoor sensors will be connected through cellular networks, and some Internet gateways (connecting mission-critical services) will also use cellular interfaces as backup for the wireline Internet connection. The access connection will frequently terminate in a local-edge cloud where real-time processing and streaming analytics will occur. The resulting data will then be routed through the wide area network to a more centralized cloud where the key IoT software platforms manage and control devices, expose APIs to web-based applications and integrate with enterprise applications and processes.

The future begins now

The automation of our physical world has been ongoing for more than 50 years, and communications networks have played an integral part in the connection of these automated systems to allow remote control over ever-increasing distances. Network operators, web companies, software and hardware developers, and enterprises in specific industry sectors are all currently investing in new technologies and capabilities and experimenting with new services and business models, proofs-of-concept and limited-scale trials, and even new commercial offerings. These industry activities span all three of the major technology areas, namely IoT network technologies, software platforms and device technologies, as well as the integration of IoT into business processes:

- Telecommunications equipment vendors have announced the availability of LTE-M technology for deployment in 2016. They have also announced the availability of a smart-band IoT device using the LTE-M technology (Le Maistre 2015).

- Worldwide, mobile network operators are partnering with automotive suppliers to create the "connected" car. In North America, a major operator has announced a simple device for the car that can turn the car into a Wi-Fi hotspot and allow Wi-Fi enabled devices in the car to connect to the Internet and consume media and other applications (Goldstein 2015). A number of IoT services such as remote car diagnostics, automatic emergency crash notification, stolen vehicle tracking, roadside assistance, locating the car in a parking lot, sending directions to the car, and starting the car remotely to warm up on a cold day are also provided to the connected car.

- A European operator and a pharmaceutical company announced a partnership to provide mobile health services to patients (Vodafone 2014).

Both drones and self-driving cars are among the new applications that call for low-latency and highly reliable mission-critical communication that will dominate our lives in the future.

The patients will be empowered to manage their conditions as a part of their daily routine using advanced technology.

- A startup network vendor acquired funding to deploy its LPWA network in the Belgian market (Sigfox 2015). In 2015, the company raised an additional 100 million euros in extra funding that is expected to finance network rollouts in the US, Latin America, Japan and South Korea following the signing of commercial agreements in these markets (Morris 2015).

- The mobile network operator forum, Next Generation Mobile Network (NGMN), has identified massive machine communication as a critical use case for 5G. In their white paper on 5G, they report that 5G should target connection densities of up to 200,000 devices per sq km in dense areas, battery life up to 15 years, low data rates of 1 to 100 kb/s and transaction latencies ranging from several seconds to hours.

- In 2015, the Federal Aviation Administration (FAA) approved over 500 companies to operate commercial drones in the United States. The drone applications include agriculture (mapping and assessment of crop damage), surveillance of pipes and other infrastructure in oil and gas, utilities and mining industries, and photography and videography for film production, real estate and mapping (Popper 2015).

- Innovation in wearables is also accelerating with numerous new devices targeting a range of applications across health, fitness, tracking and convenience. On the launch day of their range of smart watches, one of the leading device manufacturers also announced a software platform for large-scale eHealth studies enabled by the smart watch (New Scientist 2015). The instrumentation of the body will yield masses of public health data on a scale and speed never before attempted.

On software platforms, the following announcements suggest that the industry will transition to a more horizontal platform approach to building IoT solutions:

- A major vendor of IT hardware, software and services has announced a $3 billion investment in IoT technologies. Their investment is targeted at enabling the use of their data centers and sophisticated analytics software to crunch huge volumes of data streaming in from billions of connected devices (wind turbines, jet engines, building climate-monitoring systems, pacemakers and refrigerators) and encouraging application developers to use their platforms as a service upon which to build applications (Miller 2015).

- A major North American operator is opening their application development environment platform and services as part of an IoT developer program, to enable their customers to take their IoT applications from concept to production quickly and easily. The platform has been used for a wide range of vertical applications, such as farm-to-fork solutions, smart vending, mining site and equipment monitoring, building-energy management systems, medical device monitoring, and manufacturing site monitoring, demonstrating the horizontal nature of the platform (Verizon 2015).

- A leading European telecom vendor and Global M2M Association (GMA), an association of six tier-one operators worldwide, announced the deployment of the vendor's connectivity management platform for "multi-domestic" service. This platform incorporates the subscription-management solution that allows field-configurable embedded SIMs to operate in any of the operator's networks. In addition, the platform provides real-time connectivity management so that enterprises can manage, monitor, troubleshoot and support their connected devices operated globally from a single source (Allevan 2015).

Beyond individual industry segments, communities and cities are beginning to transform their infrastructure to become "smart". The term "smart city" refers to an urban area where basic infrastructure, utilities and the environment are instrumented using IoT technology and a cross-application platform is used on which local government operations, planning staff, utility providers, businesses and individual citizens can build services and applications.

- Santander in Spain is an example smart city that has taken a comprehensive approach to IoT spanning multiple dimensions of a city. As part of a European Union-supported Future Internet Research and Experimentation project named SmartSantander, an extensive network of 12,000 sensors over 8 square kilometers have been deployed with a comprehensive software platform opening up APIs to third-party application developers. The range of applications in Santander is impressive, targeting timesaving for individual citizens (car park availability), informing city planners about the environment (environmental monitoring), and managing resources by local government operations (control of street lighting).

- While SmartSantander is clearly at the level of a limited-area pilot, other cities such as Singapore are now embarking on full-scale smart city deployments, with even more ambitious plans for transforming services used on a day-to-day basis by its citizens and businesses. The city is planning to deploy a Smart Nation Platform (SNP) that will include a connectivity layer and an IoT layer to facilitate sensing and data collection within the city with a view to making the environment more intelligent and interactive. The government is working with an ecosystem of partners to ensure that devices and applications interoperate (Fai 2015, IDA 2015).

• As the world's premier sporting event, the Olympics have typically also been an opportunity for showcasing new technology, and the upcoming 2020 Olympics in Tokyo will be no exception. We can expect new IoT technology such as drones and self-driving cars equipped with five cameras, five laser sensors and an electronic map (Williams 2013) to be deployed to provide a phenomenal out-of-the-world experience for attendees. IoT technologies have already been used for several applications at the Sochi Olympics in 2014, including surveillance, facial recognition, digital signage and timekeeping (Ramsey 2014), and we can expect more at the Winter Olympics in Korea in 2018 and the Tokyo Olympics in 2020.

Summary

Humanity is on a never-ending journey to manage the world using technologies that save time and increase the ease with which we can perform tasks or acquire knowledge. In the panoply of human technological marvels, it is clear that pervasive digital automation of the physical world in the coming decade will likely loom large. The technologies needed to embed multiple dimensions of sensing and connectivity into billions of objects are maturing rapidly, with remarkable innovation around the globe. In addition, the set of networking innovations required to connect these digitally interfaced objects to the cloud, along with the software platforms and analytical techniques required to process streams (and reams) of data in real time and to turn it into valuable information and knowledge, are also being developed.

Further research and development is still needed to overcome remaining challenges in cost, massively increasing control plane capacity, dramatically increasing battery life and processing massive amounts of information in real time. But we have moved beyond the question of "if we will instrument and digitize our world?" to "how quickly?" We are entering the first phase of pervasive digitization, resulting in a sea change in how we monitor, control, live and work, using devices and things to "augment our existence" in every way imaginable. Geographic distance and economic circumstance will cease to matter and humanity will be changed in ways that make prior technological revolution waves seem more like ripples.

The technologies needed to embed multiple dimensions of sensing and connectivity into billions of objects are maturing rapidly, with remarkable innovation around the globe.

References

Allevan, M., 2015. "Ericsson, GMA to showcase 'Multi-Domestic Service' M2M platform at MWC," *Fierce Wireless Tech*, February 18 (http://www.fiercewireless.com/tech/story/ericsson-gma-showcase-multi-domestic-service-m2m-platform-mwc/2015-02-18).

Beecham 2008. *Beecham Internet of Things - Worldwide M2M Market forecast*, Beecham Research.

Fai, L. K., 2015. "Singapore's Smart Nation vision enters 'build' phase, focus on infrastructure, services," *Channel NewsAsia*, June 2 (http://www.channelnewsasia.com/news/singapore/singapore-s-smart-nation/1886696.html).

Goldstein, P., 2015. "AT&T working with connected car partners on exclusive content, games," *Fierce Wireless*, May 19 (http://www.fiercewireless.com/story/att-working-connected-car-partners-exclusive-content-games/2015-05-19).

GSK 2015. "The impact of chronic diseases on healthcare," *Triple Solution for a Healthier America* (http://www.forahealthieramerica.com/ds/impact-of-chronic-disease.html).

Guzman, Z., 2015. "The California drought is even worse than you think," *CNBC*, July 16 (http://www.cnbc.com/2015/07/16/the-california-drought-is-even-worse-than-you-think.html).

IDA Singapore 2015. "IoT and the Smart Nation," *Singapore Government iDA web site* (http://www.ida.gov.sg/Tech-Scene-News/Tech-News/Smart-Nation/2015/6/IoT-and-the-Smart-Nation).

Kim, D. -H., Kim, Ahn J.-H., Choi W.-M., Kim, H.-S., Kim, T.-H., Song, J., Huang, Y. Y., Zhuangjian, L., Chun, L. and Rogers, J. A., 2008. "Stretchable and Foldable Silicon Integrated Circuits," *Science 320*, pp. 507-511.

Le Maistre, R., 2015. "Huawei Promises 4.5G & LTE-M in 2016," *Light Reading*, February 24 (http://www.lightreading.com/data-center/cloud-strategies/huawei-promises-45g-and-lte-m-in-2016/d/d-id/713971).

Machina 2011a. M2M Comunications in Automotive 2010-2020, July.

Machina 2011b. M2M Communications in Healthcare 2010-2020, May.

Machina 2011c. M2M Communications in Consumer Electronics 2010-2020, July.

Machina 2011d. M2M Communications in Utilities 2010-2020, July.

Machina 2011e. M2M Global Forecast and Analysis 2010-20, October.

Magee, C., 2014. "VCs Look To The Future As IoT Investments Soar," *TechCrunch* November 5 (http://techcrunch.com/2014/11/05/vcs-look-to-the-future-as-iot-investments-soar).

Meeker, M., 2015. "Internet Trends 2015," KPCB web site, May 27 (http://www.kpcb.com/internet-trends).

Miller, P., 2015. "IBM Bets $3 Billion on Internet of Things Opportunity," *Forbes*, March 21 (http://www.forbes.com/sites/paulmiller/2015/03/31/ibm-bets-3-billion-on-internet-of-things-opportunity).

Morris, A., 2015. "French IoT start-up Sigfox raises $115M from Telefónica, others," *Fierce Wireless*, February 11 (http://www.fiercewireless.com/europe/story/report-french-iot-start-sigfox-raises-100m-telef-nica-others/2015-02-11).

New Scientist 2015. "Apple ResearchKit and Watch will boost health research", *New Scientist*, March 10 (https://www.newscientist.com/article/dn27123-apple-researchkit-and-watch-will-boost-health-research).

Offwat 2009. "Changes in average household bills since privatization," *Offwat web site* (http://www.ofwat.gov.uk/pricereview/pr09faqs/prs_faq_prcltssinceprivat).

Popper, B., 2015. "These are the first 500 companies allowed to fly drones over the US," *The Verge*, July 7 (http://www.theverge.com/2015/7/7/8883821/drone-search-engine-faa-approved-commercial-333-exemptions).

Postscapes 2015. "Internet of Things Investments," *Postscapes* (http://postscapes.com/internet-of-things-investment).

Pyramid Research 2011. *Mobile Data Total Region – December for NA, EMEA, CALA, APAC*.

Pyramid Research 2014. *Mobile Data Total Global – December*.

Seitz, P., 2015. "Commerical Drone Sales Set to Soar," *Investor Business Daily*, July 22 (http://news.investors.com/technology/072215-762954-drone-sales-forecast-2015-to-2025-from-tractica.htm).

Sigfox 2015. "ENGIE to Roll out SIGFOX Internet of Things Network in Belgium," *Sigfox Press Release*, June 16 (http://www.sigfox.com/en/#!/press-release/engie-to-roll-out-sigfox-internet-of-things-network-in-belgium-1).

Ramsey, R., 2014. "Seven Ways The Olympics Use M2M Technology in Sochi," *IoT Evolution*, February 21 (http://www.iotevolutionworld.com/m2m/articles/371044-seven-ways-olympics-use-m2m-technology-sochi.htm).

Telegraph 2012. "Energy costs in home rise five times faster than income," *The Telegraph* (http://www.iotevolutionworld.com/m2m/articles/371044-seven-ways-olympics-use-m2m-technology-sochi.htm).

Tractica 2015. "Commercial Drone Shipments to Surpass 2.6 Million Units Annually by 2025," *Tractica web site* (https://www.tractica.com/newsroom/press-releases/commercial-drone-shipments-to-surpass-2-6-million-units-annually-by-2025-according-to-tractica/).

UK 2014. *The Internet of Things: making the most of the Second Digital Revolution*, UK Government Chief Scientific Adviser.

Verizon 2015. "M2M Application Development Environment (M2M ADE), *Verizon web site* (https://m2mdeveloper.verizon.com/build/m2m-application-development-environment-m2m-ade-0).

Vodafone 2014. "Vodafone and AstraZeneca announce partnership for mHealth patient services," *Vodafone Press Release*, March 18 (http://www.vodafone.com/content/index/ media/vodafone-group-releases/2014/astrazeneca-mhealth.html).

Williams, M., 2013. "How Japan's Olympics will revolutionize tech," *ComputerWorld*, October 3 (http://www.computerworld.com/article/2485475/personal-technology/how-japan-s-olympics-will-revolutionize-tech.html).

CHAPTER 12

THE FUTURE
of THE HOME

Erwin Six and Jan Bouwen

The essential vision

Homes are unique. They define the occupants, their preferences and lifestyles. In addition to the purely functional aspects, homes also provide the boundary and interface to the rest of the world. They contain personal possessions, memories and familial relationships and they provide security and privacy for everything contained within them.

We are at a turning point in the history of homes. The coming proliferation of connected home devices driven by the Internet of Things (IoT), the increased intelligence of these devices and the advent of an increasing number of robotic devices and electronic assistants have the potential to change our home lives. But, only those technologies that meet a defining human need and augment and automate our existence will be adopted.

The IoT and augmented intelligence systems will create a third wave of home automation, following the information automation of the 1990s and the electromechanical automation of the 1950s that preceded it. Connected smart home objects will create more time by taking over the mundane and operational aspects of our lives, leveraging intelligence agents running in edge clouds. This will allow homes to become virtualized so they can be carried with us wherever we go. In essence, the home will become nomadic to match our increasingly nomadic digital existence.

This new home era will be defined by:

- *The ultra-sensory home:* Devices will have an improved ability to see, listen, talk, and move in natural ways, effectively transforming every home network into the central nervous system of our personal digital lives.

- *The simple secure home:* The configuration and management of the future home will be as simple and secure as the current physical and mechanical home.

- *The nomadic home:* A new virtual Home (vHome) environment will create a cyber-counterpart of our physical home that can be instantiated anywhere, with connectivity that replicates the physical space, programmable partitions between our personal and professional existence, and the ability to invite people into our vHomes.

- *New lifestyle services:* Instead of services built on islands of automation, such as home lighting, security, health monitoring applications and smart meters, lifestyle services will emerge that will take over more complex cognitive tasks to sense and adapt the home environment automatically to our needs.

This chapter outlines the major trends and the associated technologies that will have an impact on the future home, as well as the societal impact that will be enabled by an era of global digital nomadicity.

The IoT and augmented intelligence systems will make the home nomadic to match our increasingly nomadic digital existence.

The past and the present

After three decades of information-centric technology innovation in the home, the emerging IoT era promises to transform daily home life, continuing the transformation started by the electromechanical automation era in the 1950s.

The electromechanical home automation era

The history of labor-saving home innovations goes back millennia. But in the 1950s, the desire to accelerate and automate tasks led to widespread adoption of electromechanical automation. Power tools, such as drills and saws, and electromechanical machines, such as pumps, alleviated physical effort and accelerated the completion of household tasks. But, these innovations required manual operation and scheduling. Similarly, electromechanical appliances and devices, such as vacuum cleaners, washing machines and refrigerators, liberated people from many manual tasks. However, these innovations still depended on human intervention and cognition to move, guide and load/empty the device, as well as to determine when and where to perform the task. The introduction of the microwave oven in the 1960s and 1970s was the last significant innovation of the electromechanical automation era. It reduced the time needed to cook food, although with new cognition requirements: users had to know the required settings for each food type.

In effect, the result of all these electromechanical innovations was that the time spent on the physical task was saved, but cognitive time was largely unchanged or, in some cases, increased.

The information automation era

In the 1980s, a transition from electromechanical to electronic automation took place. The VCR, CD players and game consoles were the vanguards of a new era that shifted the focus away from manual housekeeping toward enhancement and automation of information and entertainment. The VCR allowed automated video recording for later viewing at a different location. The CD allowed greater portability of audio recordings because of the smaller format and increased mechanical stability during playback. And gaming systems introduced an enhanced form of playing games that could be transported easily to another place or connected to another system. Each new capability leveraged the time saved by electromechanical automation.

The adoption of broadband Internet in the 1990s moved the home into the first digital information age (figure 1). By enabling home access to the information contained on the World Wide Web, users could search for and access digital

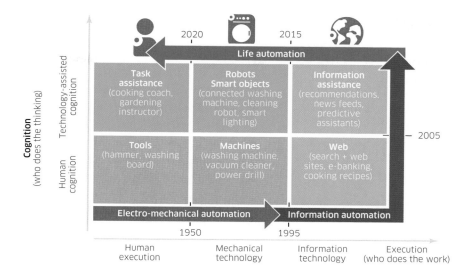

FIGURE 1: The evolution of automation eras in the home life.

information efficiently from home that previously required a physical excursion to a reference library or the purchase and shipping of a limited set of reference books. This allowed all information to be accessed from anywhere and achieved the goal of saving time in the quest for information.

In the late 1990s and early 2000s, the web evolved into a medium for commerce with the advent of online sellers, such as Amazon.com, and the emergence of online transactional services, such as banking. New online eCommerce services allowed users to save time by optimizing the search for a desired good, replacing the physical search in bricks and mortar stores, but at the expense of having to wait for the good to arrive. Online transaction services allowed users to save time and avoid the need to travel to a physical location to conduct many transactions, including financial transactions at a bank and paying bills at public utility offices. In addition, these tasks could be completed at the optimum time for the individual, rather than on a limited and fixed schedule of opening hours. And with the advent of mobile-optimized applications (apps), the optimum time was no longer location dependent, allowing users to conduct true anytime, anywhere transactions.

The information assistance era

During the pre-digital era of the previous century, the most common

Society is about to enter the era of life automation that follows naturally from the prior eras of electro-mechanical automation and information automation.

information assistance functions were directory assistance offered by telephony service providers, research support provided by librarians and personal support provided by administrative assistants. The goal of each of these information assistance functions was to accelerate the discovery of information (phone numbers, printed text, meeting and travel schedules) by use of an expert in these tasks.

Since the beginning of the web era, billions of web pages have been launched by millions of publishers who have created an unprecedented volume of accessible information to navigate. Discovery and selection has become an increasingly time-consuming task. This has driven the growth of new information assistance functions (figure 1), such as enhanced search capabilities (for example, using auto-complete based on browser cookies), recommendation engines based on buying behaviors, search terms and social-network data and, more recently, digital virtual assistants, such as Apple's Siri, Microsoft's Cortana, Google Now and IBM's Watson analytics engine. But, as discussed in chapter 10 *The future of information*, these services respond with answers in the form of data (facts) or information (the basic connection between data), but not knowledge (the meaning of the information). Therefore, they are not fully cognitive. They do not assist in the acquisition of knowledge as a true intelligent virtual assistant would, only the acquisition of the information that can be

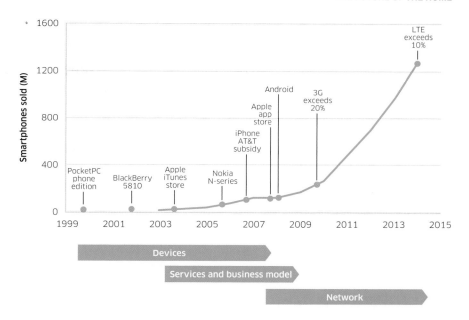

FIGURE 2: Adoption of smartphones by the mass market, along with key developments.[1]

used to glean knowledge. The acquisition of knowledge is still a human task.

But, as shown in figure 1, society is about to enter a new era – the era of life automation that follows naturally from the prior eras of electromechanical automation and information automation. In this new era, human cognition will be enhanced by new augmented intelligence functions, as described in chapter 10 *The future of information*, and by connected smart devices for the home that will lead to a significantly enhanced digitally-enabled and automated physical lifestyle.

Consider the present reality that will lead to this future nirvana.

It is estimated that smart homes will be less than 10% of the broadband connected homes in the US by 2020 (SNL Kagan 2014), based on current technological trends. To understand the factors that may limit the rate of adoption, it is instructive to look at the prior adoption of another transformative consumer technology: smartphones.

The smartphone's success and the growth of smartphone sales worldwide is built on a combination of key factors (figure 2). Although devices emerged

1 - Modeling by Bell Labs Consulting based on various data (Pyramid 2013, 2015; Infonetics 2013

in the early 2000s, the screens were hard to read, most required a stylus for interaction and memory was too limited to allow many applications to be stored on the devices. In addition, the wireless network had insufficient coverage and capacity to support mass adoption of high-bandwidth data services. Last, there was no new "life assisting" service to drive a market disruption. However, with the combined emergence of a compelling, powerful new device (the Apple iPhone), a new service and business model that made individual pieces of digital music available legally on these devices at an attractive price (the iTunes store) and the concomitant support of a new high capacity network (3G evolving to LTE) to connect the two, the adoption rate was unprecedented.

Looking ahead to the future of the consumer home, a similar set of requirements will be needed to drive change: a new service and business model, a new set of device or technological capabilities to support the service and a new networking regime to provide the required connectivity.

The future
The era of home life automation

We are now on the threshold of this future: a new era of "life automation" where the devices in the home will cooperate to provide cognitive digital assistance. In this future, the home environment will be virtualized and users will be able to recreate it anywhere to allow this digital assistance to be available everywhere.

The future home environment will leverage pervasive sensing to monitor location, status and usage of objects throughout the home and infer behavioral patterns from the collected data. Based on this information, it will proactively identify needs and coordinate the actuation of connected devices, appliances and information services, leveraging augmented intelligence tools.

The essential home assistance functions in this new era can be broadly categorized into four sets of services:

- *Infrastructure and environment management:* Energy and water consumption monitoring and automation, heating and air conditioning monitoring and control, lighting control, appliance control (including automated maintenance), object tracking and locating, access control (lock controls), security camera monitoring and control, smoke/CO_2/ Radon detector monitoring and control

- *Information and entertainment management:* Digital family scheduling and calendaring (with display on any device), contextual information delivery and control (news, social updates, alerts), universal home media control, with zero-touch sharing of information/media and controls with any defined virtual group

- *Housekeeping management:* Automated replenishment of consumables, automated cleaning, automated pet feeding, automated garden/yard management (watering and treatment automation)

- *Health management:* Personalized biometric data analysis from smart wearables, motion sensors, full spectrum sensors, medication usage sensors, child/infant monitors, accident detection, with characteristic signature analysis for each resident

In essence, the life automation era will create a *new service* – a virtual staff – comprised of a virtual housekeeper, security and maintenance service, gardener, pet sitter, IT/networking guru and media manager, the majority of which were previously unaffordable and inaccessible for the vast majority of humankind. This virtual staff will be created as virtual instances in the edge cloud and service chained together using software-defined networking (SDN) – *the new network* – as described in the following and in chapter 5 *The future of the cloud.*

And like the future enterprise network discussed in chapter 8 *The future of the enterprise LAN*, in which enterprises will embrace IoT for data-driven optimization of manufacturing, logistics and business processes, the future home will leverage the new IoT era – *the new devices* – to optimize familial and familiar daily routines and tasks. These devices will optimize execution of the mundane and assist in the learning and execution of new human tasks.

This essential shift has three constituent technological elements:

1. *A new augmented intelligence and task assistance function* that uses inference, model-based control and guidance, rather than simple human intuition and trial and error

2. *A device and network (r)evolution* to provide the basic technological infrastructure or fabric that will enable this new realm with simplicity and security

3. *A virtualization of home control* that allows interaction and recreation of the home space and digital content anywhere, rather than leaving it constrained to a single physical space, together with dynamic partitioning of resources into secure access groups for work and personal functions, rather than a monolithic fixed environment

These elements are examined in detail below.

Augmented intelligence and task assistance

Humans are creatures of habit with a strong predisposition to attempt to recognize patterns from small (or no) data sets – a behavior that is essentially the origin of what is considered intuition. However, in many cases humans see correlations where no causation exists or don't see causation where it does exist. In addition, humans are relatively poor at schedule management (too many variables to solve) and frequently in the repetition of mundane tasks, which we consider mindless (too simple to focus attention on). Consequently, a more accurate data and information-based function would be an invaluable aid to human behavioral cognition and the optimal automation of tasks.

As discussed in chapter 10 *The future of information*, this need will give rise to a new class of intelligent virtual assistant that uses inferred similarity and user feedback to create a complete set of intelligent, personalized and contextualized rules that can be executed by a task manager. For example, the hourly, daily or weekly movement vectors of each individual in the home could be stored and used as the basis of a set of recommended actions for home management of heating/cooling cycles, security, and the likely need for consumables, as well as knowledge of activity for each individual. The value of this type of augmented assistance is clearly the reason for the coupling of wearable devices to thermostats that is being pursued by some home and personal device makers (Bogard T. 2014).

An intriguing extension of this new reality is the possibility of having a physical task assisted by an augmented overlay of audio or video guidance or cues with the help of environmental and wearable sensors or augmented reality headsets. In essence, this is an extension of the current turn-by-turn directions provided by navigation applications to life at home. It would use inferred needs (the virtual staff function suggests a solution to a query) or a direct request for assistance. The new augmented messaging paradigm described in the chapter 9 *The future of communications*, could also be used to provide the controls required and communicate with the intelligent virtual assistant staff function. In addition, the media communications function of this platform will allow possible solutions to be continuously expanded by crowd-sourced and user-generated content that includes visual, audio and even sensory or tactile guidance using the appropriate set of new IoT-attached devices.

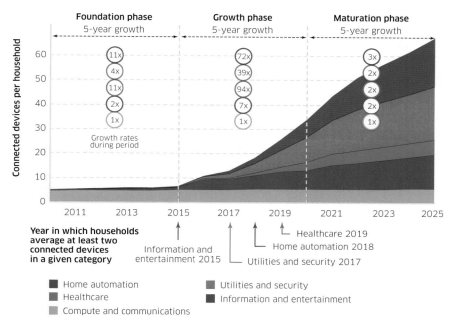

FIGURE 3: Evolution of connected home devices and the move toward synergistic value.

Device and network revolution
Device

Historically, the primary connected devices in a home have been the ones that were used for compute and communications purposes – personal computers, laptops, phones and tablets. However, this category of connected devices has already reached its near-saturation level of five to six devices per home as depicted in figure 3. But, the growth in broadband internet connectivity to the home and robust wireless coverage inside the home to support these devices has laid the foundation for seamless connections for a variety of other kinds of devices. Furthermore, once a home has more than one standalone device of a certain type, the synergistic value of connecting these devices will become evident, driving adoption of more connected devices in that category. This synergistic value is essentially a Metcalfe's Law effect for home networks, with the value of the network increasing as the square of the number of connected nodes (devices). Intelligent software can create even further synergistic value by connecting, correlating and controlling devices across various categories, which will realize the vision of a truly automated smart home. Therefore, it is likely that the growth in connected devices in a home will accelerate dramatically through 2025.

The future augmented and automated home will depend on the willingness of consumers to share private data with virtual assistants and associated services.

The initial driver of this trend will be further growth of smart media devices – homes already have more than one on average – such as smart TVs and game consoles that provide information and entertainment services. This will be followed by smart monitoring and regulating devices, like learning thermostats and surveillance systems that provide a utility and security type of service. This category is expected to cross the two devices per home threshold in 2016. Connected healthcare devices are also a potential growth driver, taking off beginning in 2019. But even as sensor-fitted disposables and smart wearables fuel part of the growth phase from 2015 to 2020, the real driver to maturity will be the gradual but steady replacement of more and more durable home appliances (for example, washing machines, refrigerators, etc.) by their smarter new versions that make fuller home automation a reality. This will start sometime around 2018.

However, device management will become a key issue. By 2020, an average home network could be comprised of 30 to 60 connected devices, effectively resembling in scale the IT networks of current small and medium enterprises. This will require network connectivity management and the use of different air interfaces by devices with varying degrees of mobility (inside and outside the home). In addition, the security and authentication of each device must be strong and verified perpetually given the critical and personal nature of digital life automation.

Network

Meanwhile, on the network side, the plethora of new devices being added to the home, each with different power, rate (bandwidth), reach and mobility requirements, has led to the proliferation of multiple networking technologies. Today, networking options include Wi-Fi access points, Bluetooth (and beacons), proprietary wireless, Li-Fi, Ethernet, external LTE, femtocells, ZiGigEe and others. Installation of each independent network and associated devices may be relatively simple for plug-and-play technologies, but several problems arise:

- Coordinating communications devices on the same network and especially over different networks is beyond the capability of the average consumer

- Managing hardware upgrades or firmware upgrades is also complex and, for some devices, requires consumer intervention and cooperation, which will create a barrier to upgrading capabilities and coordinating services over different generations and technology combinations of home networking infrastructure

- Managing the wide area network policy for each different device type and specific application to deliver the desired quality of experience (QoE) requires an understanding of the interplay of different devices and services. This must be adjusted dynamically as communications needs change and different devices are present/absent or active/inactive on the network

The solution, which is described in more detail below, is to leverage virtualization of the home device control plane functions and SDN to create a virtual home network that is managed and run in the edge cloud. With this approach, specific per-user service chains can be created based on the capabilities and composition of each user's home network infrastructure and the intra-home (LAN) and extra-home (WAN) communications requirements. This vHome approach also allows centralized management of service upgrades, as well as the creation of virtual instances of the home that can be accessed and controlled by any designated user from anywhere.

Simplicity and security

While this vision of future augmented and automated home management and task assistance is attractive, it has a critical dependency on the willingness of consumers to share private data with virtual assistants and associated services. This may be the biggest challenge in the realization of this vision because consumers have somewhat of a love-hate relationship

20%
benefits of smart
devices outweigh
any privacy concern

69%
people should
own personal
collected data

86%
afraid of
malicious use
of their data

FIGURE 4: Summary of user concerns for the smart, connected home.
(Ipsos MORI on behalf of TRUSTe 2015)

with new technology. People are eager to be assisted by technology that relieves them of repetitive tasks and assists in information discovery and entertainment delivery. But, as a recent US consumer survey showed, only 20% of consumers believe that the benefits of a smart, connected home outweigh privacy concerns (figure 4). The majority of those concerns are about the data collected by connected devices and the possibility of the malicious use of the data or infection of devices by malware.

These concerns are based on fear, uncertainty and doubt (FUD), rather than actual experience. But the following will be required to overcome this perceptual hurdle:

- *Full visibility:* The standard for data privacy for vHomes should be at least as strong as that for the physical home: only those permitted to enter the home should be allowed to "see" personal media and possessions, but no-one should be able to see personal data contained within the home or on home devices without permission. So, it will be essential to design vHome connectivity and services with full and simple to understand visibility of the data being captured and stored, where it is stored, and who has access to that data. The right to have data erased for any service should be absolute.

- *Home rule and overrule:* A simple programmable rules interface will be needed, which will provide templates users can use to determine whom they want to have access to vHome technologies, what these users will be able to see and control and under what circumstances. For example, the information gathered from health sensors might be shareable with family members or medical professionals only when an anomalous reading has occurred with no response from the individual. A similar schema might apply to security services and emergency responders. This is common today, but it must be replicated in a virtual services space and in a way

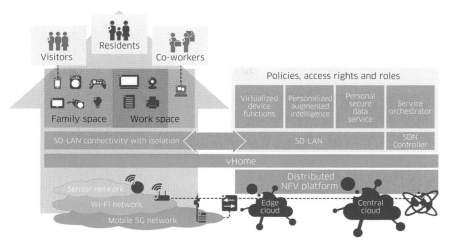

FIGURE 5: Potential vHome solution and architecture.

that is simple to program and control. There must be a separation of work and home control and device and content access schema. Users must also have the ability to allow physical guests in the space to connect their devices based on simple, selectable permissions that will enable interaction with existing home devices.

- *Trusted vHome provider:* The value of a vHome assistant that provides augmented intelligence to occupants was discussed above. When this highly personal service is combined with data privacy and security needs and support for a cloud service that allows recreation of the vHome environment in any location (discussed in more detail below), it is clear that a vHome service provider will be required to manage this service for the vast majority of people. Given the level of access and control that must be allowed in the most private of spaces, the level of trust with this service provider will be higher than that for any existing service.

The essential solution and architecture

Figure 5 illustrates the future vHome solution concept. The solution is based on the creation of virtual service instances that are instantiated in (edge) cloud infrastructure and service chained together to create the desired solution for each user, as described in chapter 5 *The future of the cloud*.

This vHome environment will host the software logic to manage and assist smart devices and servers in the physical home and make them accessible from anywhere. As a result, the home-state will be portable. The vHome service will manage and control separate instances for different user groups (work/office, home member/visitor). It will aggregate data for all connected home devices and allow sophisticated analysis to be performed to enable home life automation and augmented intelligent assistance.

The vHome solution will consist of a number of key technological components:

- *SDN connectivity:* SDN will be leveraged to create software-defined local area (SD-LAN) network connections between the home router function and the edge cloud vHome service platform (Boussard, Le Sauze, et al. 2015). It will also be used to define and enforce networking policies that determine which devices can communicate by creating service chains between the allowed elements and modifying them dynamically. Additionally, the fine-grained flow-capabilities of SDN will be used to monitor vHome service components, detect potential anomalies based on learned behavior of the constituent smart devices and provide security services, as described in chapter 3 *The future of security*.

- *Virtualized device functions:* Home devices will virtualize their software functions in a similar way as the virtualized network functions described for wide area networks in the chapters on *The future of wide area networks*, *The future of wireless access* and *The future of broadband access*. The physical layer technologies – displays, sensors, actuators, and basic Layer 1 functionality that interfaces with the physical world – will remain in the home. But the Layer 2 to 7 software logic will be hosted in the edge cloud to allow facile service creation and management (Marks 2015). In addition, device software upgrades to fix bugs and security flaws, as well as to add new functionality, will be simple to configure and manage.

- *Private secure data services:* The vHome edge cloud environment will aggregate and store the data collected by home devices, so data localization will be assured. Clear roles and access rights to data will be defined and enforced for family, visitors, work colleagues and third party services, such as healthcare, security, and virtual assistants. This will define which cloud services and devices have authorized data access and under which circumstances or contexts. The option will also exist for in-home storage where required.

- *Personalized augmented intelligence:* To simplify home operation and automation, intelligent virtual staff functions can be created in the edge cloud and added to service chains, as required. These assistants will be able to analyze the device data stored locally and provide cognitive support via complex data analysis on data aggregates and based on learned and known behaviors for each user and device type. In addition, new augmented overlays (for example, for audio or video assistance) can be instantiated in the edge cloud and connected to the vHome device service as needed.

These capabilities will address the critical network, device and security requirements of a vHome solution and provide the desired simplicity to support widespread adoption.

A key component of the overall vHome service offering will be the ability to define ad hoc vHome contexts or virtual spaces. These spaces will be spontaneously created by users and will consist of a subset of devices and associated cloud services for a specific use context, such as a family, work or e-health space. Users will have the ability to share virtual spaces with different people. For example, a family space can be used to share media content and devices (screens, media players, computing peripherals, etc.) based on different permissions granted for family and visitors. Likewise, a work space can be used to enable content devices to access an enterprise environment to allow remote working and collaboration via screen mirroring and sharing or a video conference service. And these spaces can be extended to include devices and cloud services owned by other participants. For example, a family space can include the smartphone of a visitor and a work space can include a family member's corporate laptop.

The realization of this ultimate new vHome paradigm will depend on the deployment of edge cloud facilities.

As these virtual spaces are instantiated in the cloud, they will be accessible from anywhere. The dynamic networking connectivity provided by the SDN controller will push new forwarding rules to allow such connections between the vHome and the authorized user location. When two such spaces are interconnected, the controller will also ensure that gateway functions are instantiated, both to assure seamless interworking between two devices even if protocols are different and/or to monitor the behavior of the devices in both spaces.

To understand the potential of this new vHome and vSpace reality, consider a scenario with a visitor wearing a healthcare monitoring device. As the visitor enters the home, the device is detected and identified. The vHome service discovers the device capabilities and associated services. The homeowner receives a notification on a smartphone and allows the home to grant permission for the device to access the home network with a recommended policy. The service orchestrator asks the SDN network controller to install the required networking policies to allow the device to connect to the local home infrastructure and to the device's healthcare monitoring service, as well as to any third party monitors, such as the devices of designated family members. If an event occurs, access to additional vHome functionality, such as video monitoring or conferencing using home screens and cameras, could be requested automatically by the healthcare monitoring service. If this occurs, the vHome service manages the request and alerts the homeowner to permit or deny the request for a limited time.

A myriad of other service scenarios can be imagined and will, no doubt, be created by a diverse group of home and IoT application providers. In this new reality, the trusted partner role of the vHome service provider will be absolutely vital.

The future is now

The (edge) cloud will play a critical role in the evolution of connected home devices shown in figure 3. As illustrated in figure 6, there has been an increasing reliance on cloud storage and backup of personal content and device configurations since the adoption of smartphones and tablets in 2010. In addition, with the introduction of HTML5 and web streaming technologies applications that used to run locally on devices are now moving to network-based hosted service models. For example, streaming audio services are replacing mp3 audio file downloads.

This shift is driven primarily by the desire to access more content than can be stored locally on a typical smart device, the ability to access content from any device,

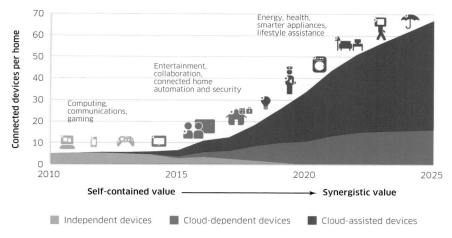

FIGURE 6: Evolution of connected home devices and the role of the (edge) cloud in assisting devices. (Machina, Beecham, Gartner, Bell Labs Consulting)

the interest in sampling and discovering new content without having to acquire a priori, and the recommendation engines that facilitate and automate this process.

Likewise, the personal computing market has moved toward cloud storage with services such as Google Drive, Dropbox and Microsoft's Office 365. And increasingly, networking and video services offered by service providers are becoming virtualized. This is an attempt to shift complex security, parental control, and Wi-Fi configuration and management into the network with virtual residential gateways (vRGWs) (Alcatel-Lucent 2015), or to minimize cost and allow facile feature evolution using a combination of smart TVs and virtual set-top-boxes (vSTBs) with cloud DVR services.

The next wave of cloudification will be driven by IoT devices, which will be designed to be small, low-cost and have long battery lives to minimize the need for proximity to charging sources, as described in chapter 11 *The future of the Internet of Things*. As a result, IoT devices will have relatively little processing power or local storage and will require cloud assistance to provide a complete service. For example, cloud support will be needed to store the data generated and support more advanced data analysis over longer timeframes. It will also be needed to support the analysis of data from different devices, such as that collected from a variety of e-health sensors, wearables and home motion sensors that will enable an application to infer user activity level and type. And it will be needed to provide guidance via intelligent assistants that support sophisticated behavioral learning algorithms.

In addition, there will be a significant growth in cloud-assisted smart home appliances, such as washing machines and dryers, refrigerators, dishwashers and heating and cooling systems. Due to their relative cost and connection to a power source, these devices will not be as limited in processing power as IoT devices. Cloud support will allow control and management of these devices remotely (value for the end user), as well as remote diagnostics and software upgrades (value for the device manufacturer and end user). And it will enable integration of the data captured by these appliances to be part of the larger vHome automation service.

As illustrated in Figure 6, this wave of IoT device and smart appliance deployments will require the majority of edge cloud services going forward. This is because of the highly personal nature of these devices, the need for low-latency and high-performance connectivity, and the need for strong data privacy and security at the per flow level. Consequently, the realization of this ultimate new vHome paradigm will depend on the rate of deployment of edge cloud facilities by web scale, IT or network service providers. However, the first phase of the smart home reality is already upon us. Table 1 shows a sample of connected home devices available in 2015, as well as how the cloud is assisting and enhancing basic device functionality.

Device	Local	Cloud assisted
Infrastructure and environment management		
Smart outlets, connected metering (for example, Belkin WeMo switches (Belkin 2015))	Electricity distribution and measurement	• Remote control – status check • Statistical consumption analysis
Washing machine (for example, LG SmartWasher (LG 2015))	Washing	• Diagnostics and troubleshooting guidance • Consumption statistics • Preventive maintenance scheduling
Security cameras (for example, Nest Cam (Nest 2015a))	Motion detection	• Storage • Advanced video analysis for security and personal use
Door/garage locks, doorbells (for example, August lock (August 2015))	Locking/unlocking User identification	• Remote control • Policy configuration • Connected to "if this, then that" rules
Thermostats (for example, Nest (Nest 2015b))	Temperature sensing Heating control	• Learning of presence patterns of residents • Energy consumption optimization

Device	Local	Cloud assisted
Light bulbs (for example, Philips Hue (Phillips 2015))	Lighting	• Programming, "if this, then that" rules • Preprogrammed configurations • Usage patterns • Remote control
vCPE vi (Alcatel-Lucent 2015b)	L1 and L2 connectivity	• NAT, firewalling, DHCP, parental control, Wi-Fi management
Information and entertainment management		
vSTB (Alcatel-Lucent 2015a)	Human computer interaction	• Cloud DVR
Media storage device (for example, DropBox (DropBox 2015), Google Drive (Google 2015))	<None>	• Cloud storage
Toys (for example, Cognitoys + IBM Watson (Cognitoys 2015))	Physical toy functionality	• Voice interaction
Smart bedroom and bathroom mirror (e.g. Samsung Mirror Displays (Business Wire 2015))	Display device (news, make-up info, virtual changing room)	• Information delivery
Housekeeping management		
Plant care (for example, Parrot Flowerpot (Parrot 2015), Edyn (Edyn 2015))	Sensors for humidity, temperature, sunlight, soil	• Statistics • Plant database • Watering prediction
Coffee machines (Amazon Dash Replenishment Services + Quirky Poppy (Amazon 2015))	Coffee	• Remote control and configuration • Coffee replenishment ordering
Vacuum cleaning (for example, iRobot Roomba (iRobot 2015))	Automated cleaning	• Indoor 3D mapping
Intelligent Pan (Pantelligent 2015)	Frying Temperature sensor	• Cooking recipes with timed temperature profiles and actions
Health management		
Smart weighing scale (for example, Withings Smart Body Analyser (Withings 2015))	Weighing, body fat measurement	• Fitness plans • Data storage and analysis
Mattress (for example, Luna Cover (Luna 2015))	Standard mattress function Personalized local temperature control	• Sleeping pattern analysis • Waking time optimization
Smart pacifier (BlueMaestro 2015)	Temperature and location sensors	• Temperature profiles • Movement alerts
Connected fitness devices (for example, SmartMat yoga mat (SmartMat 2015))	Position/movement measurement Instruction playout	• Instruction database • Progress monitoring • Social functions

TABLE 1: Summary of smart home services and devices available in 2015 and the role of the cloud in service assistance.

Summary

The future X network paradigm will extend into the home. Smart devices and augmented intelligence will bring a new wave of life automation to the home environment. This will be achieved by leveraging virtualization, SDN and the edge cloud architectural shift to enable the home to be virtualized. Ultimately, this will allow both intelligent life automation in the home and the recreation of the home anywhere. Partitioning of this vHome into virtual spaces will be enabled with different access permissions and policies. This will extend the concept of future X network slicing to be truly end-to-end. And it will enable a rich set of new services to be provided by global and local service providers.

References

Alcatel-Lucent 2015a. "IP Video Innovation: Seeing is Believing," *Alcatel-Lucent web site* (https://www.alcatel-lucent.com/solutions/ip-video).

Alcatel-Lucent 2015b. "Virtualized residential gateway,driving the delivery of enhanced residential services," *Alcatel-Lucent web site*, July (http://www2.alcatel-lucent.com/landing/virtualized-residential-gateway).

Amazon.com 2015. "Dash Replenishment Services," *Amazon web site* (https://www.amazon.com/oc/dash-replenishment-service).

August 2015. "A New Level Of Safety," *August web site* (http://august.com).

Belkin 2015. "WEMO Home Automation," *Belkin web site* http://www.belkin.com/us/Products/home-automation/c/wemo-home-automation/

BlueMaestro 2015. "Introducing Pacif-i–The Smart Pacifier." *Blue Maestro web site* (http://bluemaestro.com/pacifi-smart-pacifier).

Bogard T. 2014. "Jawbone Now Works with Nest," June (https://jawbone.com/blog/jawbone-up-works-with-nest).

Boussard M., Le Sauze N., et al. 2015. "Software-Defined LANs for Interconnected Smart Environments," 27th International Teletrafic Congress, September.

Business Wire 2015. "Samsung Display Introduces First Mirror and Transparent OLED Display Panel," June, (http://www.businesswire.com/news/home/20150609006775/en/Samsung-Display-Introduces-Mirror-Transparent-OLED-Display).

Cognitoys 2015. "Play and Learn," *Cognitoys web site* (https://cognitoys.com/#).

DropBox 2015. "Good Stuff Happens When Your Stuff Lives Here," *Dropbox web site* (https://www.dropbox.com).

Edyn 2015. "Keep a Good Thing Growing," *Edyn web site,* (https://edyn.com).

Google 2015. "A Safe Place for All Your Files," *Google web site* (https://www.google.com/drive).

Infonetics Research 2013. "Service Provider Capex, Open, Revenue, and Subscribers, Quarterly Worldwide and Regional Database: 4th Edition," December 18.

iRobot 2015. "iRobot Roomba Vacuum Cleaning Robot," *iRobot web site* (http://www.irobot.com/For-the-Home/Vacuum-Cleaning/Roomba.aspx).

LG 2015. "LG Smartwasher Saves Time and Money," *LG web site* (http://www.lg.com/au/smartwasher).

Luna 2015. "Wake Up Refreshed Every Day," *Luna web site* (http://lunasleep.com).

Marks A., 2015. "Virtual Home Gateways improve customer experience," June (http://www.alcatel-lucent.com/blog/2015/virtual-home-gateways-improve-customer-experience).

Nest 2015a. "See Your Home Away from Home," *Nest web site* (https://nest.com/camera/meet-nest-cam).

Nest 2015b. "The Brighter Way To Save Energy," *Nest web site* (https://nest.com/thermostat/meet-nest-thermostat).

Pantelligent 2015. "Cook Everything Perfectly, Every Time," *Pantelligent web site* (https://www.pantelligent.com/).

Parrot 2015. "Parrot Flower Power," *Parrot web site* (http://www.parrot.com/usa/products/flower-power).

Phillips 2015. "Lighting has Changed," *Phillips Hue web site* (http://www2.meethue.com).

Pyramid Research 2013. "Smartphone Forecast," February.

Pyramid Research 2015. "Smartphone Forecast," July.

SmartMat 2015. "The World's First Intelligent Yoga Mat," *Smart Mat web site* (https://www.smartmat.com).

SNL Kagan 2014. "Home Automation Picking up Steam," *SNL web site* (http://www.snl.com).

Withings 2015. "Smart Body Analyzer," *Withings web site* (http://www2.withings.com/eu/en/products/smart-body-analyzer).

CHAPTER 13

THE FUTURE of
NETWORK OPERATIONS

Deepak Kanwar and Danny Raz

The essential vision

For decades the evolution of network operations has been manifestly slower than the growth in electronics and network technology, which are largely driven by Moore's Law. Network operations are comprised of multiple dimensions: people, processes, platforms/tools and performance/ metrics, which we will collectively refer to as the "4Ps". The coordination and improvement of these dimensions is a complex undertaking with many human factors that are notoriously slow to evolve; so it is little surprise that the state of network operations lags far behind that of the technologically- driven evolution of the network.

Today's network operations are expected to be proactive in enabling a network service provider to cost-effectively provision and manage their network and services to meet the demand with the required economics before failures occur (Dragich, 2012). In the future, operations will need to be predictive in nature to manage the new demands of the "everything digital" era: the demands from enterprises and consumers in terms of the diversity of services, applications and connected end points (humans and machines) and the volume of data transported and processed, based on quantified future trends. We have to move from an approach in which network operations are the innovations bottleneck to one in which they are the competitive differentiator.

The following four key components of the future of network operations are needed to achieve this goal:

1. *Contextualization is the killer app* – In the emerging hyper-competitive services world every service experience will be personalized, accessible and easy to use. Consequently, operations that are network-centric today will become primarily customer-centric. Small data knowledge and machine learning will be leveraged to autonomously adapt the network to rapidly changing customer demands and service attributes.

2. *The challenge of seemingly infinite scalability* – The advent of software-defined networking (SDN), virtualization and growth of several orders of magnitude in demand for services as a result of the digitization of "everything" will require operations systems to have the following attributes:

   - Zero incremental cost to network operations while optimizing service delivery costs

   - Near instantaneous provisioning of new virtual resources and services, with network operations enabling innovation instead of being a bottleneck

   - A single seamless operations environment across network and data center (IT) operations

This shift toward agile, reliable and easy to maintain software-based services calls for a new model where operators and network equipment vendors work in much closer collaboration...

3. *Enabling global-local customer reach* – As massive-scale service providers build a global customer base, local access providers will provide complementary services critical to the global providers. New network operations will be required to enable this global-local customer reach, including:

   - The ability to dynamically interface to multiple local or global partners

   - Flexibility to adapt to local variations in language and culture, government regulations for data and privacy laws, and various local network technologies

   - A diverse partner ecosystem without compromised security

4. *Development for operations (not DevOps)* – Services will evolve from being largely hardware- to software-based. They will be deployed, updated or deleted in minutes or hours. Time to market, cost, reliability and operational simplicity of these software solutions will determine market success. This shift toward agile, reliable and easy to maintain software-based services calls for a new model where operators and network equipment vendors work in much closer collaboration, requiring a new development for operations (Dev-for-Ops) paradigm.

The underlying solutions in all four components increase the degree of automation, leveraging analytics that have been developed to improve maturity and discipline in operational processes. In the rest of the chapter, we will expand on these solutions, describing where we stand today and how to measure success on the journey to operational maturity.

The past and the present

Generic customer satisfaction kills

We are increasingly becoming an "experience economy", in which businesses that focus on customer experience financially outperform the competition (figure 1). With the barriers to entry shrinking and competition increasing, one of the primary differentiators for service providers is improving customer experience.

Service providers have invested heavily in building networks with five-nines reliability and outrank most other industries in customer care personnel, yet they have the lowest net promoter scores (NPS) across all major industries (figure 2).

It could be argued that given the complexity of building and maintaining multiple generations of (mobile and fixed) network infrastructure and the increasing dependence people place on connectivity as a "digital lifeline"

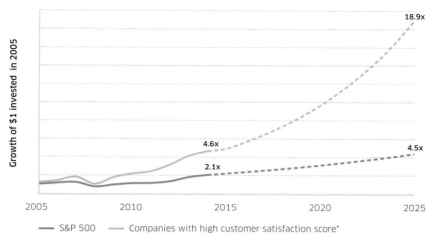

FIGURE 1. The value of customer satisfaction as judged by growth in valuation of different companies.[1]

that the telecom industry is different compared to other industries. Customers may simply have an unrealistically high expectation of service availability. Even if we focus on content services rather than basic connectivity services, we find that competitive media service providers have significantly higher NPS than similar services from connectivity and content service providers (figure 3).

The new, competitive media players such as Amazon and Netflix have done two things that may account for this disparity in NPS scores:

- Reduced the occurrence of service outages to near zero through continuous and relentless focus on operational improvements and sophisticated fail-over or self-healing mechanisms for their (unreliable) cloud infrastructure that hosts the service. This has minimized dissatisfaction with the service and nearly eliminated the need for a help desk.

- Highly personalized the service using big data analytics and sophisticated recommendation engines. Consequently, the more a consumer uses the service, the better the service becomes.

It is no doubt true that we are living in an era dominated by Millennial culture and values,[2] in which there is less intrinsic long-term loyalty and a

1 - Data analysis and modeling by Bell Labs Consulting from various sources (ACSI 2015, Morningstar 2015).
2 - Millennials are now the largest generation in the US labor force (Fry 2015b) and are set to pass the boomers as the largest generation in the US population by the end of 2015 (Fry 2015a).

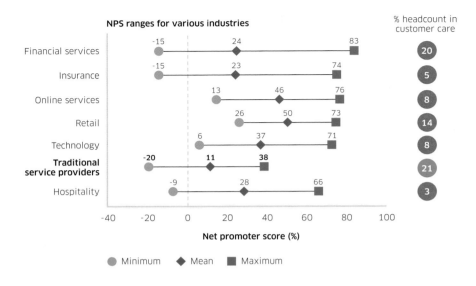

FIGURE 2: Relative customer experience versus expenditure of different industries (Satmetrix 2015, Beyond 2015).

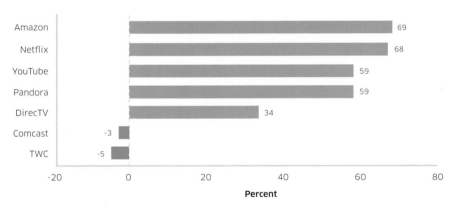

FIGURE 3: Net promoter score (NPS) benchmarks for multimedia service providers (NPS 2015).

positive mistrust for incumbency, which negatively impacts the traditional service providers. But this is unlikely to change in the coming decade(s), and so clearly a new relationship with service providers need to be established, based on a new customer experience paradigm.

The impact of finite scale

Network, traffic, services and devices have been growing at an exponential rate over the past few decades and the expectation is that this trend will

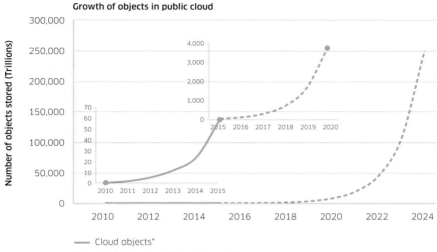

Growth of objects in public cloud

Cloud objects*

*Historical data based on AWS and Axure platforms

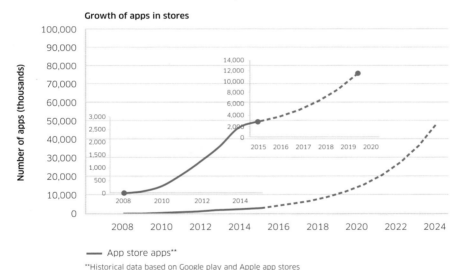

Growth of apps in stores

App store apps**

**Historical data based on Google play and Apple app stores

FIGURE 4: Successful innovation platforms have demonstrated the ability to scale by multiple orders of magnitude over a few years.[3]

continue to accelerate, as discussed in other chapters of this book – *The future of the cloud*, *The future of broadband*, *The future of wireless access*, and *The future of the Internet of Things*. Applications are being developed with increasing reliance on analytics for value differentiation. Network traffic patterns are more dynamic as the source of the traffic becomes increasingly

3 - Modeling by Bell Labs Consulting using data from various sources (Nelson 2014, Statista 2015).

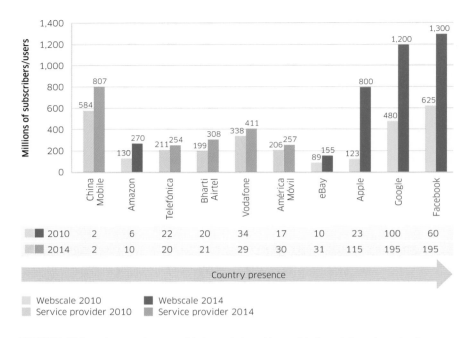

FIGURE 5: Web-scale operators are able to scale to achieve global reach based on cloud scaling rules.[4]

disparate. Service provider networks are struggling to scale due to the cost of support and operations processes that are too slow.

Figure 4 illustrates two examples of successful platforms that have supported this type of exponential growth – the scale of objects stored in Amazon Web Service and Microsoft Azure cloud infrastructure; and the scale of apps in the Apple and Google App stores. In preparation for the future where most of the network functions are virtualized and services are software-centric, service providers must demonstrate a similar ability to scale network capacity and the associated software platforms. In turn this will drive increased capability, agility and capacity requirements for network operations.

At the same time we are leaving the era of the "best effort" internet/web experience and entering an era where the network will become a digital fabric for life and business, service guarantees and service level-agreements (SLAs) will become stringent. This will drive the need for network operations systems that support dynamic scalability and service delivery optimization. Increasing levels of automation and analytics will also be required, as the

4 - Modeling by Bell Labs Consulting based data from Wikipedia, specific company annual reports and Statista (Statista 2015).

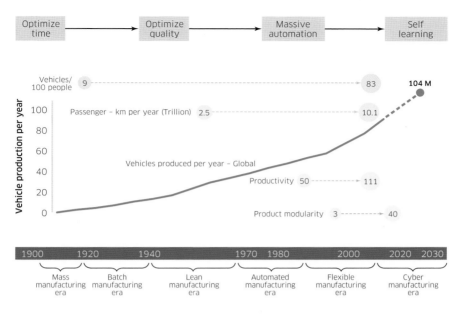

FIGURE 6: Operational simplification is the major driver of reduction in cost-to-performance ratio.[5]

complexity of the tasks required would be beyond what is possible by humans, even aided by computers.

Limited reach

Web-scale operators have leveraged their sophisticated data center scaling systems to achieve far greater global reach than traditional service providers, as shown in figure 5. Web-scale operators have typically built significant data center infrastructure and interconnect networks, but have been able to "parallelize" that task by using federated and distributed operation support systems to bring new global infrastructure online in a matter of a few years. In contrast, service providers have been extending their reach through mergers and acquisitions which intrinsically result in a complex web of dissimilar systems, interfaces and capabilities. With the need to save operational costs after the merger or acquisition, this often results in maintenance of separate systems with little unification or commonality and thus, excess operational costs and the inability to offer services with extended reach. The exceptions are the IP connectivity and TDM/voice connectivity services, which have been extensively standardized to attain a measure of "automatic" interoperability and therefore global reach.

5 - Modeling by Bell Labs Consulting base on various sources (Wikipedia 2015a, OST-R 2013, BLS 2015).

More specifically, the 4P challenges limit the ability to achieve global reach (with each challenge reflecting a lack of some form of standardization):

- *People:* Lack of standard roles and responsibilities between global and local operations, with no management imperative to unify them, and cultural opposition to doing so. Consequently, decision-making is a slow, multi-step process with an inclination to resist change and no clear accountability.

- *Process:* Deficiency in standard processes between local and global organizations. What is not standardized cannot be efficiently automated and, thus, requires large, but localized workforces without the capability to create a global service.

- *Platforms:* A need for standard platforms from country to country, and even within countries. This creates an integration tax each time there is a change in the network or service. Changes made in multiple places result in power-law increases in failure incidences.

- *Performance:* Lack of standard operational and experience metrics leads to incompatible management across different countries and areas. Operational metrics are often manually generated and computed but, without alignment on the methodology, the impact and imperative to improve is lost.

As we look toward the network of 2020 and beyond, we see the emergence of a global-local paradigm, which is described throughout this book. It is imperative for the new set of global service providers (GSPs), but also for local service providers (LSPs), that these limitations are removed (LSPS will at minimum require standard interfaces and metrics that they share with GSPs).

Separate development and operations

It is instructive to look at how industries have simplified, standardized and automated their processes over decades to achieve substantial gains in productivity. In figure 6 we show data for the automotive industry, whose focus has shifted from initial innovation to mass adoption. With standardization and automation, vehicle production per person has increased nearly 10 fold, even with more than a 10 fold increase in the number of components (modularity).

Continuous improvement in development, operations and maintenance processes and practices has played a key role in this overall transformation, as well as the sophisticated handling of an extensive network of parts suppliers and partners. The financial struggles of the automotive industry are well documented and served to catalyze this transformation; it truly was a case of adapt or die.

Since the break-up of the monopolies in the telecommunication industry in the 1990s, service providers have similarly relied on their suppliers – hardware and software vendors – to produce the essential network elements and associated software platforms, such as business and operations support systems (B/OSS). These same vendors (and others) provided network design, installation and integration services, and then the service providers customized the systems to their needs (according to the local market) and offered the relevant services to end customers. This separation of roles and responsibilities can work relatively well and lead to efficiency, as evidenced by the automotive industry.

However, a primary difference in the service provider case is that multiple legacy systems and networks (hundreds) have been deployed over the last 100 years and continue to provide a useful service. Each legacy system has to be maintained and integrated with each new additional system. This creates a massive complexity problem that is unlike almost any other industry. In most industries with similar lifetimes, the new systems quickly replace the old. Support for the old systems and products typically ends within a decade or two (at most), and even where older models persist (for example, cars), there is no interworking required between the older models and the newer models. They operate and can be maintained completely independently.

The persistence of interdependent systems for prolonged periods has resulted in the

A primary difference in the service provider case is that multiple legacy systems and networks (hundreds) have been deployed ... Each legacy system has to be maintained and integrated with each new additional system. This creates a massive complexity problem that is unlike almost any other industry.

cumulative complexity that has developed in service provider networks. This is exacerbated by the bespoke nature of the equipment systems provided by the vendors, each striving to differentiate itself in order to win market share and lock-in their service provider customers. Although the vendor solutions are standardized in terms of interfaces and protocols, their capabilities and roadmaps all differ sufficiently that integration of new vendors or new equipment into the network operations systems is onerous and often not worth the effort required, at least until a real technological disruption appears, typically once every decade or so in each networking domain. Even then, it is likely that a new management or operations system will be required for the new technology, given the history of the industry.

This is about to change with the advent of SDN and NFV. These new management and control systems have been architected with specifications that require support for the integration of elements using common methods and management, as well as a common orchestration framework. They rely on open source code (for example, OpenStack) and open interfaces, protocols and APIs (for example, protocols such as Yang, NetConf and BGP). These open standards allow federation of multiple controllers and simplified integration with network operations systems. The success of this open standards approach to SDN and NFV opens the door for a new approach to standardizing operations, which will require deep collaboration between vendors and network operator development teams, which we will discuss below.

The future

In the previous section, we discussed the historical and current state of network operations and the significant challenges that exist in network operations systems – in short it is an area ripe for disruption. Indeed, one could argue that without such change, the entire vision of the Future X Network will be left unrealized, as it will simply be too operationally complex and costly to implement. The change will require optimization of the 4Ps identified earlier, with the dominant P being people, as it is ultimately people that define the other Ps: processes, performance and platforms. The key change on the people front will be increasing the level of automation, taking people out of the minute-by-minute decision process and relying on a high degree of analytics with self/machine-learning and optimization to increase agility and scale. This will leave people more time to deal with complex exceptions and deal with increasingly improving the other Ps: process, platforms and performance.

Contextualization is the killer app

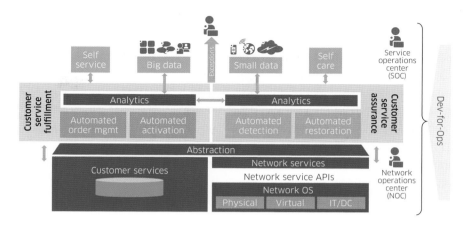

FIGURE 7: Network operations high-level architecture driven by automation and analytics.

To successfully compete in the new digital era, network service providers will need to support a high level of personalization and autonomy. From a network user perspective this will require that:

- Both enterprise users and consumers are able to rapidly explore new offers and modify their current services with an ability to choose any mix of pay-for-use or bulk-usage models depending on which payment solution best matches the contextualized needs of the user or enterprise.

- There should be no perceived service disruption or impairment as the network adapts to changes in demand, capabilities, services requirements, usage patterns or impairments and failures. Any issue that arises should be self-diagnosable and resolvable without the need for operator intervention. This is accomplished by providing tools and intelligent virtual assistants to solve problems without needing to engage a customer support technician (Amazon 2015).

This level of personalization and autonomy will require a new network operations architecture as shown in figure 7.

Consider a use case where an enterprise is experiencing a capacity overload (congestion) due to unanticipated demand for a service. They need to mitigate this immediately in order not to impact the service uptake and their external reputation or internal efficiency. The automated detection capability within the *customer service assurance* functions will be continuously analyzing the network state provided by the network operating system (OS) and will therefore be able to predict the congestion state before it occurs.

As described in chapter 4, *The future of wide area networks*, the network OS will provide an abstracted view of the network state that is relevant to the service in question. This allows an alert to be sent to the enterprise that a capacity threshold is being approached, together with potential remediation options. If the alert is sent as an augmented message (as described in chapter 9, *The future of communications*), the potential actions can be sent as "controls" that allow the enterprise user to directly re-program the network. The *customer service fulfillment* set of functions is then responsible for implementation of the network service modification and the monitoring of adherence to the new SLA.

This level of automation of network operations will be essential to support dynamic service creation, management and value optimization. A key component of this capability will be the use of data from publically available information sources, such as news, weather, infrastructure, entertainment or other large-scale events – often reflected in social media, for instance – as an input to the predictive analytics engine. This will allow historical, predictable trend data to be augmented with stochastic current event data to predict the likely network load in each location and each context.

Standard big data techniques and machine or deep-learning approaches are unlikely to be well-suited to the design and implementation of such automated, predictive network operations. As described in chapter 10 *The future of information*, correlation-based big data analytics do not contain or produce any insight on root causes or consequences. They do not have an underlying model, only statistics, and as a result the potential for incorrect predictions is too large (unbounded, in fact). Similarly, although deep learning has some potential for specific scenarios on which a system can be trained to create a heuristic model, the level of dynamism in the Future X Network makes it unlikely that such an approach will be broadly applicable.

Instead network operations needs to adopt an augmented intelligence approach, based on inference rather than correlation or model-based training, also described in chapter 10. This approach uses probabilistic outcomes that are mapped to policy-based rules, which, with human expert oversight to select the recommended action or option, represents the optimal approach. In addition, observed outcomes and user feedback will be incorporated to allow constant refinement of the predicted/inferred probabilities.

Looking forward, there will be one constant in network operations: change. The Future X Network will be subject to it constantly:

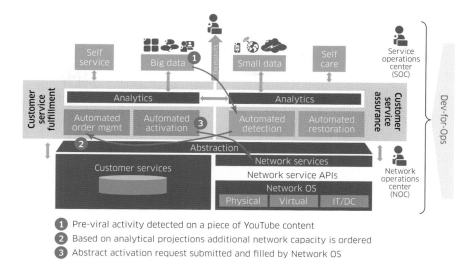

1. Pre-viral activity detected on a piece of YouTube content
2. Based on analytical projections additional network capacity is ordered
3. Abstract activation request submitted and filled by Network OS

FIGURE 8: Capacity for viral event driven by big data analysis.

- In network capabilities, as new virtualized functions and capabilities are added
- In services, as new applications are on-boarded and service-chained to create new service mash-ups
- In end points, as GSP and LSP customers (and connected things) are added or modified, or they move from one location to another

The entire network operations architecture will therefore need to be programmable to allow simple adaptation to change, but also to present a consistent set of abstracted interfaces that can be consumed and understood by the requesting user, service or system. The entire architecture also needs to be driven by a new development for operations paradigm, which is discussed below.

Seemingly infinite scalability

As discussed earlier, network service providers will need to emulate the capabilities of web-scale operators and implement platforms that provide seemingly infinite scalability. Figure 8 depicts a use case to illustrate how the proposed architecture provides network operations with infinite scale.

1. The automated detection capability and the corresponding analytics identifies that a video clip is starting to get a significant number of views; although the absolute numbers are low, the growth trajectory is very steep. Based on this trajectory, the predictive analytics engine generates

a set of high probabilities for a congestion event, which then generates an alert that more capacity will likely be needed in a specific region of the metro area network.

2. The automated order management function in the service fulfillment domain uses a policy rule to determine that action must be taken to alleviate the potential congestion.

3. The automated activation module generates a capacity request to the network OS that returns a set of options. The activation function selects the path based on the defined SLA and cost profile, and requests the path set-up from the network OS.

This simple use case can be expanded to include multiple cycles of capacity increases based on the actual usage and a subsequent reduction in capacity as the (viral) video interest diminishes.

This use case illustrates the three key elements network operators need to support seemingly infinite scale:

1. *The network operation centers (NOCs) will be unified* – With the advent of large-scale network virtualization and the seamless connectivity and integration of the wide-area and data-center networks the new cloud-integrated network (CIN) data center will need to cover both, as well as network and IT operations. The network state and service dependencies between the physical and virtualized network elements and IT infrastructure will also require seamless management across these domains. The development of the network OS is essential to create a unified, abstracted view of virtualized and physical resources and an end-to-end view of network state and resource availability.

   The unified NOC will manage both physical and virtual resources using the network OS to provide an end-to-end view of the network and across all resources, enabling the NOC to determine the root cause of events to avoid triggering actions from erroneous alarm signatures. The NOC will also take into consideration both the current and predicted state of the network, the physical equipment and virtual functions in the service path, as well as the virtual resources that can be utilized or instantiated to adapt the service as needed to address dynamically changing service needs.

   Through this end-to-end view and coordination with the NFV management and orchestration function, the unified NOC will be able to dynamically manage function and capacity, rapidly scaling virtual functions based on demand and spinning up new virtual functions to augment physical network functions that are approaching capacity limits. Tight coordination between the physical and virtual function domains will allow the location of network functions to be dynamically optimized, shifting traffic away from sub-optimally located functions (both physical and virtual) to virtual functions at locations in the network where they can be most efficiently utilized, for example, due to time of day variations in density of mobile users at various points in the network. It is also possible for a service to be maintained with the requisite quality of service as a user moves between physical and virtual resources, and both legacy and new infrastructure domains, which is critical both for user satisfaction but also for ensuring maximum use of prior infrastructure investments.

2. *The service operations center (SOC) will be the single customer interface* – Customer and service awareness becomes the focal point of operations, rather than just simple network connectivity or operator-owned service availability. Therefore, the SOC also becomes the primary interface into network operations.

3. *Trust in predictive automation will be uniquely important* – The Future X Network operates on a timescale and with a number of tunable parameters

Through this end-to-end view and coordination with the NFV management and orchestration function, the unified NOC will be able to dynamically manage function and capacity, rapidly scaling virtual functions based on demand and spinning up new virtual functions to augment physical network functions that are approaching capacity limits.

that exceeds what can be supported by human cognition. The basic performance and scaling of virtualized network systems has been verified, there will be a "transition of trust" to support increasing levels of automation, initially with human supervision, but with increasing levels of autonomy over time. Even the addition of new capabilities and services will be automated using a Dev-for-Ops approach, as described below.

Global-local customer scale

Given the challenges and opportunities discussed previously, GSPs will need to have the ability to seamlessly offer services though multiple local partners, as described in chapter 7, T*he future of broadband access*. Web-scale players are proving the global-local model with new services available instantly wherever they have local partners and the right to offer service. Similarly, LSPs will need to partner with one or more global providers in order to offer their subscriber base a competitive array of services.

To make this new global-local alliance paradigm a reality, service providers will need to operate in a secure, "open garden" environment. Standardized interactions in processes and platforms through abstract, open and secure APIs will ensure that service automations execute fluidly across partnership boundaries. SLAs on process and performance metrics will provide clear information on partner performance.

In addition, GSPs will need to define the aspects of a service that must be enforced or can be modified by an LSP through the use of defined policies that, for example, support:

- Local legal requirements, including regulation, content rights, privacy or obscenity laws. For example, video content offered by the GSP has a set of viewing rights that must be respected. As well, local obscenity laws must be respected and enforced by the LSP based on restrictions in each location.

- Local languages for user interfaces (including for self-care and self-provisioning). For example, a GSP will offer a suite of services to its LSP partner; the self-service interface used to explore these new services must be translated into the local language by the LSP.

- Local pricing incentives. For example, the GSP provides a base pricing structure for its service package to the LSP partner but with the option to offer a promotional discount in certain regions based on business aspirations and current market reality.

- Local variations of network technology capabilities. For example, the GSP offers ultra-high-definition video content streams to its subscribers as an option. The

LSP will flag whether this can be supported for subscribers in their region, which will allow adaptation of the offered content quality to the end user.

Many such scenarios can be imagined in the future that will require a flexible, federated set of service policies to be defined and implemented using a standard schema and APIs between GSPs and LSPs.

More generally, federation among network operators will become a business necessity and technical implementation of federated services over the complex virtual and physical infrastructure will be required. This will be done based on a well-defined network and services definition schema with associated sets of APIs and policies that define the capabilities and services supported by each provider's network. Service federation will be used as an additional tool to create seemingly infinite scalability by allowing the combination of resources between GSP and LSP networks. This new federated reality will also require a new approach to development of new services. The creation of new hybrid services will require higher levels of interaction between the development teams and the (multiple) operations teams, which can be facilitated by the approach described in the next section.

Dev-for-Ops (not DevOps)

In the software industry, DevOps is seen as a key methodology to simplify and accelerate production and innovation within a complex and dynamic software production environment by integrating the development and operations teams. However, the service provider environment has unique characteristics that require a different solution.

- Network service providers must provide services with high reliability and therefore be extremely fault tolerant. Historically, the recognition of communications services as a "lifeline" that was vital in emergency situations was the main driving force behind regulations that require high availability (99.999% uptime) of the services anywhere anytime. This is very different from many web services that are "nice to have" and thus can tolerate service downtime, although as discussed in chapter 9, *The future of communications*, this is also changing for some messaging platforms due to the central role they now play in human communications.

- Network services are dependent almost entirely on equipment and supporting software provided by third party vendors, which effectively prevents the full unification of the development team and the operations team – an essential requirement of the DevOps model.

At a high level, the Dev-for-Ops concept resembles the "design for manufacturing" engineering concept, where the design facilitates the manufacturing process to improve the quality and reduce cost.

Therefore a new innovative approach will be needed. Dev-for-Ops is an alternative innovative framework being defined that delivers agility, scalability, cost and reliability to deploying and operating network services in which the service is developed by the vendor but with a deep understanding of the operational environment and its life cycle management requirements based on operator-specific knowledge. As we described above, the increasing standardization and unification of network operations systems and interfaces will allow better collaboration among the operators. However, this will not eliminate the need for differentiation among the operators, very similar to the need for personalization of the end-user services.

At a high level, the Dev-for-Ops concept resembles the "design for manufacturing" engineering concept, where the design facilitates the manufacturing process to improve the quality and reduce cost. Design for manufacturing and its predecessors, such as Six Sigma, were adopted in many engineering disciplines including software engineering and have delivered valuable outcomes (Goss 2010).

In the current DevOps methodology (shown in the left side of figure 9), the application software is delivered from the development team to the operations team, which then deploys and operates it. The development team uses feedback from the operations team to generate the next software release using agile methodology, implementing the prioritized changes that are requested by operations.

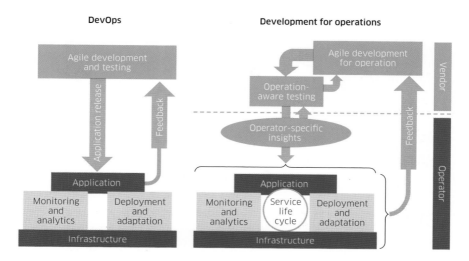

FIGURE 9. The difference between DevOps and Dev-for-Ops, illustrating how the collaboration between vendors and service providers is critical to Dev-for-Ops (Bell Labs).

The right side of figure 9 shows the Dev-for-Ops framework, including the dependencies between the vendor and the network service provider, and how availability, visibility, adaptability and scalability are provided. Note that this is the same functionality described in the previous subsection as service fulfillments and service assurance but the focus here is on the development of the overall solution. These characteristics must be embedded in the solutions delivered by the vendor based on a synergistic, trusted relationship between vendor and operator.

The essential differences between the two approaches are:

- DevOps delivers the application; Dev-for-Ops delivers a fully realized service including the core application, monitoring and analytics and deployment and adaptation capabilities. A Dev-for-Ops service is ready to meet the demands of contextualization, scale and global-local integration.

- In many cases, operators have important insights regarding their networks and their customers that influence methods and techniques of deploying, managing and operating the services and infrastructure. Dev-for-Ops maintains this by allowing the specific adaptation of the overall solution according to the operator specific insights.

- The feedback mechanism takes on a different character in Dev-for-Ops. In order to address the separation between vendor and operator, an automatic or semi-automatic mechanism (similar to Microsoft's "report a bug" feature)

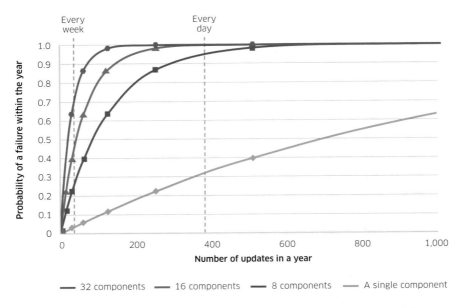

FIGURE 10: Ensuring reliability with a component failure probability of 0.1% (Bell Labs).

is needed to provide the feedback in a timely manner, without revealing sensitive operator business information (Microsoft 2015).

- The insights and feedback contribute to automated, operation-aware testing, which is crucial to maintain high reliability with rapid, complex releases. Failure probabilities grow exponentially in this environment driving the need for very high-quality, bulletproof software releases.

As in Six Sigma, the ability to maintain reliability when reusing components heavily depends on the ability to detect and fix small defects in any of the sub-components early in the development process. This effect is even more important today, as we not only deal with many subcomponents but also with the high-frequency of software updates.

To illustrate this point, in figure 10 we show the case of the service failure for a service based on independent components each with a very small failure probability (0.001). If weekly updates were performed, there is almost a 50% probability of a service failure within a year. This underscores the importance of testing and improving software quality as a critical building block of Dev-for-Ops to deliver required reliability. The testing procedure must be automated with the ability to detect defects in both the application and its life cycle operational processes.

The Dev-for-Ops framework will further require business and human interaction between the vendors and network service providers, and the use of cutting edge testing and agile development techniques to develop flexible, agile, easy-to-provision, end-to-end solutions. This capability will be a crucial component of the future network operation architecture.

The future begins now

Network operators and service providers are beginning to evolve their networks to support unified, scalable and more dynamic network operations in concert with their move to network virtualization, cloud-based platforms and global partnerships to increase their reach beyond their traditional markets (Ovum 2014). The following are a few such examples:

1. A North American operator is offering a new digital life service for the consumer market that offers home security and home automation with user-specific configuration and control (Rose 2015).

2. An APAC operator acquired a video streaming analytics company and a health analytics company to support enhanced user-specific service awareness and optimization, as well as potential new revenue growth (Telecoms 2014, Pearce 2015).

3. A European operator is enhancing the capabilities of their customer experience platform to include analytics and allow users more control of their broadband service, including self-help tools to reduce the need for customer care intervention and optimize the time to problem resolution (Alcatel-Lucent 2015).

4. Two North American service providers use user-specific contextual recommendation engines for their video service offers, similar to the web-scale video service providers (Baumgartner 2010). Another provider has recently announced plans to allow their video subscribers to customize their channel packages (Ramachandran 2015).

5. New standards are being defined by the ETSI Industry Specification Group (ETSI ISG) with a significant emphasis on defining a new architecture for network and service operations (ETSI 2014).

6. There is an increasing move to all-IP networks, with the retirement of legacy service infrastructure (for example, the PSTN network), which is driving a critical first phase of simplification of network operations (Brodkin 2013).

7. Two large operators in the US, as well as several global service providers are offering global enterprise services targeting the multinational enterprise market, with global and local partners. These services include security, managed work space, communications-enabled applications, 360-degree CEM and service management and control. They are being augmented with new vertical specific services in vehicular management, healthcare device management and smart city management (Ovum 2014).

8. Many operators are moving forward with integration of their IT and network operations organizations to mirror the evolutionary path of the underlying technologies they manage (Goldstein 2015, Donegan 2015).

9. A leading web-scale company is optimizing their network operations to support the explosive growth in demand through heavy reliance on automation (Andreyev 2014).

10. Network analytics are being leveraged by a Tier 1 European mobile operator to optimize the customer's experience. They are using big data analytics to anticipate evolutionary trends and address anomalies proactively (SAS 2015).

11. Software-defined networking has been embraced as a key network optimization tool by leading web-scale companies (Levine 2015, Duffy 2015, Han 2015, ACG 2014).

Summary

The competitive landscape of network service providers is changing rapidly and their network operations must be an integral part of the solution. Support for user choice and service personalization, as well as advanced automated customer care, allows companies in many industries to earn significant advantages in customer satisfaction and profitable growth. Adapting to this new reality, and the attendant massive growth in the volume and complexity of services, will place increasingly high-demands on network operations, for both LSPs (access and aggregation) and GSPs (core network).

The new architecture for the future of network operations must support high levels of operational automation and predictive analytics to support the required speed of innovation, with high levels of abstraction of network systems. Contextual awareness of the network, service and customer experience will also be critical. This new operations architecture will allow the near-complete automation of fulfillment and assurance activities, seemingly infinite scaling and the connection of virtualized (NFV) and software-defined networks (SDN). Adopting a new Dev-for-Ops methodology will also provide the tools for service providers to innovate rapidly while maintaining the carrier-grade reliability their users require.

The largest barrier will then be the "people" barrier, discussed in the introduction to this chapter. It will be critical for network (and IT) operations to adapt to the customer-experience reality and adopt new processes, a new architecture and, critically to trust in and coexist with increasingly automated systems. More than any technology, this human component of the future equation will likely determine the success or failure of the Future X Network.

References

ACG Research 2014. "Business Case for Cisco SDN for the WAN," Cisco web site, (http://www.cisco.com/c/dam/en/us/products/collateral/routers/wan-automation-engine/business-case-for-cisco-sdn-for-the-wan.pdf).

ACSI 2015. *American Customer Satisfaction Index* (http://www.theacsi.org).

Alcatel-Lucent 2015. "Accenture and Alcatel-Lucent helping Telefonica enhance experience for consumers in Europe and Latin America," *Alcatel-Lucent press release*, April 7 (https://www.alcatel-lucent.com/press/2015/accenture-and-alcatel-lucent-helping-telefonica-enhance-experience-consumers-europe-and-latin#sthash.fQKCditL.dpuf).

Amazon 2015. "Amazon Help Community, Self-Service," *Amazon* (http://www.amazon.com/gp/help/customer/forums).

Andreyev, A., 2014. "Introducing data center fabric, the next-generation Facebook data center network," *Facebook web site*, November 14 (https://code.facebook.com/posts/ 360346274145943/introducing-data-center-fabric-the-next-generation-facebook-data-center-network).

Baumgartner, J., 2010. "Jinni Puts Cable Guys in the Mood," *LightReading*, February 8 (http://www.lightreading.com/jinni-puts-cable-guys-in-the-mood/a/d-id/674432).

Beyond Philosophy 2011. "The 2011 Beyond Philosophy Global Customer Experience Management Survey", *Beyond Philosophy web site* (http://beyondphilosophy.com/2011-beyond-philosophy-global-customer-experience-management-survey).

BLS 2015. "Labor Productivity and Costs by Industry Tables," *U.S. Dept. of Labor, Bureau of Labor Statistics web site* (http://www.bls.gov/lpc).

Brodkin, J., 2013. "'The telephone network is obsolete': Get ready for the all-IP telco," *ArsTechnica*, January 7 (http://arstechnica.com/information-technology/2013/01/the-telephone-network-is-obsolete-get-ready-for-the-all-ip-telco).

Donegan, M., 2015. "Integrating Networks & IT the Tele2 Way," *LightReading*, April 27 (http://www.lightreading.com/business-employment/business-transformation/integrating-networks-and-it-the-tele2-way-/d/d-id/715318).

Dragich, L., 2012. "Event Management: Reactive, Proactive or Predictive?", *APMdigest*, August 1 (http://www.apmdigest.com/event-management-reactive-proactive-or-predictive).

Duffy, J., 2015. "Alcatel-Lucent uses SDN to meld IP, optical services," *NetworkWorld*, May 20 (http://www.networkworld.com/article/2924956/sdn/alcatel-lucent-uses-sdn-to-meld-ip-optical-services.html).

ETSI 2014. "Network Functions Virtualisation – White Paper #3," ETSI web site, presented at SDN & OpenFlow World Congress", Dusseldorf-Germany, October 14–17 (https://portal.etsi.org/Portals/0/TBpages/NFV/Docs/NFV_White_Paper3.pdf).

Fry, R., 2015a. "This year, Millennials will overtake Baby Boomers," *PewResearchCenter: FactTank*, January 16 (http://www.pewresearch.org/fact-tank/2015/01/16/this-year-millennials-will-overtake-baby-boomers).

Fry, R., 2015b. "Millennials surpass Gen Xers as the largest generation in U.S. labor force," *PewResearchCenter: FactTank*, May 11 (http://www.pewresearch.org/fact-tank/2015/05/11/millennials-surpass-gen-xers-as-the-largest-generation-in-u-s-labor-force).

Goldstein, P., 2015. "Verizon's technology chief Tony Melone to retire, as carrier merges network and IT functions," *FierceWireless*, March 10 (http://www.fiercewireless.com/story/verizons-technology-chief-tony-melone-retire-carrier-merges-network-and-it/2015-03-10).

Goss, J., 2010. "How Lean Six Sigma affirms agile programming practices," *TechRepublic*, February 8 (http://www.techrepublic.com/blog/tech-decision-maker/how-lean-six-sigma-affirms-agile-programming-practices).

Had, F., et al 2014. "Path computation for IP network optimization," *Alcatel-Lucent TechZine* June 16 (https://techzine.alcatel-lucent.com/path-computation-ip-network-optimization).

Jobs, S., 2011. "Apple Worldwide Developer's Conference Keynote Address," *Apple Developer web site* (https://developer.apple.com/videos/wwdc/2011).

Levine, H., 2015. "The 2020 WAN takes shape – SDN, virtualization, and hybrid WANs," *NetworkWorld*, August 13 (http://www.networkworld.com/article/2971214/wan-optimization/the-2020-wan-takes-whape-sdn-virtualization-and-hybrid-wans.html).

Microsoft 2015. "Search Products accepting bugs or suggestions," *Microsoft Connect* (https://connect.microsoft.com).

Morningstar 2015. *Morningstar web site* (http://performance.morningstar.com).

Nelson, F., 2014. "Microsoft Azure's 44 New Enhancements, 20 Trillion Objects", *Tom's IT Pro* April 4 (http://www.tomsitpro.com/articles/microsoft-azure-paas-iaas-cloud-computing,1-1841.html).

NPS 2015. *NPS Benchmarks* (http://www.npsbenchmarks.com).

Ovum 2014. "Ovum Decision Matrix: Selecting a Global Telco Managed Security Services Provider," *Ovum web site*, 17 September (http://www.ovum.com/research/ovum-decision-matrix-selecting-a-global-telco-managed-security-services-provider).

OST-R 2013. "Table 1-35M: U.S. Vehicle-Kilometers (Millions)," *U.S. Dept. of Transportation, Office of the Assistant Secretary for Research and Technology web site* (http://www.rita.dot.gov/ bts/sites/rita.dot.gov.bts/files/publications/national_transportation_statistics/html/table_01_35_m.html).

Pearce, R., 2015. "Telstra acquires healthcare analytics company," *Computerworld* (Australia), March 27 (http://www.computerworld.com.au/article/571439/telstra-acquires-healthcare-analytics-company).

Ramachandran, S. and Knutson, R., 2015. "Verizon Breaks Pay-TV Bundle as Competition Mounts," *WSJ web site*, April 16 (http://www.wsj.com/articles/verizon-breaks-pay-tv-bundle-as-competition-mounts-1429240054).

Rose, 2015. "Digital Life by AT&T Review," SecurityGem, August 8 (http://securitygem.com/digital-life-att).

SAS 2015. "Building customer satisfaction with better network monitoring," *SAS Customer Stories web site* (http://www.sas.com/en_us/customers/telecom-italia-analytics.html).

Satmetrix 2015. "2015 Net Promoter Score Benchmarks by Industry: US Consumer Benchmarks," *Satmetrix web site* (http://www.satmetrix.com/expertise/benchmarks/benchmarks-by-industry).

Statista 2015. Statista (http://www.statista.com).

Telecoms 2014. "Telstra acquires video analytics company Ooyala," *Telecoms.com*, August 13 (http://telecoms.com/277881/telstra-acquires-video-analytics-company-ooyala).

Wikipedia 2015a. "List of Countries by Motor Vehicle Production," *Wikipedia* (https://en.wikipedia.org/wiki/List_of_countries_by_motor_vehicle_production#cite).

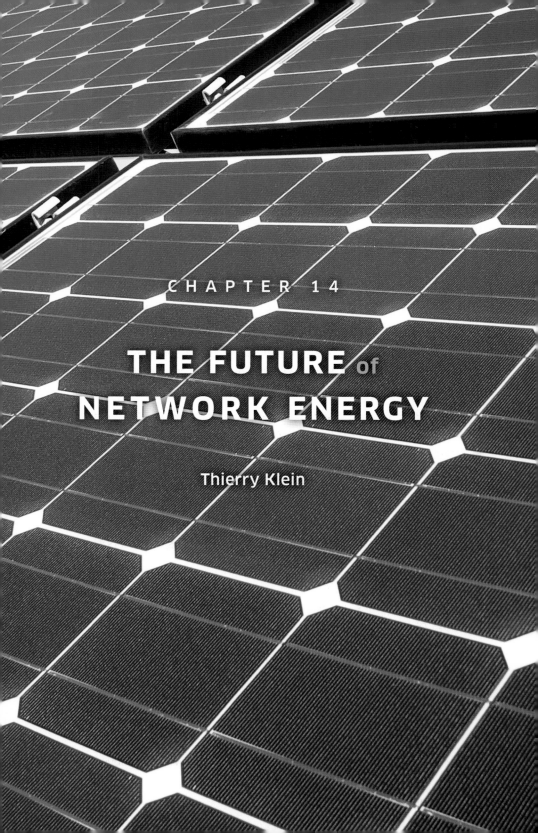

CHAPTER 14

THE FUTURE of
NETWORK ENERGY

Thierry Klein

The essential vision

Climate change is one of the defining topics of our generation. At the heart of the issue is a growing understanding of the impact of human activities on our planet and the desire to balance the evolution of human technology to meet our needs with the preservation of natural resources for current and future generations. Global energy consumption, especially the use of fossil fuels, has been dramatically increasing since the industrial revolution and increases with each technological revolution. Every single industry sector is facing this challenge and the information and communications technologies (ICT) sector is no different. The ICT sector, however, is in a very unique position since it both contributes to carbon emissions and enables reductions of carbon emissions in other industry sectors.

The Global e-Sustainability Initiative (GeSI) published its landmark SMART2020 report in 2008. In this report both the direct impact of the ICT sector and the enabling abatement effect of ICT are assessed. The most recently published SMARTer2030 report, issued in June 2015, has emphasized that the ICT sector's greenhouse gas emissions are expected to account for 1.97% of the global greenhouse gas emissions by 2030, or 1.25 Gigatonnes (Gt) of carbon dioxide equivalents (CO_2e) (GeSI 2015). However, in the same time frame,

ICT can enable a 20% reduction of global CO2e emissions. In other words, the emissions avoided by other industry sectors, through the use of ICT, are nearly 10 times larger than the emissions generated by deploying these same ICT solutions. Tremendous additional economic benefits are also generated, demonstrating that, in ICT at least, it is possible to avoid having to choose between economic prosperity and environmental protection.

We are witnessing an explosion of the traffic volumes in communications and data networks. The expansion of the worldwide broadband subscriber base and the increasing number and diversity of connected applications, services, devices and objects account for most of the increase. The increased traffic volumes and connectivity requirements necessitate continuous development and deployment of new technologies and infrastructures to deliver the expected performance and user experiences. The resulting energy consumption and energy cost of the broader communications infrastructure (including communication networks, connected devices and data centers) are quickly becoming one of the major challenges for the ICT industry. The benefits of ICT are undisputed. It enables a more prosperous, more accessible and more sustainable future for all. It is nevertheless the responsibility of the ICT sector to also effectively manage its own impact on the environment.

In this chapter we evaluate the key drivers and opportunities for the future of network energy and find the following:

1. *Energy consumption of networks is at the precipice of economic unsustainability* – The anticipated increase in traffic will lead to a substantial increase in energy consumption across all parts of the network. When coupled with rising energy costs, this increasing energy consumption will inevitably lead to prohibitive energy costs for network providers, seriously threatening their economic viability.

2. *The new digital era will require "zero power networking"* – A fully connected world poses immense challenges from an overall energy consumption and energy cost perspective. The scale and the sheer number of devices and objects that will be connected create critical operational, technical and business challenges in providing access to reliable power sources in all markets. This drives the need for unprecedented energy efficiency and energy autonomy, particularly in wireless access, devices and the cloud, which are the dominant energy consumers in end-to-end networking.

3. *New disruptive technologies will be required to fundamentally change the energy equation* – We outline the different facets of this energy challenge

and describe a portfolio of key technologies that will form the essential solution, considering everything from experimental approaches still in research, to technologies already commercialized and deployed.

The past and present

Unsustainable energy consumption and cost

Over the past decades, the energy consumption of telecommunication networks has grown significantly to reach 350 TWh worldwide in 2012 (Lamber 2013). This represents about 2% of the worldwide electricity consumption, whereas the entire ICT sector (including data centers, devices, computers and peripherals) accounts for about 6% of global electricity consumption. The network energy bill typically represents between 7–15% of the operational expense of telecommunication service providers in developed countries and up to 40–50% in some developing countries (Intelligent 2012, GSMA 2013, Kim 2014). A major European network operator recently stated that its energy bill would reach the $1 billion mark by 2020 (Le Maistre 2014), whereas the energy bill of some of the large operators in the US had already topped the $1 billion mark by 2012. In the UK and Italy, for instance, the telecommunications operators are the largest consumers of electricity, consuming about 1% of the total electricity generation of their countries.

The rapid adoption of smartphones and tablets is dramatically driving up daily Internet traffic, and forecasts indicate that traffic will increase up to 85 times by 2017, as compared to 2010. The population of people online is expected to grow from 2.3 billion in 2012 to 3.6 billion in 2017, representing about half of the world's population. By 2017 more than 5 trillion gigabytes of data will pass through the global communications network every year, which is the equivalent of everyone on the planet tweeting non-stop for more than 100 years. This dramatic increase is driving significant increases in energy consumption and taking us to the precipice of economic unsustainability.

Figure 1 shows the alarming trend of the annual energy cost of the global telecommunication networks between 2011 and 2025. If we continue on the current trajectory, the global energy cost is projected to increase from $40B to $343B annually. The energy consumptions for wireless and wireline access traffic are calculated for end-to-end services, including the energy consumption of the access networks, the metro and edge networks, the core networks and the associated data centers. Clearly current technology

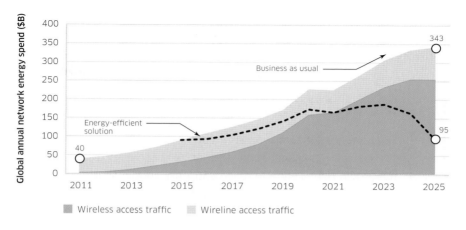

FIGURE 1: Global annual network energy spend by operators (Bell Labs Consulting).

improvements are not sufficient to keep up with the rising energy costs and this is the imperative we address in this chapter.

Powering the fully connected world

Network and data-center energy consumption is just one side of the energy equation; we also need to focus on the device side. The energy challenges for connected devices point to different opportunities and require different solutions than networks. Mobile and desktop computing devices (personal computers, laptops, printers, feature phones, smartphones and tablets) consumed about 39 GW of power globally in 2013, which is equivalent to the annual electricity consumption of seven cities the size of New York (GWATT 2015).

In figure 2a, we show a high-level analysis of the energy consumption of connected devices, including all consumer devices with network connections, as well as devices for intelligent buildings, public safety, surveillance cameras and smart homes, and including all sensors, machines and connected objects generally part of Internet of Things (IoT) applications. The global energy consumption of these connected devices is expected to grow from about 200 TWh in 2011 to 1,400 TWh in 2025. Such a level of global energy consumption would cost the world almost $400 billion annually in 2025 and represents a 14.6x increase from 2011. Clearly this level of energy increase between 2011 and 2025 is undesirable.

To really understand the magnitude of the issue, it is also important to look at the energy supply options for the different devices. IoT devices and sensors individually tend to consume relatively small amounts of energy in absolute

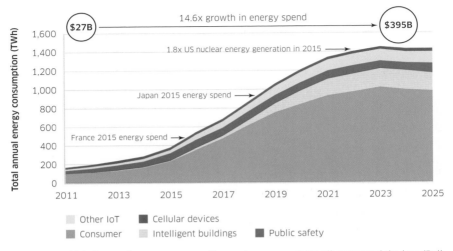

FIGURE 2A: Global annual energy consumption and energy cost for all connected devices (Bell Labs Consulting).

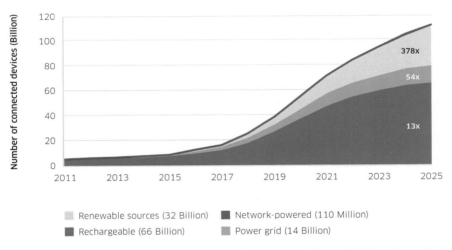

FIGURE 2B: Different power supply options for global connected devices (Bell Labs Consulting).

terms, but there will be huge numbers of such devices deployed. In addition, the relevant energy challenge is not just the actual energy consumed, but more importantly, the energy supply and energy autonomy of these devices.

In figure 2b, we have considered four different energy sources and classified devices accordingly:

1. *Power grid* devices are always connected to and powered by the grid, such as home appliances

2. *Rechargeable devices* rely on batteries but are periodically recharged from the grid, such as smartphones, tablets, certain smart home or industrial monitoring sensors

3. *Renewable energy devices* are off-grid for their entire lifetime, being so low-powered that they can rely solely on their batteries (many IoT devices), or harvest local energy sources, such as solar- or wind-powered devices

4. *Network-powered* devices can be reverse-powered through the communications network, such as video surveillance cameras

For each category of device shown in figure 2a, we have estimated the percentage of devices that could be powered by the different energy sources and assumed a gradual increase in the use of renewable energy sources between 2011 and 2025.

Figure 2b makes apparent the need to find new low power solutions and rely to a greater degree on renewable energy sources. With over 100 billion devices connecting to the network in 2025, it would be neither practical nor economically viable to repeatedly change and/or recharge batteries or run power lines to that many devices. The full potential of a connected knowledge world and IoT will not be realized at a large scale unless the energy supply challenge is appropriately addressed. We need to find a balance in the energy equation between greater and more efficient use of renewable energy sources, high-density energy storage and lower power consumption by devices.

Energy hot spots in networks

Figure 1 shows the trend on the total network energy consumption, but a closer look is required to understand the energy "hot spots" in networks, because this is where future technologies will have their greatest impact. This analysis is possible using the Bell Labs GWATT (www.gwatt.net) application, an interactive tool that provides insights into the energy consumption of different network sub-domains, as a function of traffic growth and taking into consideration the potential benefits of new technologies and network architectures.

In figure 3a, we breakdown network energy consumption into six main network domains: home/enterprise, access/aggregation, metro, edge and core networks, and service core/data centers. Consumption for each category is shown for 2013 and projected for 2020 in terms of absolute power consumption.

Figure 3b shows percentages, with a slightly different take on the categories, separating radio and fixed access, combining metro, edge and core, and

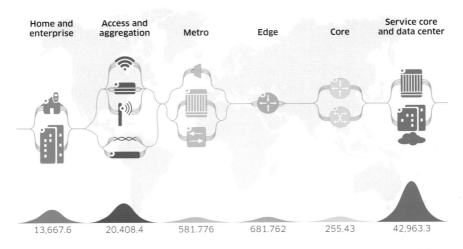

FIGURE 3A: 2013 global network power consumption in MW (Bell Labs GWATT application).

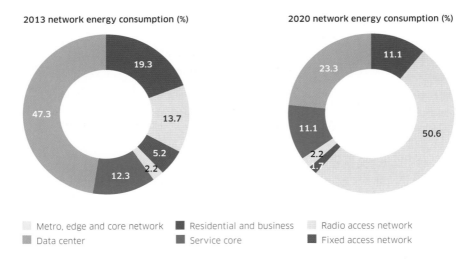

FIGURE 3B: Evolution of network energy consumption puts the focus on wireless access and the cloud — represented by data centers (Bell Labs GWATT application).

separating service core and data centers. Importantly, it also shows changes in relative power consumption by 2020.

The analysis shows that the critical energy consumption in networks occurs in three places: the wireless radio access network, the cloud/ data centers, and also the end devices. Key technology advancements are required in each of these domains to ensure that future traffic growth can be supported in a sustainable and economically viable way.

We are at a critical junction with respect to current and future energy consumption and energy costs of telecommunication networks and devices. Energy considerations have typically been an after-thought in network design, but we have reached a point where energy has truly become the new frontier for innovation and competitive differentiation. It is no longer a nice-to-have feature but a make-or-break requirement for the network of the new digital era.

The future

To address the looming energy challenges over the next decade, as outlined in the preceding section, we believe that a new approach, which we call *zero-power networking*, is required. This approach is comprised of three different but related dimensions, which will form a foundational block of networks in 2020 and beyond.

- Zero grid power: It is imperative for scenarios when access to power is not available, not reliable or cost prohibitive. Today's renewable energy technologies – based on solar, wind, vibration and RF harvesting, with batteries for storage – are often not practical due to the size of energy collectors, such as solar panels and wind turbines, and the weight of batteries. Current research is investigating ways to provide low-energy/high-efficiency solutions that enable practical, off-grid network deployments.

- Zero wasted power: Up to 50% of the energy consumed by communications equipment does not provide any directly useful work. This includes overhead for cooling and thermal management, inefficiencies in power supplies and AC-DC power conversion, and the idle power consumption, which can be up to 90% of the peak power consumption even without any traffic load. The goal must be to eliminate or, at least, significantly reduce this energy waste.

- Zero energy per bit: An energy-optimized network would process, store, and transport as much data as possible for the smallest amount of energy spent. To achieve this, we need to improve the functional energy efficiency of communications equipment and networks. Our goal is zero energy per bit above the absolute minimum required energy.

Several major architectural and technology disruptions are happening that will collectively drive the zero-power networking vision. An entire portfolio of technologies is required – spanning architectures, hardware components and materials, software, algorithms and protocols – to minimize the energy consumption of networks and enable full energy autonomy. There is no silver bullet; we will have to marshal a broad range of ideas to realize the full benefit of each technology.

Paving the road to maximum network energy efficiency

In 2010, the GreenTouch consortium[1] was formed with the mission to provide architectures, technologies and solutions to improve network energy efficiency by a factor 1,000, based on a state-of-the-art 2010 reference scenario. Built as a precompetitive research consortium, GreenTouch brought together more than 40 organizations worldwide, including equipment vendors, service providers, research institutes and universities. This consortium has demonstrated a number of technologies that will transform network energy consumption. We describe several of those technologies below, focusing on wireless access networks, since this is the area with the greatest opportunities for energy efficiency gains.

Ultra-dense small cells with intelligent resource management

Small cells and heterogeneous networks are an energy efficient solution for satisfying large and time-varying capacity demands in wireless networks (Zeller 2013). To achieve the maximum energy efficiency benefits, ultra-dense networks are required. We need, first, to increase the number of small cells by at least a factor 10 beyond currently imagined deployment numbers and, second, reduce the coverage area of a single small cell to about 50–100m. Deploying such a large number of small cells will present installation and operation challenges. Also, because capacity use varies significantly with time, power control techniques and smart algorithms must be able to turn off the small cells when not needed, to avoid unnecessary interference and power consumption.

Researchers working together with telecom industry leaders are pioneering novel wireless network architectures that separate the control and data planes and reduce or even eliminate the permanent emissions of pilot signals, used to determine the presence and location of users. They also reduce or eliminate broadcast signaling by allowing user terminals to request connectivity through an asynchronous wake-up channel that does not require continuous signaling and response. In these architectures, the data connectivity is provided by small cells that are dynamically controlled and powered on and off, while the network control and mobility management

1 - Further information on the GreenTouch consortium can be found at www.greentouch.org.

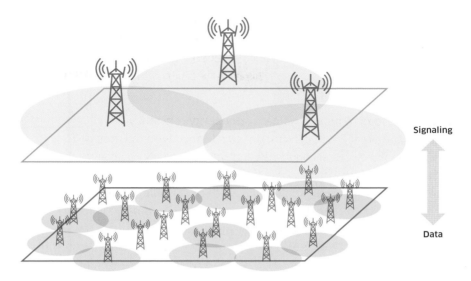

FIGURE 4: Multi-tiered wireless network architecture with separate control and data planes. It uses macro base stations for signaling and small cells for ultra broadband transmission (Capone 2012).

functions are provided by the overlaying macro cells, which, although less efficient in terms of energy per bit, transmit a relatively small amount of signaling traffic. This multi-tiered wireless architecture is depicted in figure 4. The full benefit can be realized when combined with lower-power, energy-efficient wireless equipment that can quickly be put into sleep mode and has a power profile that is proportional to the traffic load.

Ultra-energy efficiency: Massive MIMO

As discussed in chapter 6, *The future of wireless access*, Massive MIMO is a promising and disruptive wireless air interface technology that achieves large gains in spectral and energy efficiency. Originally invented by Bell Labs in 2005 (Marzetta 2015), Massive MIMO uses a large excess of service antennas at the base station to serve a relatively smaller number of active user terminals, as shown in figure 5. It enables selective transmission and reception of simultaneous beams to and from terminals using every time and frequency resource simultaneously. Essentially, the power is directed only where the terminals are located. It is highly scalable, with additional service antennas continuously improving the spectral efficiency. It also has the desirable property that all the complexity (for example, channel estimation, pre-coding and signaling processing) is at the base station, allowing for low-power and low-complexity terminals. Massive MIMO is broadly applicable to a number of

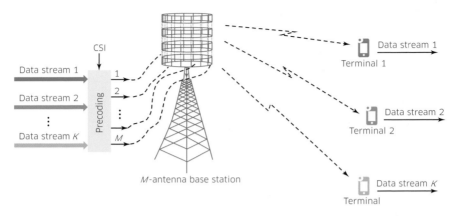

FIGURE 5: Massive MIMO architecture and transmission principle (Marzetta, T., GreenTouch Large-Scale Antenna Systems Project).

deployment scenarios, including mobile access to user terminals, fixed wireless access to stationary terminals, as well as wireless backhaul for small cells.

The potential performance gains are impressive, with 4–250x spectral efficiency gains and 1,000-14,000x total energy efficiency gains compared to LTE, with uniformly good service throughout the cell and at the cell edge. A solid, information-theoretic foundation for this new air-interface technology has been developed and is also supported by realistic system-level simulations to validate both the performance gains in terms of spectral efficiency, as well as total energy efficiency. However, a major challenge remains in the system design, a fundamental redesign of the network architecture, the signaling mechanisms, algorithms and protocols. Also, a new base station architecture is required to accommodate the large number of antennas and the associated number of RF signal-processing chains.

It takes a portfolio of technologies
In order to maximize the energy efficiency of future networks, a complete portfolio of technologies, architectures, algorithms and protocols is needed (from GreenTouch):

- *New power models for wireless access, fixed access and core networking equipment* – energy-efficient hardware is required with tighter integration of components, reduced peak power and idle power levels, optical interconnects, deep sleep modes and power profiles that are proportional to traffic loads

- *Algorithms for dynamic allocation of resources at the equipment and network levels* – intelligent allocation of wireless air interface resources, including antenna muting, sleep modes, adaptive packet processing, energy optimized mixed line rates, dynamic routing and topology optimizations and real-time allocation of protection and restoration resources

- *Optimizing Shannon's tradeoff between energy efficiency and spectral efficiency*[2] – the use of single-user and multi-user MIMO, coordinated transmissions, beamforming, bandwidth expansion and interference alignment will enable efficient allocation of wireless air interface resources (power, time, bandwidth, frequency, space) and operation at the "sweet spot" of Shannon's tradeoff – giving maximum energy efficiency while still supporting wireless system performance targets

- *Virtualization of home gateway functionalities* – power-hungry, unshared and always-on home gateways will be replaced by virtualized service functions hosted by the network access provider (as described in chapter 12, *The future of the home*). The general-purpose servers hosting these virtualized device functions will be shared dynamically between hundreds or even thousands of virtual home gateways, significantly improving energy efficiency.

- *Energy-optimized content distribution and network function virtualization* – content distribution and caching – traditionally used to improve content

2 - Shannon's theory highlights a fundamental tradeoff between energy efficiency and spectral efficiency.

It is possible through a combination of technologies, architectures, components, algorithms and protocols to reduce the net energy consumption in end-to-end communications networks by up to 98% by 2020.

access time and reliability – can be better optimized for energy use by re-designing the number and location of edge cloud data centers, and the content placement and replacement policies, taking into account the total energy cost of transporting, processing and storing information in end-to-end networks

Measuring the gain: Green Meter results

GreenTouch has developed a methodology, called the *Green Meter*, to evaluate end-to-end network energy efficiency and quantify the individual and combined impact of different technologies, providing a roadmap for future technology developments (GreenTouch 2015). The study concluded that it is possible through a combination of technologies, architectures, components, algorithms and protocols to reduce the net energy consumption in end-to-end communications networks by up to 98% by 2020 compared to the 2010 reference scenario. This impressive reduction in network energy consumption takes into account the dramatic increase in traffic predicted between 2010 and 2020, and is based on significant improvements in the energy efficiencies of mobile access networks (up to 10,000x), residential fixed access networks (up to 254x) and core networks (up to 316x) compared to the lightly loaded and relatively energy inefficient 2010 network reference scenario.

Although the reference case can be debated in terms of its applicability to networks in 2015, energy efficiency gains of even one or two orders of magnitude in end-to-end, wireless access networks would reduce operational expenses and the environmental impact for network operators, and provide significant economic gains to consumers and businesses. In the following section, we examine a few solutions that combine some of the aforementioned technologies and offer promise in moving us toward the ultimate goal, zero-power networking.

Wireless, wireless small cells with full energy autonomy

Using ultra-dense deployments of small cells, we could meet future traffic demands and at the same time provide significant network energy efficiency improvements. However, deploying hundreds of thousands of small cells creates operational challenges, such as connecting them to both the backhaul network and the power grid. It is frequently not practical or economically affordable to run backhaul and power wires to each small cell. The cost of connection could represent as much as 50% of the total cost of deployment. This is a problem for dense urban and rural areas, as well as developing countries and remote regions where access to reliable power is not guaranteed. It is therefore critical to find innovative ways to connect small

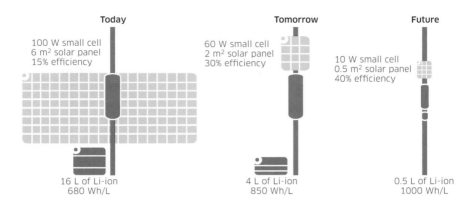

FIGURE 6: State of the art technologies and vision for future energy-autonomous small cells (Bell Labs Research).

cells wirelessly. A completely wireless solution (power and network) would also increase network survivability, providing access to communications services during power outages, emergencies and disasters.

The solution will require a wireless backhaul solution and access to local renewable energy sources. Among the potential renewable energy sources (wind, solar, RF harvesting, vibration), the only viable alternative is solar power since, except for wind, the amount of energy harvested by other means is not sufficient for wireless applications. Unfortunately, for safety reasons, small wind turbines would not normally be practical for such small units, which are often deployed at a height of only a few meters for safety reasons, and due to the required wind start-up speed.

Unfortunately, deploying small cells with solar panels is not practical today at large scale since powering a wireless cell consuming 100 W would typically require a solar panel of about 4m², based on current commercially available technologies and solar panel efficiencies of around 30%. Improvements in solar cell efficiency are possible, although the state of the art cells already operate at 46% efficiency (Renew 2014). Even with a maximum theoretical limit of 85% efficiency (Byrnes 2014), the greater opportunity is to reduce the energy consumption of the system and, thus, the required size of the solar panels by at least a factor of 10 to 20x, as shown in figure 6.

In order to accomplish energy autonomy for small cells a completely integrated network solution will be required, focusing on:

- End-to-end system architecture, integrating wireless access and wireless backhaul
- Ultra-low-power, small cells hardware design
- Optimized signal processing, dynamic power control and management of small cells
- Utilization of highly efficient energy harvesting and energy storage technologies

With sufficient focus on these four areas, a small cell that would be completely standalone without any wires for the power supply or the backhaul, could be supported by a solar panel that would fit within the footprint of the product. Such an autonomous small cell would allow the creation of wireless networks anywhere, at any time. This would completely change the future of wireless networks and open up new possibilities for network deployment, installation and even network ownership.

Virtualization and distributed cloud networks

Future Internet traffic will likely be dominated by interaction-intensive cloud services and applications running on massive numbers of resource-limited communication end points. The reduced capabilities of the end devices and the requirement for battery lifetimes of ~10 years, pushes more data storage, processing, analytics and control functionality to the cloud – thus deeper into the network. In contrast, the network throughput and low-latency required to transport this data tends to push functionality in the opposite direction, toward the end device.

The result is a fundamental transformation of today's centralized network and cloud architectures toward a highly distributed, converged cloud integrated network platform (CIN), as discussed in several earlier chapters. The CIN is composed of a large number of edge cloud nodes distributed across a massive, meshed metro-core network (Bell Labs 2013, Weldon 2014). Virtual cloud service functions are dynamically and elastically instantiated over shared, commodity-type servers at multiple cloud locations close to end users and interconnected by a programmable network.

Figure 7 shows an overall network architecture with both physical and virtualized network cloud architectures. From an energy standpoint, there are several important questions on how to optimally allocate resources. For example, taking into account the resources required for processing, storing and transmitting information in the virtual infrastructure, where should the network and data plane processing and storage functions be implemented to

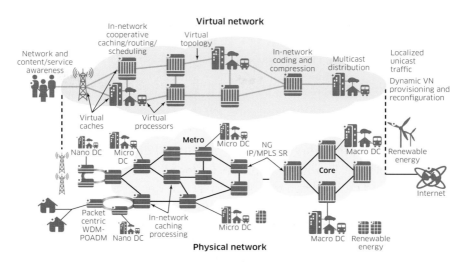

FIGURE 7: Vision of the distributed cloud network architecture comprising both the physical network and the virtualized network as a communications and computing platform (Bell Labs Research).

minimize the use of the physical infrastructure and therefore the total energy consumption? Part of the answer involves the energy consumption of data center networks, which we consider in the next section.

Data center architectures and energy consumption

Data centers are a major element of ICT energy consumption. While the mega data centers run by Google, Facebook, Amazon and others receive a lot of media attention, the vast majority of the energy consumed by data centers (90–95%) is actually consumed by small, medium and large corporate and multi-tenant data centers. A majority (52%) of the energy in a typical data center is consumed by IT equipment: servers, top-of-rack switches, interconnection and aggregation routers. The rest is consumed by building-related infrastructure, with the lion's share (32%) by cooling and thermal management infrastructure (DataCenter 2013).

Optimization of energy dissipation using innovative thermal management technologies is thus critical for lowering that 32% share while maintaining the data center in its operating range and within its reliability target. Convective air-cooling is the traditional approach and works quite well for low to moderate server heat densities. Recently, a number of data center operators have published impressive power utilization efficiencies (PUE) using innovative air flow management techniques, such as hot- and cold-aisle segregation, higher room operating temperatures and free cooling, whenever possible.

More recently there has been growing interest in liquid-cooling, which offers significant benefits relative to conventional air-based approaches, especially with respect to energy efficiency and equipment heat density. Liquid-cooling can be implemented at a range of equipment scales, including the rack, the server and even the component level, and will be especially important for ultra-dense systems, such as those expected in future data centers. Although liquid-cooling offers 10–40x improvements in heat transfer efficiency compared to air-cooling, some significant challenges remain, including the need for high reliability, low acoustic noise, physical footprint constraints and the ability to easily maintain and upgrade server equipment.

New research directions into hybrid liquid- and air-cooling have emerged that address these challenges and provide the benefits of liquid-cooling while being invisible to the data center end-user. For example, approaches based on hybrid air- and pumped-refrigerant-based cooling have shown some very promising results, with up to 40x energy efficiency improvement compared to air-cooling demonstrated in lab experiments and real-world data center environments.

A classical, hierarchical intra-data-center system architecture can be split in three parts: the racks of physical servers (disaggregated or not) and their top of the rack switches (TOR); the interconnection layer, including some aggregation stages; and the load balancers interconnected to a border router. Since the major part of the energy consumption is in the racks of servers (that could reach up to 80% of the total energy consumption of the IT equipment), maximizing the utilization of the racks of servers is a critical objective. Figure 8 illustrates a future architecture using low-cost wavelength division multiplexing (WDM) approaches and modular, photonic cross-connects (PXC) (discussed in chapter 4, *The future of wide area networks*). This architecture optimizes energy utilization and lowers energy cost per bit with inter-server defragmentation, smart placement of workloads, regrouping of data processing to enable server sleep mode strategies, and establishment of optical bypass to manage elephant flows and preserve QoS on the interconnection layer.

In the future, greater utilization and energy efficiency could be achieved by further disaggregation of the computing architecture, embedding optics directly onto the server card and increasing the use of all-optical switching between banks of compute, memory and storage servers. To support the required memory bandwidth, however, would require terabit optical interfaces; thus, it remains a challenging techno-economic goal.

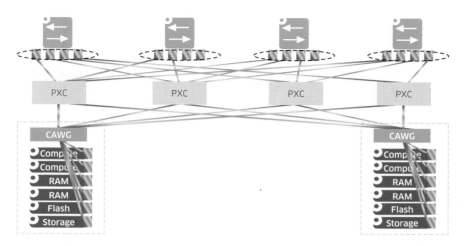

FIGURE 8: Future data center architecture leveraging wavelength division multiplexing approaches and photonic cross-connects to optimize server resource utilization (Bell Labs Research).

Stepping back from the precipice

The potential of the new technologies, architectures and solutions we have discussed in this chapter could reverse the dramatic trend in energy consumption (shown as the dotted line in figure 9). A number of the technologies described are disruptive and will require further research and standardization before being commercially deployed. Large scale deployment will not likely happen until 2020 and beyond, when 5G networks are deployed. Nevertheless, there are significant opportunities for improving network energy consumption today, as outlined below (and captured on the dotted line in figure 9).

Wireless access networks – 2G and 3G networks are gradually being decommissioned and replaced by more efficient 4G LTE networks. The rate of 3G decommissioning has a significant impact on overall network energy consumption. By 2020, we assume 98% deployment of LTE in North America, and 95%, for the rest of the world. Within these LTE networks, we consider the following technology evolution paths:

- LTE will enjoy Moore's law improvements in hardware and see limited deployment of small cells, the use of remote radio heads and improved macro base stations to support 20 MHz bandwidth with a single transceiver. These technologies will be gradually deployed with an assumed penetration rate of 100% by 2017.

- Also coming, optimized networking, architecture and technologies that are more disruptive and require standards changes, including the use of Massive

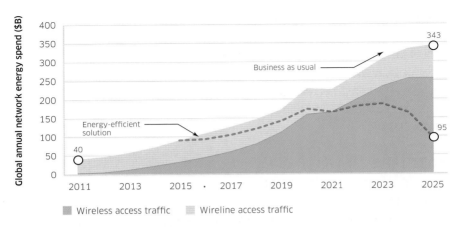

FIGURE 9: Global annual network energy spend by operators (Bell Labs Consulting).

MIMO for dense urban scenarios, advanced scheduling algorithms, separation of control and data plane functionalities and aggressive sleep modes in ultra-dense small cell networks. These are already foreseen in the future 5G specifications and will be deployed starting in 2020, reaching a 20% penetration rate by 2025.

Fixed access networks: There will be a gradual increase in the deployment rate of passive optical networks (PON) and decommissioning of copper-based access networks:

- PON evolution will include Gigabit-capable PON access and IP/MPLS metro aggregation with Moore's law hardware improvements, power shedding, more efficient hardware design and more efficient cooling. Since these solutions are part of the European Broadband Code of Conduct for 2020, we assume they will be gradually deployed with 100% penetration by 2020.

- Also deployed will be virtualized home gateway functions, optimized optical transceivers, sleep modes and improvements in electronics and optics hardware with an assumed penetration rate of 30% by 2025.

As can be seen in figure 9, the network energy cost can almost be reduced by a factor of four by 2025, through the deployment of these more efficient and energy-optimized technologies. Starting in 2020, the energy cost flattens and then decreases, despite the significant increase in traffic that continues beyond 2020. This is a consequence of energy efficiency improvements outpacing the predicted traffic growth. Note that compared to 2011, the energy cost has more than doubled from $40 to $95 billion in 2025, reflecting the fact that energy optimized technologies will not yet be universally deployed. As their network penetration increases beyond 2025, further reductions in energy consumption and cost will occur, reducing energy consumption and cost below 2011 levels.

The future begins now

Optimizing communications networks for energy efficiency is a huge undertaking requiring an equally large shift in network design thinking. This shift is necessary if we are to optimize networks for energy efficiency and maintain our focus on performance, throughput, latency, quality of service and reliability. It will take a full network generation, roughly 10+ years, for the opportunities and the new technologies to fully take hold. Fortunately, a lot of attention is already being paid to the energy challenges confronting the ICT industry today. The following are examples of recent and current energy initiatives:

- The European FP7 project, EARTH (Energy Aware Radio and Network Technologies), investigated the energy efficiency of mobile communications and demonstrated that the energy consumption of LTE networks can be reduced by more than 50% while staying within the LTE standards (Auer 2011, Zeller 2013, Parsons 2015).

- A major Asian operator launched its "Green Action Plan" in 2007 with a strong focus on energy saving and carbon emission reduction (Green IT 2009).

Optimizing communications networks for energy efficiency is a huge undertaking... It will take a full network generation, roughly 10+ years, for the opportunities and the new technologies to fully take hold.

- A major European operator is putting power consumption management at the heart of its corporate strategy to prevent its annual power bill reaching the $1 billion mark by 2020 (Le Maistre 2014).

- A major North American operator announced in 2014 that it would deploy new solar power systems at some of its network facilities bringing its total investment in solar technologies to $140M to date. This deployment is part of a strategy to have a 50% reduction by 2020 of the carbon emissions per terabyte of data flowing through its networks (Mearian 2014).

- Network equipment vendors have energy features in their product roadmap. For example, the functional energy efficiency of a typical optical transport platform increased by up to 100% (depending on configurations) from 2012 to 2014. The energy efficiency of a routing platform increased by 25%. Similarly the functional energy efficiency of radio access technology is on track to exceed the 75% targeted improvement by 2015 compared to 2008 (Alcatel-Lucent 2014).

- Equipment vendors and service providers are also very much concerned about their own carbon footprint, with commitments to reduce by 50% the absolute carbon footprint from operations by 2020 (compared to 2008 baseline) and a 40% reduction had already been achieved by 2014 (Alcatel-Lucent 2014).

- The European Commission has included energy efficiency as one of the key performance indicators for 5G networks in its 5G Infrastructure Private Public Partnership (5G PPP) with targeted energy savings of 90% per service indicated (5g-ppp.eu).

Summary

The energy consumption required to power our current and future communications networks is dramatically increasing. One of the greatest challenges facing the ICT industry in the next decade is to address the rising energy bills for network operators and the environmental impact of their operations.

In this chapter, we have highlighted the key trends in network energy consumption, energy cost and energy supply. Fortunately new architecture and technology breakthroughs have emerged that pave the way to a more efficient and a more sustainable network energy future. From our analysis, it is evident that future networks can support substantial growth in applications, services and connected devices and operate with lower energy consumption than today.

Without a doubt, a lot of work remains to fully realize these potential energy efficiency gains and truly achieve the vision of a zero-power network. The initial research results need to be further validated; technologies need to be positioned and standardized appropriately; product development and commercialization efforts need to be launched; and technology roll-out plans need to be developed.

The challenges and opportunities are clear. The benefits are tremendous and span beyond the ICT sector and potentially touch all industry sectors across the globe, helping them to reduce their own energy footprints. Networking technologies are vital to bridge the digital divide and connect the unconnected in the world. If we are to successfully launch the next digital era, energy efficiency will be key to a more productive and sustainable future for all of us.

References

Alcatel-Lucent 2014. *Alcatel-Lucent Sustainability Report 2014* (https://www.alcatel-lucent.com/sustainability).

Auer. G, 2011. "How Much Energy is Needed to Run a Wireless Network?" *IEEE Wireless Communications Magazine: Special Issue on Technologies for Green Radio Communication Networks* 18(5), Oct 2011, pp. 40-49.

Bell Labs 2013. "Metro Network Traffic Growth: An Architecture Impact Study," *Bell Labs Strategic White Paper*, December (http://resources.alcatel-lucent.com/asset/171568).

Byrnes, S., 2013. "Why Are Solar Panels So Inefficient?", Quora, November 4 (http://www.forbes.com/sites/quora/2013/11/04/why-are-solar-panels-so-inefficient).

Capone, A., Fonseca dos Santos, A., Filippini, I. and Gloss, B., 2012. "Looking beyond green cellular networks," *Conference on Wireless On-demand Network Systems and Services (WONS)*, January, pp. 127–130 (archived at http://www.greentouch.org/uploads/documents/wons2012.pdf).

DataCenter Dynamics 2013. "Data Center Market Trend Report 2012," June 2 (statistics reported on LeGrande web site: http://datacenter.legrand.com/EN/efficiency).

GeSI 2015. "#Smarter2030 ICT Solutions for the 21st Century challenges," *Global e-Sustainability Initiative* (GeSI), June (http://smarter2030.gesi.org/downloads/Full_report2.pdf).

Green IT Review 2009. "China Mobile - Green Action Plan," 29 June (http://www.thegreenitreview.com/2009/06/china-mobile-green-action-plan.html).

GreenTouch 2015. "Final Results from Green Meter Research Study: Reducing the Net Energy Consumption in Communications Networks by up to 98% by 2020," *GreenTouch web site*, June 2015 (http://www.greentouch.org/uploads/documents/GreenTouch_Green_Meter_Final_Results_18_June_2015.pdf).

GSMA 2013. "Greening Telecoms: Pakistan and Aghanistan Market Analysis," *Green Power for Mobile*.

GWATT 2015. "Global What-if Analyzer of Network Energy Consumption," *Bell Labs Research web site* (www.gwatt.net).

Intelligent Energy, 2012. "The True Cost of Providing Energy to Telecom Towers in India," *White Paper*.

Kim, 2014. "Profitability of the Mobile Business Model ... The Rise! & Inevitable Fall?" *Techneconomy Blog*, July 21 (http://techneconomyblog. com/2014/07/21/profitability-of-the-mobile-business-model).

Lambert, S., et al, 2013. "Worldwide electricity consumption of communication networks," *Optical Society of America*.

Le Maistre, R., 2014. "Energy Bill Shocks Orange Into Action," *Light Reading*, September 24 (http://www.lightreading.com/energy-efficiency/energy-bill-shocks-orange-into-action/d/d-id/711038).

Marzetta, T., 2015. "Massive MIMO: An Introduction," *Bell Labs Technical Journal*, March.

Mearian, L., 2014. "Verizon to become solar-power leader in the U.S. telecom industry," *Computerworld*, August 26 (http://www.computerworld.com/article/2599182/sustainable-it/verizon-to-become-solar-power-leader-in-the-u-s-telecom-industry.html).

Parsons, N., 2015. "SDN-Enabled Optical Circuit Technology for Efficient and Scalable Data Center Networks," *EuCNC 2015*.

Renew 2014. "New World Record for Solar Cell Efficiency Set at 46%", *ReNew Economy* (web site), December 2014 (http://reneweconomy.com.au/2014/new-world-record-for-solar-cell-efficiency-set-at-46-73582).

Weldon, M., 2014. "Defining the Future of Networks", *CommsDay Summit*, April (http://www.slideshare.net/CommsDay/alcatellucents-marcus-weldon-at-commsday-summit-2014).

Zeller, D. et al 2013. "Sustainable Wireless Broadband Access to the Future Internet - The EARTH Project," Galis, A. and Gavras, A. (eds), *The Future Internet - Future Internet Assembly 2013: Validated Results and New Horizons*, pp. 249–271 (article archived at http://rd.springer.com/content/pdf/10.1007%2F978-3-642-38082-2_21.pdf).

Glossary / Abbreviations

2G	second generation
3Cs	capture, compute and communicate (data)
3GPP	3rd Generation Partnership Project
4G LTE	fourth generation long term evolution
4Ps	people, processes, platforms/tools and performance/metrics
5G LTE	5th generation long term evolution
ABR	adaptive bit rate
ADSL	asymmetric digital subscriber line
AES	Advanced Encryption Standard
AI	artificial intelligence
ANN	artificial neural networks
APAC	Asia-Pacific
APN	access point name
APT	advanced persistent threat
AR	augmented reality
ARPANET	Advanced Research Projects Agency Network
ATM	asynchronous transfer mode
AugI	augmented intelligence
AWS	Amazon web services
BBU	baseband unit
bit-AES	binary digit Advanced Encryption Standard
B/OSS	business and operations support systems
BPON	broadband PON
BYOD	bring your own device
CAPEX	capital expenditures
CATV	community antenna television
CBR	constant bit rate
C&C	content and control
CCN	content-centric networking
CDMA	Code Division Multiple Access
CDN	Content Delivery Network
CEC	converged enterprise cells
CEM	customer experience management
CIDR	classless inter-domain routing
CIGI	Centre for International Governance Innovation

CIN	cloud integrated network
CMOS	complementary metal oxide semiconductor
CO	central office
CoMP	coherent multipoint processing
CORD	central office re-architected as data center
CP-OFDM	cyclic prefix - orthogonal frequency division multiplexing
CPRI	common public radio interface
CPU	central processing unit
CRM	customer relationship management
CSMA	collision sense multiple access
CSPs	communications service providers
CSPF	constrained-shortest-path-first
CTSS	compatible time sharing systems
DAS	distributed antenna system
DC	data center
DCI	data center interconnect
DDoS	distributed denial of service
DÉCOR	dedicated core networks
DMIPS	Dhrystone million instructions per second
DMZs	Demilitarized Zones
DNS	domain name system
DOCSIS	data over cable service interface specification
DOF	degree-of-freedom
DoS	denial of service
DPI	deep packet inspection
DPU	distribution point unit
DRM	dynamic risk management
DSL	digital subscriber line
DTLS	datagram transport layer security
DWDM	dense wavelength division multiplexing
EARTH	energy aware radio and network technologies
ECC	elliptic-curve cryptography
EDGE	Enhanced Data rates for GSM Evolution
EDR	endpoint detection and response
EPC	evolved packet core
EPON	Ethernet passive optical network
ERM	enterprise resource management
ETSI	European Telecommunications Standards Institute

ETSI ISG NFV	European Telecommunications Standards Institute Industry Specification Group for Network Functions Virtualization
EV-DO	Evolution data optimized
FANS	fixed access network sharing
FCC	Federal Communications Commission (US)
FCF	free cash flow
FEC	forward error correction
FFT	Fast Fourier Transform
FIDO	fast identity online
finFET	Fin-based Field Effect Transistor
FPGAs	field-programmable gate arrays
FTTH	fiber to the home
Gb/s	gigabits per second
GeSI	Global E-sustainability initiative
GHz	gigahertz
GPON	gigabit passive optical network
GPP	general purpose processors
GSM	Global system for mobile communications
GSPs	global service providers
GWATT	Global "What if" Analyzer of neTwork energy consumpTion
HetNet	Heterogeneous network
HFC	hybrid fiber-coaxial
HFT	high-frequency trading
HSPA+	High-speed download/upload packet access (+ designates the evolved newer spec)
HSS	home subscriber server
HTTPS	hypertext transfer protocol secure
IaaS	infrastructure as a service
IBO	in-building operator
ICN	information-centric networking
ICT	information and communications technology
IDS	intrusion detection system
IEEE	Institute of Electrical and Electronics Engineers
IETF	Internet Engineering Task Force
IGP	interior gateway protocol
IMS	IP multimedia subsystem
iOS	iPhone Operating System (Apple)
IoT	Internet of Things

IP	internet protocol
IP/MPLS	internet protocol/multi-protocol label switching
IPSec	internet protocol security
ISAC	industry-specific information sharing and analysis centers
ISM	industrial, scientific and medical
IVA	intelligent virtual assistant
IVR	interactive voice response
KPIs	key performance indicators
LAA	license-assisted access
LAN	local area network
Li-ion	lithium ion battery
LoRa	long range radio
LPWA	low-power wireless access
LSPs	local service providers
LT	line termination
LTE	long term evolution
LTE Adv CA	long term evolution advanced carrier aggregation
LTE Adv MIMO	long term evolution advanced multiple input and multiple output
LTE RRH	long term evolution remote radio head
LTE-U	unlicensed LTE carriers
LWA	LTE and Wi-Fi Link Aggregation
MAC	Media Access Control
MAUs	monthly average users
MEMS	micromechanical devices
MIMO	multiple input and multiple output
MIMs	machine identification modules
MIPS	million instructions per second
mmWave	millimeter wave bands
MME	mobility management entity
MOCN	multi-operator core networks
MP-TCP	multipath transmission control protocol
MSO	multi-service operator
Multi-RAT	multi-remote administration tool
MVNO	mobile virtual network operator
M2M	machine-to-machine
NETCONF	network configuration
NFV	network functions virtualization
NG CPE	next generation customer premises equipment

NGMN	next generation mobile network
NIC	network interface cards
NICT	National Institute of Information and Communications Technology (Japan)
NIST	National Institute of Standards and Technology
NOCs	network operation centers
NP	network policies
NPS	net promoter scores
NSA	National Security Agency
NSFnet	National Science Foundation network
NSO	network service orchestrator
NTIA	National Telecommunications & Information Administration
OCAs	open connect appliances
OFDM	orthogonal frequency division multiplexing
OLT	optical line terminal
ONF	open networking foundation
ONOS	open network operating system
OPEX	operating expenditures
OS	operating system
OSA	open system architecture
OSS	operations support system
OT	operational technology
OTT	over-the-top
PaaS	platform as a service
PBXs	private branch exchanges
PCRF	policy and charging rules function
PDN	packet data network
PGW	packet network gateway
PID	proportional-integral-derivative
PLMN	public land mobile network
PON	passive optical network
PoPs	points of presence
PPP	private public partnership
PSTN	public switched telephone network
PUE	power utilization efficiencies
PXC	photonic cross connects
QoS	quality of service
Qubit	quantum bit
RANs	radio access networks

RCS	rich communication services
REsT	representational state transfer
ROI	return on investment
RRM	radio resource management
RSA	Rivest-Shamir-Adleman cryptosystem
RTB	real-time bidding
RTSP	real-time streaming protocol
RTT	round trip time
SaaS	software as a service
SAUs	simultaneously active users
SCADA	supervisory control and data acquisition
SCP	signal control point
SD-VPN	software-defined virtual private network
SDH	synchronous digital hierarchy
SDN	software-defined networking
SDN/NFV	software defined networking/network functions virtualization
SGW	serving gateway
SIEMs	security information and event management systems
SiGe	silicon-germanium
SIP	session initiation protocol
SLAs	service level agreements
SMS	short message service
SMTP	simple mail transfer protocol
SNP	smart nation platform
SoC	system-on-a-chip
SOC	service operations center
SON	self-optimizing network
SONET	synchronous optical network
SS7	signaling system No. 7
SSH	secure shell
SSL/TLS	secure sockets layer/transport layer security
STP	signal transfer point
T-CAM	ternary CAM
TCP	transmission control protocol
TCP/IP	transmission control protocol/internet protocol
TDM	time division multiplexing
TOR	top of the rack
TOT	triangle of truth
TriGate	3D transistors

UE	user equipment
UF-OFDM	universal filtered - orthogonal frequency division multiplexing
USB	universal serial bus
UGC	user generated content
URLs	uniform resource locators
vANC	virtual access network controller
vBBU	virtualized baseband unit
vCDN	virtualized content distribution network
VDSL	very-high-bit-rate digital subscriber line
vEPC	virtualized evolved packet core
vHome	virtual home
VMs	virtual machines
VoIP	Voice over internet protocol
VoLTE	voice-over-LTE
VOR	vestibule-ocular reflex
vP-GW	virtualized packet network gateway
VPN	virtual private network
VPWNs	virtual private wireless networks
vRAN	virtualized radio access network
vRGWs	virtual residential gateways
vSpace	virtual space
vSTBs	virtual set-top-boxes
VXLAN	virtual extensible local area network
WAN	wide area network
WDM	wavelength division multiplexing
WEP	wired equivalent privacy
Wi-Gig	Wireless gigabit
WLAN	wireless local area network
WPA	Wi-Fi protected access
WPS	Wi-Fi protected setup
WWS	world wide streams
WWW	World Wide Web